AF478646

BOLYAI SOCIETY
MATHEMATICAL STUDIES

17

BOLYAI SOCIETY MATHEMATICAL STUDIES

Ervin Győri
Gyula O. H. Katona
László Lovász (Eds.)

Horizons of Combinatorics

 Springer

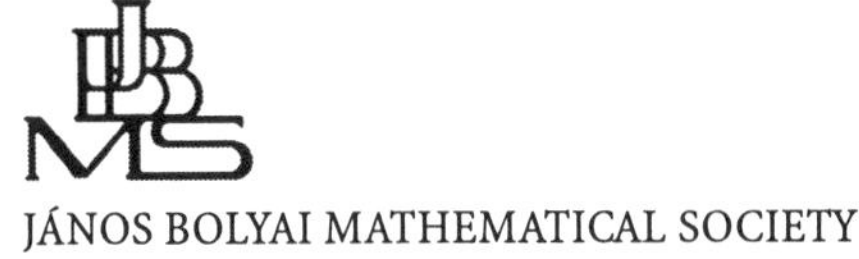

Ervin Győri
Hungarian Academy of Sciences
Alfréd Rényi Inst. of Mathematics
Reáltanoda u. 13–15
Budapest 1053, Hungary
e-mail: gyori@renyi.hu

Gyula O. H. Katona
Hungarian Academy of Sciences
Alfréd Rényi Inst. of Mathematics
Reáltanoda u. 13–15
Budapest 1053, Hungary
e-mail: ohkatona@renyi.hu

László Lovász
Eötvös University
Institute of Mathematics
Pázmány Péter sétány 1/C
Budapest 1117, Hungary
e-mail: lovasz@cs.elte.hu

Managing Editor:
Gábor Sági
Hungarian Academy of Sciences
Alfréd Rényi Inst. of Mathematics
Reáltanoda u. 13–15
Budapest 1053, Hungary
e-mail: sagi@renyi.hu

The publication of the recent volume was supported
by the Hungarian Academy of Sciences.

Mathematics Subject Classification (2000): 03XX, 05XX, 52C10, 60C5, 62XX

Library of Congress Control Number: 2007941335

ISSN 1217-4696
ISBN 3-540-77199-9 Springer Berlin Heidelberg New York
ISBN 978-963-9453-09-8 János Bolyai Mathematical Society, Budapest

Springer is a part of Springer Science+Business Media
springer.com

Cover design: Erich Kirchner, Heidelberg

Printed on acid-free paper   44/3142/db – 5 4 3 2 1 0

Készült: Regiszter Kiadó és Nyomda Kft.

CONTENTS

Preface

The János Bolyai Mathematical Society and the Alfréd Rényi Institute of Mathematics organized the conference Horizons of Combinatorics during the period July 17–21, 2006 at Balatonalmádi (Lake Balaton, Hungary). The Hungarian conferences in combinatorics have the "tradition" not to be organized with regular frequency, and having all different names. Yet, this conference was, in a certain sense, a continuation of the conferences organized in January, 1996, and January, 2001.

The present volume is strongly related to this conference. We have asked our main speakers to summarize their recent works in survey papers. Since many of them reacted positively, we are able to present the reader with this collection of papers written by excellent authors. Unlike many of our previous volumes that needed several years of preparation the current volume appears 18 months after the conference.

Let us briefly introduce the content.

The paper of Addario-Berry and Reed draws a nice picture from an observation of Bertrand (which is called the First Ballot Theorem) to recently obtained results on sums of identically distributed random variables and to analyzing random permutations of sets of real numbers.

V. Csiszár, Rejtő and Tusnády study some aspects of stochastic methods in modern combinatorics, from a rather new perspective.

The survey of Egawa illustrates three different types of proofs for theorems establishing the existence of a 2-factor.

The paper of Fox and Pach deals with special classes of graphs defined by geometric methods. For these classes, the authors answer in the affirmative the following question of Erdős and Hajnal: "Is it true that for every graph G there exists a constant $c = c(G)$ such that if a graph H on n vertices does not contain an isomorphic copy of G (as an induced subgraph) then H has a complete or empty subgraph of size n^c?"

Ron Graham, the leading expert in Ramsey theory has collected a variety of problems and recently obtained related results in the theory which make progress on some of the presented problems.

Katona surveys (mostly quite recent) results in Sperner theory. The maximum number of subsets is searched under conditions excluding configurations which can be expressed by merely inclusions. A new method, which is actually an extension of Lubell's chain method, is illustrated in detail.

Miklós discusses the results and relations between the (maximum) number of subsums of a finite sum with some additional properties assumed and extremal sets of vertices of the hypercube in the sense that their span (either over $GF(2)$ or over $\mathbb{R}$) does not contain certain (forbidden) configurations from the hypercube.

Recski's paper surveys some, partly new, combinatorial results concerning the rigidity of tensegrity frameworks. Issues related to computational complexity are also emphasized.

Seress presents several constructions of polygonal and near-polygonal graphs. Possible classifications of these graphs are also discussed.

The paper of Soukup presents generalizations of several well known theorems in the theory of finite graphs, finite partially ordered sets, etc. to graphs with infinitely many vertices, partially ordered sets with infinitely many elements, and so on. The paper accurately illustrates, that such generalizations are sometimes straightforward, sometimes hard to obtain, sometimes true in "small" infinite sets but fail in the higher infinity, or sometimes simply not true.

Tokushige surveys Frankl's random walk method in the theory of intersecting families and explains its usage with many examples.

Vu's survey discusses some basic problems concerning random matrices with discrete distributions. Several new results, tools and conjectures have been presented.

December, 2007   The editors

BOLYAI SOCIETY
MATHEMATICAL STUDIES, 17

Horizons of Combinatorics
Balatonalmádi
pp. 9–35.

BALLOT THEOREMS, OLD AND NEW

L. ADDARIO-BERRY and B. A. REED

"There is a big difference between a fair game and a game it's wise to play."
– [7].

1. A BRIEF HISTORY OF BALLOT THEOREMS

1.1. Discrete time ballot theorems

We begin by sketching the development of the classical ballot theorem as it first appeared in the Comptes Rendus de l'Academie des Sciences. The statement that is fairly called the first Ballot Theorem was due to Bertrand:

Theorem 1 ([8]). *We suppose that two candidates have been submitted to a vote in which the number of voters is μ. Candidate A obtains n votes and is elected; candidate B obtains $m = \mu - n$ votes. We ask for the probability that during the counting of the votes, the number of votes for A is at all times greater than the number of votes for B. This probability is $(2n - \mu)/\mu = (n - m)/(n + m)$.*

Bertrand's "proof" of this theorem consists only of the observation that if $P_{n,m}$ counts the number of "favourable" voting records in which A obtains n votes, B obtains m votes and A always leads during counting of the votes, then

$$P_{n+1,m+1} = P_{n+1,m} + P_{n,m+1},$$

the two terms on the right-hand side corresponding to whether the last vote counted is for candidate B or candidate A, respectively. This "proof" can be easily formalized as follows. We first note that the binomial coefficient

$B_{n,m} = (n+m)!/n!m!$ counts the total number of possible voting records in which A receives n votes and B receives m, Thus, the theorem equivalently states that for any $n \geq m$, $B_{n,m} - P_{n,m}$, which we denote by $\Delta_{n,m}$, equals $2mB_{n,m}/(n+m)$. This certainly holds in the case $m = 0$ as $B_{n,0} = 1 = P_{n,0}$, and in the case $m = n$, as $P_{n,n} = 0$. The binomial coefficients also satisfy the recurrence $B_{n+1,m+1} = B_{n+1,m} + B_{n,m+1}$, thus so does the difference $\Delta_{n,m}$. By induction,

$$\Delta_{n+1,m+1} = \Delta_{n+1,m} + \Delta_{n,m+1}$$

$$= \frac{2m}{n+m+1}B_{n+1,m} + \frac{2(m+1)}{n+m+1}B_{n,m+1}$$

$$= \frac{2(m+1)}{n+m+2}B_{n+1,m+1},$$

as is easily checked; it is likely that this is the proof Bertrand had in mind.

After Bertrand announced his result, there was a brief flurry of research into ballot theorems and coin-tossing games by the probabilists at the Academie des Sciences. The first formal proof of Bertrand's Ballot Theorem was due to André and appeared only weeks later [3]. André exhibited a bijection between unfavourable voting records starting with a vote for A and unfavourable voting records starting with a vote for B. As the latter number is clearly $B_{n,m-1}$, this immediately establishes that $B_{n,m} - P_{n,m} = 2B_{n,m-1} = 2mB_{n,m}/(n+m)$.

A little later, [5] asserted but did not prove the following generalization of the classical Ballot Theorem: if $n > km$ for some integer k, then the probability candidate A always has more than k-times as many votes as B is precisely $(n - km)/(n + m)$. In response to the work of André and Barbier, Bertrand had this to say:

> "Though I proposed this curious question as an exercise in reason and calculation, in fact it is of great importance. It is linked to the important question of duration of games, previously considered by Huygens, [de] Moivre, Laplace, Lagrange, and Ampere. The problem is this: a gambler plays a game of chance in which in each round he wagers $\frac{1}{n}$'th of his initial fortune. What is the probability he is eventually ruined and that he spends his last coin in the $(n + 2\mu)$'th round?" [6]

He notes that by considering the rounds in reverse order and applying Theorem 1 it is clear that the probability that ruin occurs in the $(n + 2\mu)$'th round is nothing but $\frac{n}{n+2\mu}\binom{n+2\mu}{\mu}2^{-(2\mu+n)}$. By informal but basic computations, he then derives that the probability ruin occurs *before* the $(n+2\mu)$'th round is approximately $1 - \frac{\sqrt{2/\pi n}}{\sqrt{n+2\mu}}$, so for this probability to be large, μ must be large compared to n^2. (Bertrand might have added Pascal, Fermat, and the Bernoullis [16, pp. 226–228] to his list of notable mathematicians who had considered the game of ruin; [4, pp. 98–114] gives an overview of prior work on ruin with an eye to its connections to the ballot theorem.)

Later in the same year, he proved that in a *fair game* (a game in which, at each step, the average change in fortunes is nil) where at each step, one coin changes hands, the expected number of rounds before ruin is infinite. He did so using the fact that by the above formula, the probability of ruin in the t'th round (for t large) is of the order $1/t^{3/2}$, so the expected time to ruin behaves as the sum of $1/t^{1/2}$, which is divergent. He also stated that in a fair game in which player A starts with a dollars and player B starts with b dollars, the expected time until the game ends (until one is ruined) is precisely ab [7]. This fact is easily proved by letting $e_{a,b}$ denote the expected time until the game ends and using the recurrence $e_{a,b} = 1 + (e_{a-1,b} + e_{a,b-1})/2$ (with boundary conditions $e_{a+b,0} = e_{0,a+b} = 0$). Expanding on Bertrand's work, Rouché provided an alternate proof of the above formula for the probability of ruin [24]. He also provided an exact formula for the expected number of rounds in a biased game in which player A has a dollars and bets a_0 dollars each round, player B has b dollars and bets b_0 dollars each round, and in each round player A wins with probability p satisfying $a_0 p > b_0(1-p)$ [25].

All the above questions and results can be restated in terms of a *random walk* on the set of integers $\mathbb{Z}$. For example, let $S_0 = 0$ and, for $i \geq 0$, $S_{i+1} = S_i \pm 1$, each with probability $1/2$ and independently of the other steps – this is called a *symmetric simple random walk*. (For the remainder of this section, we will phrase our discussion in terms of random walks instead of votes, with $X_{i+1} = S_{i+1} - S_i$ constituting a step of the random walk.) Then Theorem 1 simply states that given that $S_t = h > 0$, the probability that $S_i > 0$ for all $i = 1, 2, \ldots, t$ (i.e. the random walk is favourable for A) is precisely h/t. Furthermore, the time to ruin when player A has a dollars and player B has b dollars is the *exit time* of the random walk S from the interval $[a, -b]$. The research sketched above constitutes the first detailed examination of the properties of a random walk $S_0, S_1, \ldots, S_n$ conditioned on the value S_n, and the use of such information to study the asymptotic properties of such a walk.

In 1923, Aeppli proved Barbier's generalized ballot theorem by an argument similar to that used by André's. This proof is presented in [4, pp. 101–102], where it is also observed that Barbier's theorem can be proved using Bertrand's original recurrence in the same fashion as above. A simple and elegant technique was used in [9] to prove Barbier's theorem; we use it to prove Bertrand's theorem as an example of its application, as it highlights an interesting perspective on ballot-style results.

We think of $\mathcal{X} = (X_1, \ldots, X_{n+m}, X_1)$ as being arranged clockwise around a cycle (so that $X_{n+m+1} = X_1$). There are precisely $n + m$ walks corresponding to this set, obtained by choosing a first step X_i, so to establish Bertrand's theorem it suffices to show that however $X_1, \ldots, X_{n+m}$ are chosen such that $S_n = n - m > 0$, precisely $n - m$ of the walks $X_{i+1}, \ldots, X_{n+m}, X_1, \ldots, X_i$ are favourable for A. Let $S_{ij} = X_{i+1} + \cdots + X_j$ (this sum includes X_{n+m} if $i < j$). We say that $X_i, \ldots, X_j$ is a *bad run* if $S_{ij} = 0$ and $S_{i'j} < 0$ for all $i' \in \{i+1, \ldots, j\}$ (this set includes $n + m$ if $i > j$). In words, this condition states that i is the first index for which the reversed walk starting with X_j and ending with X_{i+1} is nonnegative. It is immediate that if two bad runs intersect then one is contained in the other, so the maximal bad runs are pairwise disjoint. (An example of a random walk and its bad runs is shown in Figure 1).

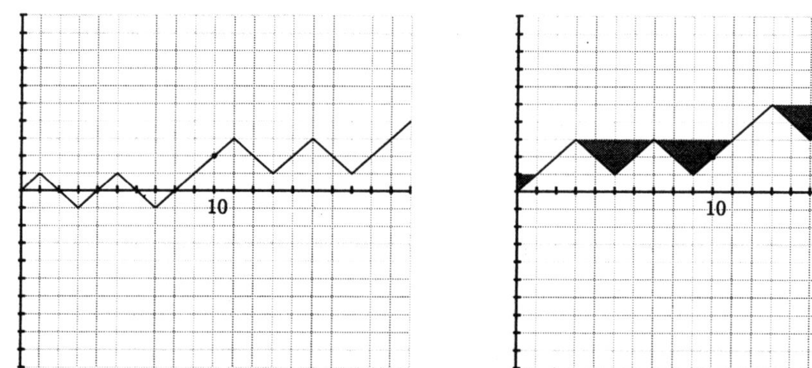

Fig. 1. On the left appears the random walk corresponding to the voting sequence $(1, -1, -1, 1, 1, -1, -1, 1, 1, 1)$, doubled to indicate the cyclic nature of the argument. On the right is the reversal of the random walk; the maximal bad runs are shaded grey

If $X_i = 1$ and $S_{ij} = 0$ for some j then X_i begins a bad run, and since $S_n = \sum_{i=1}^{n} X_i > 0$, if $X_i = -1$ then X_i ends a bad run. As $S_{ij} = 0$ for a maximal bad run and $X_i = 1$ for every X_i not in a bad run, there must be precisely $n - m$ values of i for which X_i is not in a bad run. If the walk starting with X_i is favourable for A then for all $i \neq j$, S_{ij} is positive, so X_i

is not in a bad run. Conversely, if X_i is not in a bad run then $X_i = 1$ and for all $j \neq i$, $S_{ij} > 0$, so the walk starting with X_i is favourable for A. Thus there are precisely $(n - m)$ favourable walks corresponding to $\mathcal{X}$, which is what we set out to prove.

With this technique, the proof of Barbier's theorem requires nothing more than letting the positive steps have value $1/k$ instead of 1. This proof is notable as it is the first time the idea of cyclic permutations was applied to prove a ballot-style result. This "rotation principle" is closely related to the *strong Markov property* of the random walk: for any integer $t \geq 0$, the random walk $S_t - S_t, S_{t+1} - S_t, S_{t+2} - S_t, \ldots$ has identical behavior to the walk S_0, S_1, S_2 and is independent of $S_0, S_1, \ldots, S_t$. (Informally, if we have examined the behavior of the walk up to time S, we may think of *restarting* the random walk at time t, starting from a height of S_t; this will be important in the generalized ballot theorems to be presented later in the paper.) This proof can be rewritten in terms of *lattice paths* by letting votes for A be unit steps in the positive x-direction and votes for B be unit steps in the positive y-direction. When conceived of in this manner, this proof immediately yields several natural generalizations [9, 15, 23].

Starting in 1962, Lajos Takács proved a sequence of increasingly general ballot-style results and statements about the distribution of the maxima when the ballot is viewed as a random walk [28, 29, 30, 31, 32, 33, 36]. We highlight two of these theorems below; we have not chosen the most general statements possible, but rather theorems which we believe capture key properties of ballot-style results.

A family of random variables $X_i, \ldots, X_n$ is *interchangeable* if for all $(r_1, \ldots, r_n) \in \mathbb{R}^n$ and all permutations σ of $\{1, \ldots, n\}$, $\mathbf{P}\{X_i \leq r_i \forall 1 \leq i \leq n\} = \mathbf{P}\{X_i \leq r_{\sigma(i)} \forall 1 \leq i \leq n\}$. We say $X_1, \ldots, X_n$ is *cyclically interchangeable* if this equality holds for all *cyclic* permutations σ. A family of interchangeable random variables is cyclically interchangeable, but the converse is not always true. The first theorem states:

Theorem 2. *Suppose that $X_1, \ldots, X_n$ are integer-valued, cyclically interchangeable random variables with maximum value 1, and for $1 \leq i \leq n$, let $S_i = X_1 + \cdots + X_i$. Then for any integer $0 \leq k \leq n$,*

$$\mathbf{P}\{S_i > 0 \ \forall 1 \leq i \leq n \mid S_n = k\} = \frac{k}{n}.$$

This theorem was proved independently in [10] and [39] – we note that it can also be proved by Dvoretzky and Motzkin's approach. (As a point of historical curiosity, Takács' proof of this result in the special case of interchangeable random variables, and Dwass' proof of the more general result above, appeared in the same issue of Annals of Mathematical Statistics.) Theorem 2 and the "bad run" proof of Barbier's ballot theorem both suggest that the notion of cyclic interchangeability or something similar may lie at the heart of all ballot-style results.

Theorem 3 ([36], p. 12). *Let $X_1, X_2, \ldots$ be an infinite sequence of iid integer random variables with mean μ and maximum value 1 and for any $i \geq 1$, let $S_i = X_1 + \cdots + X_i$. Then*

$$\mathbf{P}\{S_n > 0 \text{ for } n = 1, 2, \ldots\} = \begin{cases} \mu & \text{if } \mu > 0, \\ 0 & \text{if } \mu \leq 0. \end{cases}$$

The proof of Theorem 3 proceeds by applying Theorem 2 to finite subsequences $X_1, X_2, \ldots, X_n$, so this theorem also seems to be based on cyclic interchangeability. Takács has generalized these theorems even further, proving similar statements for step functions with countably many discontinuities and in many cases finding the exact distribution of $\max_{i=1}^{n}(S_i - i)$.

(Takács originally stated his results in terms of non-negative integer random variables – his original formulation results if we consider the variables $(1 - X_1), (1 - X_2), \ldots$ and the corresponding random walk.) Theorem 3 implies the following classical result about the probability of ever returning to zero in a biased simple random walk:

Theorem 4 ([11], p. 274). *In a biased simple random walk $0 = R_0, R_1, \ldots$ in which $\mathbf{P}\{R_{i+1} - R_i = 1\} = p > 1/2$, $\mathbf{P}\{R_{i+1} - R_i = -1\} = 1 - p$, the probability that there is no $n \geq 1$ for which $R_n = 0$ is $2p - 1$.*

Proof. Observe that the expected value of $R_i - R_{i-1}$ is $2p - 1 > 0$, so if $R_1 = -1$ then with probability 1, $R_i = 0$ for some $i \geq 2$. Thus,

$$\mathbf{P}\{R_n \neq 0 \text{ for all } n \geq 1\} = \mathbf{P}\{R_n > 0 \text{ for all } n \geq 1\}.$$

The latter probability is equal to $2p - 1$ by Theorem 3.   ■

We close this section by presenting the beautiful "reflection principle" proof of Bertrand's theorem. We think of representing the symmetric simple random walk as a sequence of points $(0, S_0), (1, S_1), \ldots, (n, S_n)$ and connecting neighbouring points. If $S_1 = 1$ and the walk is unfavourable, then letting

k be the smallest value for which $S_k = 0$ and "reflecting" the random walk $S_0, \ldots, S_k$ in the x-axis yields a walk from $(0,0)$ to (n,t) whose first step is negative – this is shown in Figure 2. This yields a bijection between walks that are unfavourable for A and start with a positive step, and walks that are unfavourable for A and start with a negative step. As all walks starting with a negative step are unfavourable for A, all that remains is rote calculation. This proof is often incorrectly attributed to [3], which established the same bijection in a different way – its true origins remain unknown.

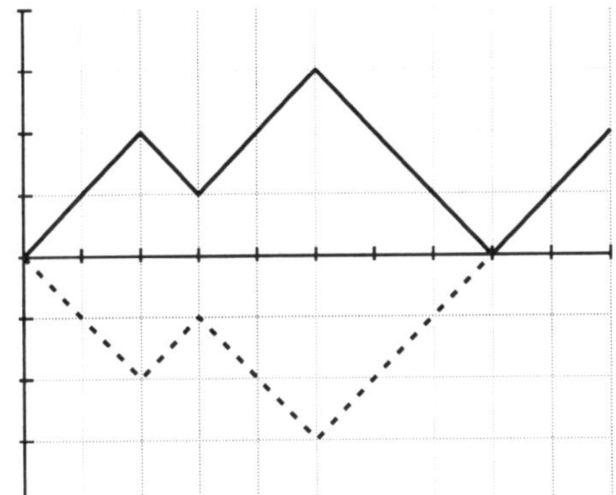

Fig. 2. The dashed line is the reflection of the random walk from $(0,0)$ to the first visit of the x-axis

1.2. Continuous time ballot theorems

The theorems which follow are natural given the results presented in Section 1.1; however, their statements require slightly more preliminaries. A *stochastic process* is simply a nonempty set of real numbers T and a collection of random variables $\{X_t, \ t \in T\}$ defined on some probability space. The collection of random variables $\{X_1, \ldots, X_n\}$ seen in Section 1.1 is an example of a stochastic process for which $T = \{1, 2, \ldots, n\}$. In this section we are concerned with stochastic processes for which $T = [0, r]$ for some $0 < r < \infty$ or else $T = [0, \infty)$.

A stochastic process $\{X_t, \ 0 \le t \le r\}$ has *(cyclically) interchangeable increments* if for all $n = 2, 3, \ldots$, the finite collection of random variables $\{X_{rt/n} - X_{r(t-1)/n}, \ t = 1, 2, \ldots, n\}$ is (cyclically) intechangeable. A process $\{X_t, \ t \ge 0\}$ has *interchangeable increments* if for all $r > 0$ and $n > 0$, $\{X_{rt/n} - X_{r(t-1)/n}, \ t = 1, 2, \ldots, n\}$ is interchangeable, and is *stationary* if this latter collection is composed of independent identically distributed random variables. As in the discrete case, these are natural sorts of prerequisites for a ballot-style theorem to apply.

There is an unfortunate technical restriction which applies to all the ballot-style results we will see in this section. The stochastic process $\{X_t,\ t \in T\}$ is said to be *separable* if there are almost-everywhere-unique measurable functions X^+, X_- such that almost surely $X_- \leq X_t \leq X^+$ for all $t \in T$, and there are countable subsets S_-, S^+ of T such that almost surely $X^+ = \sup_{t \in S^+} X_t$ and $X_- = \inf_{t \in S_-} X_t$. The results of this section only hold for separable stochastic processes. In defense of the results, we note that there are nonseparable stochastic processes $\{X_t,\ 0 \leq t \leq r\}$ for which $\sup\{X_t - t, 0 \leq t \leq r\}$ is non-measurable. As the distribution of this random variable is one of the key issues with which we are concerned, the assumption of separability is natural and in some sense necessary in order for the results to be meaningful. Moreover, in very general settings it is possible to construct a separable stochastic process $\{Y_t \mid t \in T\}$ such that for all $t \in T$, Y_t and X_t are almost surely equal (see, e.g., [12, Sec. IV.2]); in this case it can be fairly said that the assumption of separability is no loss.

The following theorem is the first example of a continuous-time ballot theorem. A *sample function* of a stochastic process is a function $x_\omega : T \to \mathbb{R}$ given by fixing some $\omega \in \Omega$ and letting $x_\omega(t) = X_t(\omega)$.

Theorem 5 ([34]). *If $\{X_t,\ 0 \leq t \leq r\}$ is a separable stochastic process with cyclically interchangeable increments whose sample functions are almost surely nondecreasing step functions, then*

$$\mathbf{P}\{X_t - X_0 \leq t \text{ for } 0 \leq t \leq r \mid X_r - X_0 = s\} = \begin{cases} \dfrac{t - s}{t} & \text{if } 0 \leq s \leq t, \\ 0 & \text{otherwise.} \end{cases}$$

This theorem is a natural continuous equivalent of Theorem 2 of Section 1.1; it can be used to prove a theorem in the vein of Theorem 3 which applies to stochastic processes $\{X_t,\ t \geq 0\}$. Takács' other ballot-style results for continuous stochastic processes are also essentially step-by-step extensions of his results from the discrete setting; see [34, 35, 36, 38].

In 1957, Baxter and Donsker derived a double integral representation for $\sup\{X_t - t, t \geq 0\}$ when this process has stationary independent increments. Their proof proceeds by analyzing the zeros of a complex-valued function associated to the process. They are able to use their representation to explicitly derive its distribution when the process is a Gaussian process, a coin-tossing process, or a Poisson process. This result was rediscovered by Takács, who also derived the joint distribution of X_r and $\sup\{X_t - t, 0 \leq t \leq r\}$ for r finite, using a generating function approach [37]. Though these results are clearly related to the continuous ballot theorems, they are not as elegant, and neither their statements nor their proofs bring to mind the ballot theorem. It seems that considering separable stationary processes in their full generality does not impose enough structure for it to be possible to prove these results via straightforward generalization of the discrete equivalents.

A beautiful perspective on the ballot theorem appears by considering *random measures* instead of stochastic processes. Given an almost surely nondecreasing separable stochastic process $\{X_t, 0 \leq t \leq r\}$, fixing any element ω of the underlying probability space Ω yields a sample function x_ω. By our assumptions on the stochastic process, almost every sample function x_ω yields a measure μ_ω on $[0, r]$, where $\mu_\omega[0, b] = x_\omega(b) - x_\omega(a)$. This allows us to define a "random" measure μ on $[0, r]$; μ is a function with domain Ω, $\mu(\omega) = \mu_\omega$, and for almost all $\omega \in \Omega$, $\mu(\omega)$ is a measure on $[0, r]$. If x_ω is a nondecreasing step function, then μ_ω has countable support, so is *singular* with respect to the Lebesgue measure (i.e. the set of points which have positive μ_ω-measure has *Lebesgue* measure 0); if this holds for almost all ω then μ is an "almost surely singular" random measure.

We have just seen an example of a random measure; we now turn to a more precise definition. Given a probability space $\mathcal{S} = (\Omega, \Sigma, \mathbf{P})$, a random measure on a possibly infinite interval $T \subset \mathbb{R}$ is a function μ with domain $\Omega \times T$ satisfying that for all $r \in T$, $\mu(\cdot, r)$ is a random variable in $\mathcal{S}$, and for almost all $\omega \in \Omega$, $\mu(\omega, \cdot)$ is a measure on T. A random measure μ is almost surely singular if for almost all $\omega \in \Omega$, $\mu(\omega, \cdot)$ is a measure on T singular with respect to the Lebesgue measure. (This definition hides some technicality; in particular, for the definition to be useful it is key that the set of ω for which μ is singular is itself a measurable set! See [18] for details.) A random measure μ on $\mathbb{R}^+$, say, is stationary if for all t, letting $X_{t,i} = \mu(\cdot, (i + 1)/t) - \mu(\cdot, i/t)$, the family $\{X_{t,i} \mid i \in \mathbb{N}\}$ is composed of independent identically distributed random variables; stationarity for finite intervals is defined similarly.

This perspective can be used to generalize Theorem 5. Konstantopoulos has done so, as well as providing a beautiful proof using a continuous analog of the reflection principle [22]. The most powerful theorem along these lines to date is due to Kallenberg. To a given stationary random measure μ

defined on $T \subseteq \mathbb{R}^+$ we associate a random variable I called the *sample intensity* of μ. (Intuitively, I is the random average number of points in an arbitrary measurable set $B \subset T$ of positive finite measure, normalized by the measure of B. For a formal definition, see [17, Chapter 11].)

Theorem 6 ([18]). *Let μ be an almost surely singular, stationary random measure on $T = \mathbb{R}^+$ or $T = (0, 1]$ with sample intensity I and let $X_t = \mu(\cdot, t) - \mu(\cdot, 0)$ for $t \in T$. Then there exists a uniform $[0, 1]$ random variable U independent from I such that*

$$\sup_{t \in T} \frac{X_t}{t} = \frac{I}{U} \quad almost~surely.$$

It turns out that if $T = (0, 1]$ then conditional upon the event that $X_1 = m$, $I = m$ almost surely. It follows that in this case $\mathbf{P}\big\{ \sup_{t \in T} \frac{X_t}{t} \leq 1 \mid X_1 \big\} = \max\{1 - X_1, 0\}$. Similarly, if $T = \mathbb{R}^+$ and $\frac{X_t}{t} \to m$ almost surely as $t \to \infty$, then $I = m$ almost surely, so in this case $\mathbf{P}\big\{ \sup_{t \in T} \frac{X_t}{t} \leq 1 \big\} = \max\{1 - m, 0\}$. This theorem can thus be seen to include continuous generalizations of both Theorem 2 and Theorem 3.

Kallenberg has also proved the following as a corollary of Theorem 6 (this is a slight reformulation of his original statement, which applied to infinite sequences):

Theorem 7. *If X is a real random variable with maximum value 1 and $\{X_1, X_2, \ldots, X_n\}$ are iid copies of X with corresponding partial sums $\{0 = S_0, S_1, \ldots, S_n\}$, then*

$$\mathbf{P}\{S_i > 0 \forall 1 \leq i \leq n \mid S_n\} \geq \frac{S_n}{n}.$$

It is worth comparing this theorem with Theorem 2; the theorems are almost identical, but Theorem 7 relaxes the integrality restriction at the cost of replacing the equality of Theorem 2 with an inequality.

1.3. Outline

To date, Theorem 7 is the only ballot-style result which has been proved for random walks that may take non-integer values. Paraphrasing Harry Kesten [20], the goal of our research is to move towards making ballot theorems part of "the general theory of random walks" – part of the body of results that hold for *all* random walks (with independent identically distributed steps), regardless of the precise distribution of their steps. We succeed in proving ballot-style theorems that hold for a broad class of random walks, including all random walks that can be renormalized to converge in distribution to a normal random variable. A *truly* general ballot theorem, however, remains beyond our grasp.

In Section 2 we discuss in what sense existing ballot theorems such as those presented in Section 1 are optimal, and what sorts of "general ballot theorems" it makes sense to search for in light of this optimality. In Section 3 we demonstrate our approach in a restricted setting and prove a weakening of our main result. This allows us to highlight the key ideas behind our general ballot theorems without too much notational and technical burden. In Section 4, we sketch the main ideas required to strengthen the presented result. Finally, in Section 5 we address the limits of our approach and suggest some avenues for future research.

2. General Ballot Theorems

The aim of our research is to prove analogs of the discrete-time ballot theorems of Section 1 for more general random variables. The Theorems of Section 1.1 all have two restrictions: (1) they apply only to integer-valued random variables, and (2) they apply only to random variables that are bounded from one or both sides. (In the continuous-time setting, the restriction that the stochastic processes are almost surely integer-valued, increasing step functions is of the same flavour.) In this section we investigate what ballot-style theorems can be proved when such restrictions are removed.

The restrictions (1) and (2) are necessary for the results of Section 1.1 to hold. Suppose, for example, that we relax the condition of Theorem 2 requiring that the variables are bounded from above by $+1$. If X takes every value in $\mathbb{N}$ with positive probability, then $\mathbf{P}\{S_i > 0 \forall 1 \leq i \leq n \mid S_n = n\} < 1$, so the conclusion of the theorem fails to hold. For a more explicit example, let X be any random variable taking values ± 1, ± 4 and define the corresponding cyclically interchangeable sequence and random walk. For $S_3 = 2$ to occur, we must have $\{X_1, X_2, X_3\} = \{4, -1, -1\}$. In this case, for $S_i > 0$, $i = 1, 2, 3$ to occur, X_1 must equal 4. By cyclic interchangeability, this occurs with probability 1/3, and not 2/3, as Theorem 2 would suggest. This shows that the boundedness condition (2) is required. If we relax the integrality condition (1), we can construct a similar example where the conclusions of Theorem 2 do not hold.

Since the results of Section 1.1 can not be directly generalized to a broader class of random variables, we seek conditions on the distribution of X so that the bounds of that section have the correct order, i.e., so that $\mathbf{P}\{S_i > 0\ \forall\ 1 \le i \le n \mid S_n = k\} = \Theta(k/n)$. (When we consider random variables that are not necessarily integer-valued, the right conditioning will in fact be on an event such as $\{k \le S_n < k+1\}$ or something similar.) How close we can come to this conclusion will depend on what restrictions on X we are willing to accept. It turns out that a statement of this flavour holds for the *mean zero* random walk $S_n^0 = S_n - n\mathbf{E}X$ as long as there is a sequence $\{a_n\}_{n \ge 0}$ for which $(S_n - n\mathbf{E}X)/a_n$ converges to a non-degenerate normal distribution (in this case, we say that X is in the *range of attraction of the normal distribution* and write $X \in \mathcal{D}$; for example, the classical central limit theorem states that if $\mathbf{E}\{X^2\} < \infty$ then we may take $a_n = \sqrt{n}$ for all n.) For the purposes of this expository article, however, we shall impose a slightly stronger condition than that stated above.

From this point on, we restrict our attention to sums of mean zero random variables. We note this condition is in some sense necessary in order for the results we are hoping for to hold. If $\mathbf{E}X \ne 0$ – say $\mathbf{E}X > 0$ – then it is possible that X is non-negative, so the only way for $S_n = 0$ to occur is that $X_1 = \cdots = X_n = 0$, and so $\mathbf{P}\{S_i > 0\ \forall\ 1 \le i \le n \mid S_n = 0\} = 0$, and not $\Theta(1/n)$ as we would hope from the results of Section 1.

3. Ballot Theorems for Closely Fought Elections

One of the most basic questions a ballot theorem can be said to answer is: given that an election resulted in a *tie,* what is the probability that one of the candidates had the lead at every point aside from the very beginning and the very end. In the language of random walks, the question is: given that $S_n = 0$, what is the probability that S does not return to 0 or change sign between time 0 and time n? Erik Sparre Andersen has studied the conditional behavior of random walks given that $S_n = 0$ in great detail, in particular deriving beautiful results on the distribution of the maximum, the minimum, and the amount of time spent above zero. Much of the next five paragraphs can be found in [1], for example, in slightly altered terminology.

We call the event that S_n does not return to zero or change sign before time n, $Lead_n$. We can easily bound $\mathbf{P}\{Lead_n \mid S_n = 0\}$ using the fact that $X_1, \ldots, X_n$ are interchangeable. If we condition on the multiset of outcomes $\{X_1, \ldots, X_n\} = \{x_{\sigma(1)}, \ldots, x_{\sigma(n)}\}$, and then choose a uniformly random cyclic permutation σ and a uniform element i of $\{1, \ldots, n\}$, then the interchangeability of $X_1, \ldots, X_n$ implies that $\big(x_{\sigma(i)}, \ldots, x_{\sigma(n)}, x_{\sigma(1)}, \ldots, x_{\sigma(i-1)}\big)$ has the same distribution as if we had sampled directly from $(X_1, \ldots, X_n)$.

Letting $s_j = \sum_{k=1}^{j-1} x_{\sigma(k)}$, in order for $Lead_n$ to occur given that $S_n = 0$, it must be the case that s_i is either the unique maximum or the unique minimum among $\{s_1, \ldots, s_n\}$. The probability that this occurs is at most $2/n$ as it is exactly $2/n$ if there are unique maxima and minima, and less if either the maximum or minimum is not unique. Therefore,

$$(1) \qquad\qquad \mathbf{P}\{Lead_n \mid S_n = 0\} \le \frac{2}{n}.$$

On the other hand, the sequence certainly has *some* maximum (resp. minimum) s_i, and if $X_1 = x_i$ then S_j is always non-positive (resp. non-negative). Denoting this event by $Nonpos_n$ (resp. $Nonneg_n$), we therefore have

$$(2) \qquad \mathbf{P}\{Nonpos_n \mid S_n = 0\} \ge \frac{1}{n} \quad \text{and} \quad \mathbf{P}\{Nonneg_n \mid S_n = 0\} \ge \frac{1}{n}.$$

If $S_n = 0$ then the $(n-1)$ renormalized random variables given by $X_i' = X_{i+1} + X_1/(n-1)$ satisfy $(n-1)S_{n-1}' = (n-1)\sum_{i=1}^{n-1} X_i' = (n-1)\sum_{i=1}^{n} X_i = 0$. If $X_1 > 0$ and none of the *renormalized* partial sums are negative, then $Lead_n$ occurs. The renormalized random variables are still interchangeable (see [1, Lemma 2] for a proof of this easy fact), so we may apply the second bound of (2) to obtain

$$\mathbf{P}\{Lead_n \mid S_n = 0, \ X_1 > 0\} \ge \frac{1}{n-1}.$$

An identical argument yields the same bound for $\mathbf{P}\{Lead_n \mid S_n = 0, X_1 < 0\}$, and combining these bounds yields

$$\mathbf{P}\{Lead_n \mid S_n = 0\} \ge \mathbf{P}\{Lead_n \mid S_n = 0, \ X_1 \ne 0\}\mathbf{P}\{X_1 \ne 0 \mid S_n = 0\}$$

$$\ge \frac{1 - \mathbf{P}\{X_1 = 0 \mid S_n = 0\}}{n-1}.$$

As long as $\mathbf{P}\{X_1 = 0 \mid S_n = 0\} < 1$, this yields that $\mathbf{P}\{Lead_n \mid S_n = 0\} \geq \alpha/n$ for some $\alpha > 0$. By interchangeability, it is easy to see that $\mathbf{P}\{X_1 = 0 \mid S_n = 0\}$ is bounded uniformly away from 1 for large n, as long as $S_n = 0$ does not imply that $X_1 = \cdots = X_n = 0$ almost surely. (Note, however, that there *are* cases where $\mathbf{P}\{X_1 = 0 \mid S_n = 0\} = 1$, for example if the X_i only take values in the non-negative integers and in the negative multiples of $\sqrt{2}$.)

Sparre Andersen's approach gives a necessary and sufficient, though not terribly explicit, condition for $\mathbf{P}\{Lead_n \mid S_n = 0\} = \Theta(1/n)$ to hold. Philosophically, in order to make ballot theorems part of the "general theory of random walks", we would like necessary and sufficient conditions on the distribution of X_1 for $\mathbf{P}\{Lead_n \mid S_n = k\} = \Theta(k/n)$ for *all* $k = O(n)$. Even more generally, we may ask: what are sufficient conditions on the structure of a multiset S of n numbers to ensure that if the elements of the multiset sum to k, then in a uniformly random permutation of the set, all partial sums are positive with probability of order k/n? In the remainder of the section, we focus our attention on sets S whose elements are sampled independently from a mean-zero probability distribution, i.e., they are the steps of a mean-zero random walk. (We remark that it is possible to apply parts of our analysis to sets S that do not obey this restriction, but we will not pursue such an investigation here.) We will derive *sufficient* conditions for such bounds to hold in the case that $k = O(\sqrt{n})$; it turns out that for our approach to succeed it suffices that the step size X is in the range of attraction of the normal distribution, though our best result requires slightly stronger moment conditions on X than those of the classical central limit theorem.

Before stating our generalized ballot theorems, we need one additional definition. We say a variable X has *period* $d > 0$ if dX is an integer random variable and d is the smallest positive real number for which this holds; in this case X is called a *lattice* random variable, otherwise X is *non-lattice*. We can prove the following:

Theorem 8. *Suppose X satisfies $\mathbf{E}X = 0$, $\mathbf{Var}\{X\} > 0$, $\mathbf{E}\{X^{2+\alpha}\} < \infty$ for some $\alpha > 0$, and X is non-lattice. Then for any fixed $A > 0$, given independent random variables $X_1, X_2, \ldots$ distributed as X with associated partial sums $S_i = \sum_{j=1}^{i} X_j$, for all k such that $0 \leq k = O(\sqrt{n})$,*

$$\mathbf{P}\{k \leq S_n \leq k + A, \ S_i > 0 \ \forall \ 0 < i < n\} = \Theta\left(\frac{k+1}{n^{3/2}}\right).$$

Theorem 9. *Suppose X satisfies $\mathbf{E}X = 0$, $\mathbf{Var}\{X\} > 0$, $\mathbf{E}\{X^{2+\alpha}\} < \infty$ for some $\alpha > 0$, and X is a lattice random variable with period d. Then given independent random variables $X_1, X_2, \ldots$ distributed as X with associated partial sums $S_i = \sum_{j=1}^{i} X_j$, for all k such that $0 \le k = O(\sqrt{n})$ and such that k is a multiple of $1/d$,*

$$\mathbf{P}\{S_n = k, \ S_i > 0 \ \forall \ 0 < i < n\} = \Theta\left(\frac{k+1}{n^{3/2}}\right).$$

From these theorems, we may derive "true" (conditional) ballot theorems as corollaries, at least in the case that $k = O(\sqrt{n})$. The following result was proved in [27], and is the tip of an iceberg of related results. Let Φ be the density function a $\mathcal{N}(0,1)$ random variable.

Theorem 10. *Suppose S_n is a sum of independent, identically distributed random variables distributed as X with $\mathbf{E}X = 0$, and there is a constant a such that $S_n/a\sqrt{n}$ converges to a $\mathcal{N}(0,1)$ random variable. If X is non-lattice let B be any bounded set; then for any $h \in B$ and $x \in R$*

$$\mathbf{P}\{|S_n - x| \le h/2\} = \frac{h\Phi(x/a\sqrt{n})}{a\sqrt{n}} + o(1/\sqrt{n}).$$

Furthermore, if X is a lattice random variable with period d, then for any $x \in \{n/d \mid n \in \mathbb{Z}\}$,

$$\mathbf{P}\{S_n = x\} = \frac{\Phi(x/a\sqrt{n})}{a\sqrt{n}} + o(1/\sqrt{n}).$$

In both cases, $\sqrt{n}o(1/\sqrt{n}) \to 0$ as $n \to \infty$ uniformly over all $x \in \mathbb{R}$ and $h \in B$.

Together with Theorem 8 this immediately yields:

Corollary 11. *Under the conditions of Theorem 8,*

$$\mathbf{P}\{S_i > 0 \ \forall \ 0 < i < n \mid k \le S_n \le k + A\} = \Theta\left(\frac{k+1}{n}\right).$$

Similarly, combining Theorem 9 with Theorem 10, we have

Corollary 12. *Under the conditions of Theorem 9,*

$$\mathbf{P}\{S_i > 0 \ \forall \ 0 < i < n \mid S_n = k\} = \Theta\left(\frac{k+1}{n}\right).$$

As we remarked above, the approach we are about to sketch can also be used to prove a ballot theorem under the weaker restriction that X is in the range of attraction of the normal distribution, at the cost of replacing the bound $\Theta\left(\frac{k+1}{n}\right)$ by the bound $\frac{k+1}{n^{1+o(1)}}$; for the sake of brevity and clarity we will not discuss the rather minor modifications to our approach that are needed to handle this case. Furthermore, for the purposes of this expository article, we shall not prove Theorems 8 or 9 in their full generality or strength, instead restricting our attention to a special case which allows us to highlight the key elements of our proofs. Finally, we shall provide a detailed explanation of only the upper bound, after which we shall briefly discuss our approach to the lower bound. We will prove:

Theorem 13. *Suppose X satisfies $\mathbf{E}X = 0$, $\mathbf{Var}\{X\} > 0$, $|X| < C$, and X is non-lattice. Then for any fixed $A > 0$, given independent random variables $X_1, X_2, \ldots$, distributed as X with associated partial sums $S_i = \sum_{j=1}^{i} X_j$, for all $0 \le k = o\left(\sqrt{n/\log n}\right)$,*

$$\mathbf{P}\{k \le S_n \le k + A, S_i > 0 \ \forall \ 0 < i < n\} = O\left(\frac{(k+1)\log n}{n^{3/2}}\right).$$

Of course, a conditional ballot theorem that is correspondingly weaker than Corollary 11 follows by combining Theorem 13 with Theorem 10. We remark that in cases where Theorem 7 applies, it provides a lower bound on $\mathbf{P}\{k \le S_n \le k + A, S_i > 0 \ \forall \ 0 < i < n\}$ of the same order as the upper bound of Theorem 13. *From this point forward, X will always be a random variable satisfying the conditions in Theorem 13, and $X_1, X_2, \ldots$, will be independent copies of X with corresponding partial sums $S_1, S_2, \ldots$.*

To begin providing an intuition of our approach, we first remark that if $S_i > 0 \ \forall 0 < i < n$ is to occur, then for any r, letting T be the first time $t \ge 1$ that $S_t > r$ or $S_t \le 0$, we have either $S_T > 0$ or $T > n$. (We will end up choosing the value r so that $T = o(n)$ except with negligibly small probability, so to bound the previous probability we shall essentially need to bound the probability that $S_T > 0$, i.e., that the walk "stays positive". We will see shortly that Wald's identity implies that $\mathbf{P}\{S_T > 0\} = O(1/r)$.

We may impose a similar constraint on the "other end" of the random walk S, by letting S' be the *negative reversed* random walk given by $S'_0 = 0$, and for $i > 0$, $S'_{i+1} = S'_i - X_{n-i}$ (it will be useful to think of S'_i as being defined even for $i > n$, which we may do by letting $X_0, X_{-1}, \ldots$ be independent copies of X). If $S_i > 0 \ \forall 0 < i < n$ and $k \leq S_n \leq k + C$ are to occur, then letting T' be the first time t that $S'_t \leq -(k + A)$ or $S'_t > r - (k + A)$, it must be the case that either $S'_{T'} > 0$ or $T' > n$. (Again, we will choose r so that $T' = o(n)$ with extremely high probability.)

Finally, in order for $k \leq S_n \leq k + A$ to occur, the two ends of the random walk must "match up". We may make this mathematically precise by noting that as long as $T < n - T'$, we may write S_n as $S_T + (S_{n-T'} - S_T) - S'_{T'}$, and may thus write the condition $k \leq S_n \leq k + A$ as

$$k + S'_{T'} - S_T \leq (S_{n-T'} - S_T) \leq k + A + S'_{T'} - S_T.$$

If $T + T'$ is at most $n/2$, say, then $S_{n-T'} - S_T$ is the sum of at least $n/2$ random variables. In this case, the classical central limit theorem suggests that $S_{n-T'} - S_T$ should "spread itself out" over a range of order $\sqrt{n}$, and essentially this fact will allow us to show that the two ends "meet up" with probability $O(1/\sqrt{n})$.

3.1. Staying positive

To begin formalizing the above sketch, let us first turn to bounds on the probabilities of the events $S_T > 0$ and $S'_{T'} > 0$.

Lemma 14. *Fix $r > 0$ and $s \geq 0$, and let $T_{r,s}$ be the first time $t > 0$ that either $S_t > r$ or $S_t \leq -s$. Then $\mathbf{P}\{S_{T_{r,s}} > 0\} \leq (s + C)/(r + s + C)$.*

Proof. We first remark that $\mathbf{E}T_{r,s}$ is finite; this is a standard result that can be found in, e.g., [11, Chapter 14.4], and we shall also rederive this result a little later. Thus, by Wald's identity, we have that $\mathbf{E}S_{T_{r,s}} = \mathbf{E}T_{r,s}\mathbf{E}X_1 = 0$, and letting Pos_r denote the event $\{S_{T_{r,s}} > 0\}$; we may therefore write

$$(3) \quad 0 = \mathbf{E}S_{T_{r,s}} = \mathbf{E}\{S_{T_{r,s}} \mid Pos_r\}\mathbf{P}\{Pos_r\} + \mathbf{E}\{S_{T_{r,s}} \mid \overline{Pos_r}\}\mathbf{P}\{\overline{Pos_r}\}.$$

By definition $\mathbf{E}\{S_T \mid Pos_r\} \geq r$, and by our assumption that X has absolute value at most C, we have $\mathbf{E}\{S_T \mid \overline{Pos_r}\} \geq -(s + C)$. Therefore

$$0 \geq r\mathbf{P}\{Pos_r\} - (s + C)\mathbf{P}\{\overline{Pos_r}\} = r\mathbf{P}\{Pos_r\} - (s + C)\big(1 - \mathbf{P}\{Pos_r\}\big),$$

and rearranging the latter inequality yields that $\mathbf{P}\{Pos_r\} \leq (s + C)/(r + s + C)$. $\blacksquare$

As an aside, we note that may easily derive a lower bound of the same order for $\mathbf{P}\{Pos_r\}$ in a similar fashion; we first observe that $\mathbf{E}\{S_{T_{r,s}} \mid Pos_r\} < r + C$. Similarly, $\mathbf{E}\{S_{T_{r,s}} \mid \overline{Pos_r}\} \leq -s$, and using the fact that X has zero mean and positive variance, it is also easy to see that there is $\varepsilon > 0$ such that in fact $\mathbf{E}\{S_{T_{r,s}} \mid \overline{Pos_r}\} \leq -\max\{\varepsilon, s\}$. Combining (3) these two bounds, we thus have

$$0 < (r + C)\mathbf{P}\{Pos_r\} - \max\{\varepsilon, s\}\mathbf{P}\{\overline{Pos_r}\}$$

$$= (r + C)\mathbf{P}\{Pos_r\} - \max\{\varepsilon, s\}\big(1 - \mathbf{P}\{Pos_r\}\big),$$

so $\mathbf{P}\{Pos_r\} \geq \max\{\varepsilon, s\}/\big(r + C + \max\{\varepsilon, s\}\big)$. Lemma 14 immediately yields the bounds we require for $\mathbf{P}\{S_T > 0\}$ and $\mathbf{P}\{S'_{T'} > 0\}$; next we show that for a suitable choice of r, with extremely high probability, both T and T' are $o(n)$.

3.2. The time to exit a strip

For $r \geq 0$, we consider the first time t for which $|S_t| \geq r$, denoting this time T_r. We prove

Lemma 15. *There is $B > 0$ such that for all $r \geq 1$, $\mathbf{E}T_r \leq Br^2$ and for all integers $k \geq 1$, $\mathbf{P}\{T_r \geq kBr^2\} \leq 1/2^k$.*

This is an easy consequence of a classical result on how "spread out" sums of independent identically distributed random variables become (which we will also use later when bounding the probability that the two ends of the random walk "match up"). The version we present can be found in [19]:

Theorem 16. *For any family of independent identically distributed real random variables $X_1, X_2, \ldots$ with positive, possibly infinite variance and associated partial sums $S_1, S_2, \ldots$, there is a constant c depending only on the distribution of X_1 such that for all n,*

$$\sup_{x \in \mathbb{R}} \mathbf{P}\{x \leq S_n \leq x + 1\} \leq c/\sqrt{n}.$$

Proof of Lemma 15. Observe that the expectation bound follows directly from the probability bound, since if the probability bound holds then we have

$$\mathbf{E}T_r \leq \sum_{j=0}^{\infty} \mathbf{P}\{T_r \geq j\} \leq \sum_{i=0}^{\infty} \lceil Br^2 \rceil \mathbf{P}\big\{T_r > i\lceil Br^2 \rceil\big\} \leq \sum_{i=0}^{\infty} \frac{\lceil Br^2 \rceil}{2^i} = 2\lceil Br^2 \rceil,$$

which establishes the expectation bound with a slightly changed value of B. It thus remains to prove the probability bound. By Theorem 16, there is $c > 0$ (and we can and will assume $c > 1$) such that

$$(4) \qquad \mathbf{P}\big\{|S_{\lceil 128c^2r^2\rceil}| \le 2r\big\} \le \sum_{i=\lfloor -2r\rfloor}^{\lfloor 2r\rfloor} \mathbf{P}\big\{i \le S_{\lceil 128c^2r^2\rceil} \le i+1\big\}$$

$$\le (4r+1)\frac{c}{\sqrt{\lceil 128c^2r^2\rceil}} < \frac{1}{2},$$

the last inequality holding as $c > 1$ and $r > 1$. Let $t^* = \lceil 128c^2r^2\rceil$ – then $\mathbf{P}\{T_r > t^*\} \le 1/2$. We use this fact to show that for any positive integer k, $\mathbf{P}\{T_r > kt^*\} \le 1/2^k$, which will establish the claim with $B = 128c^2 + 1$, for example. We proceed by induction on k, having just proved the claim for $k = 1$. We have

$$\mathbf{P}\big\{T_r > (k+1)t^*\big\} = \mathbf{P}\big\{T_r > (k+1)t^* \cap T > kt\big\}$$

$$= \mathbf{P}\big\{T_r > (k+1)t^* \mid T_r > kt^*\big\}\mathbf{P}\big\{T_r > kt\big\}$$

$$= \frac{1}{2^k}\cdot\mathbf{P}\big\{T_r > (k+1)t^* \mid T_r > kt^*\big\},$$

by induction. It remains to show that $\mathbf{P}\big\{T_r > (k+1)t^* \mid T_r > kt^*\big\} \le 1/2$. If $T_r > kt^*$ then by the strong Markov property we may think of restarting the random walk at time kt^*. Whatever the value of S_{kt^*}, if the restarted random walk exits $[-2r, 2r]$ then the original random walk exits $[-r, r]$, so this inequality holds by (4). This proves the lemma. ∎

This bound on the time to exit a strip is the last ingredient we need; we now turn to the proof of Theorem 13.

3.3. Proof of Theorem 13

Fix $A > 0$ as in the statement of the theorem. For $r \ge 1$ we denote by T_r the first time t that $|S_t| \ge r$. We let S' be the negative reversed random walk given by $S'_0 = 0$, and for $i > 0$, $S'_{i+1} = S'_i - X_{n-i}$ (again as above, we define S'_i for $i > n$ by letting $X_0, X_{-1}, \ldots$ be independent copies of X), and let T'_r be the first time t that $|S'_t| \ge r$. We choose B such that for all $r \ge 1$ and

and for all integers $k \geq 1$, $\mathbf{P}\{T_r \geq kBr^2\} \leq 1/2^k$ and $\mathbf{P}\{T'_r \geq kBr^2\} \leq 1/2^k$ – such a choice exists by Lemma 15.

Choose $r^* = \left\lfloor \sqrt{n/9B \log n} \right\rfloor$ – then with $k = \lceil 2 \log n \rceil < 2 \log n + 1$, it is the case that

$$kB(r^*)^2 \leq \frac{kBn}{9B \log n} < \frac{(2 \log n + 1)n}{9 \log n} < \frac{n}{4},$$

so $\mathbf{P}\{T_{r^*} \geq n/4\} \leq 1/2^k \leq 1/n^2$, and similarly $\mathbf{P}\{T'_{r^*} \geq n/4\} \leq 1/n^2$.

Next let T be the first time t that $S_t > r^*$ or $S_t \leq 0$, and let T' be the first time t that $S'_t > r^* - (k + A)$ or $S'_t \leq -(k + A)$. It is immediate that $T < T_{r^*}$. Furthermore, since $k = o(\sqrt{n/\log n})$, $(k + A) < r^*$ for n large enough, so $r^* > r^* - (k + A) > 0 > -(k + A) > -r^*$; it follows that $T' < T'_{r^*}$. These two inequalities, combined with the bounds for T_{r^*} and T'_{r^*}, yield

$$(5) \qquad \mathbf{P}\left\{T \geq \frac{n}{4}\right\} \leq \frac{1}{n^2} \qquad \text{and} \qquad \mathbf{P}\left\{T' \geq \frac{n}{4}\right\} \leq \frac{1}{n^2}$$

Let E be the event that $k \leq S_n \leq k + A$, and $S_i > 0$ for all $0 < i < n$ – we aim to show that $\mathbf{P}\{E\} = O\big((k + 1) \log n/n^{3/2}\big)$. In order that E occur, it is necessary that either $T \geq n/4$ or $T' \geq n/4$ (we denote the union of these two events by D), or that the following three events occur (these events control the behavior of the beginning, end, and middle of the random walk, respectively):

E_1: $S_T > 0$ and $T < n/4$,

E_2: $S'_{T'} > 0$ and $T' < n/4$,

E_3: letting $\Delta = S'_{\lfloor n/4 \rfloor} - S_{\lfloor n/4 \rfloor}$, we have $k + \Delta \leq S_{n - \lfloor n/4 \rfloor} - S_{\lfloor n/4 \rfloor} \leq k + \Delta + A$.

It follows that

$$\mathbf{P}\{E\} \leq \mathbf{P}\{D\} + \mathbf{P}\{E_1, E_2, E_3\}.$$

Furthermore, $\mathbf{P}\{D\} \leq \mathbf{P}\{T \geq n/4\} + \mathbf{P}\{T' \geq n/4\} \leq 2/n^2$ by (5), so to show that $\mathbf{P}\{E\} = O(\log n/n^{3/2})$, it suffices to show that $\mathbf{P}\{E_1, E_2, E_3\} = O(\log n/n^{3/2})$; we now demonstrate that this latter bound holds, which will complete the proof.

The events E_1 and E_2 are independent, as E_1 is determined by the random variables $X_1, \ldots, X_{\lfloor n/4 \rfloor}$, and E_2 is determined by the random variables $X_{n - \lfloor n/4 \rfloor + 1}, \ldots, X_n$. Furthermore, in the notation of Lemma 14, T is an event of the form $T_{r,s}$ with $r = r^*$, $s = 0$; it follows that $\mathbf{P}\{S_T > 0\} \leq$

$C/(r^* + C)$. Since S' has step size $-X$ and $|-X| < C$, we may also apply Lemma 14 to the walk S' with the choice $r = r^* - k + C$, $s = k + C$, to obtain the bound $\mathbf{P}\{S'_{T'} > 0\} \le (k + 2C)/(r + k + 2C)$. Therefore

$$
\begin{aligned}
(6) \qquad \mathbf{P}\{E_1, E_2, E_3\} &= \mathbf{P}\{E_3 \mid E_1, E_2\}\mathbf{P}\{E_1\}\mathbf{P}\{E_2\} \\[2mm]
&\le \mathbf{P}\{E_3 \mid E_1, E_2\}\mathbf{P}\{S_T > 0\}\mathbf{P}\{S'_{T'} > 0\} \\[2mm]
&\le \mathbf{P}\{E_3 \mid E_1, E_2\} \cdot \frac{C(k + 2C)}{r^*(r^* + k + 2C)} \\[2mm]
&< \mathbf{P}\{E_3 \mid E_1, E_2\} \cdot \frac{2C^2(k + 1)}{(r^*)^2}.
\end{aligned}
$$

To bound $\mathbf{P}\{E_3 \mid E_1, E_2\}$, we observe that

$$
\begin{aligned}
(7) \qquad \mathbf{P}\{E_3 \mid E_1, E_2\} &\le \sup_{x \in \mathbb{R}} \mathbf{P}\{E_3 \mid E_1, E_2, \Delta = x\} \\[2mm]
&= \sup_{x \in \mathbb{R}} \mathbf{P}\big\{k + x \le S_{n - \lfloor n/4 \rfloor} - S_{\lfloor n/4 \rfloor} \le k + x + A \mid E_1, E_2, \ \Delta = x\big\}.
\end{aligned}
$$

Furthermore, the event that $k + x \le S_{n - \lfloor n/4 \rfloor} - S_{\lfloor n/4 \rfloor} \le k + x + A$ is independent from E_1, E_2, and from the event that $\Delta = x$, as the former event is determined by the random variables $X_{\lfloor n/4 \rfloor + 1}, \ldots, X_{n - \lfloor n/4 \rfloor}$, and the latter events are determined by the random variables $X_1, \ldots, X_{\lfloor n/4 \rfloor}, X_{n - \lfloor n/4 \rfloor + 1}, \ldots, X_n$. It follows from this independence, (7), and the strong Markov property that

$$
\begin{aligned}
(8) \quad \mathbf{P}\{E_3 \mid E_1, E_2\} &\le \sup_{x \in \mathbb{R}} \mathbf{P}\big\{k + x \le S_{n - \lfloor n/4 \rfloor} - S_{\lfloor n/4 \rfloor} \le k + x + A\big\} \\[2mm]
&= \sup_{x \in \mathbb{R}} \mathbf{P}\big\{k + x \le S_{n - 2\lfloor n/4 \rfloor} \le k + x + A\big\}. \\[2mm]
&\le (A + 1)\sup_{x \in \mathbb{R}} \mathbf{P}\big\{k + x \le S_{n - 2\lfloor n/4 \rfloor} \le k + x + 1\big\},
\end{aligned}
$$

the last inequality holding by a union bound. By Theorem 16, there is $c > 0$ depending only on X, such that

$$
\sup_{x \in \mathbb{R}} \mathbf{P}\big\{x \le S_{n - 2\lfloor n/4 \rfloor} \le x + 1\big\} \le \frac{c}{\sqrt{n - 2\lfloor n/4 \rfloor}} \le \frac{\sqrt{2}c}{\sqrt{n}},
$$

and it follows from this fact and from (8) that

$$
\mathbf{P}\{E_3 \mid E_1, E_2\} \le \frac{\sqrt{2}c(A + 1)}{\sqrt{n}}.
$$

Combining this bound with (6) yields

$$\mathbf{P}\{E_1, E_1, E_3\} \leq \frac{\sqrt{2}c(A+1)}{\sqrt{n}} \cdot \frac{2C^2(k+1)}{(r^*)^2} = \frac{2\sqrt{2}c(A+1)C^2(k+1)}{(r^*)^2\sqrt{n}}.$$

Since $r^* = \lfloor \sqrt{n/9B\log n} \rfloor$, $(r^*)^2 \geq n/10B\log n$ for n large enough, so letting $a = 2\sqrt{2}(A+1)cC^2 \cdot 10B = O(1)$, we have

$$(9) \qquad \mathbf{P}\{E_1, E_1, E_3\} < \frac{a(k+1)\log n}{n^{3/2}} = O\left(\frac{(k+1)\log n}{n^{3/2}}\right),$$

as claimed.

4. Strengthening Theorem 13

There are two key ingredients needed to move from the upper bound in Theorem 13 to the stronger and more general upper bound in Theorem 8. The first concerns stopping times $T_{r,s}$ of the form seen in Lemma 14. Without the assumption that the step size X is bounded, we have no *a priori* bound on $\mathbf{E}\{S_{T_{r,s}} \mid S_{T_{r,s}} > r\}$ or on $\mathbf{E}\{S_{T_{r,s}} \mid S_{T_{r,s}} \leq -s\}$, so we can not straightforwardly apply Wald's identity to bound $\mathbf{P}\{S_{T_{r,s}} > 0\}$ as we did above.

Griffin and McConnell have proven bounds on $\mathbf{E}\{|S_{T_{r,r}}| - r\}$ (a quantity they call the *overshoot* at r), for random walks with step size X in the domain of attraction of the normal distribution; their results are the best possible in the setting they consider [13]. Their bounds do not directly imply the bounds we need, but we are able to use their results to obtain such bounds using a bootstrapping technique we refer to as a "doubling argument". The key idea behind this argument can be seen by considering a symmetric simple random walk S and a stopping time $T_{3k,0}$, for some positive integer k. Let T be the first time $t > 0$ that $|S_t| = k$. If the event $S_{T_{3k,0}} > 0$ is to occur, it must be the case that $S_T = k$. Next, let T' be the first time $t > T$ that $|S_t - S_T| \geq 2k$ – if $S_{T_{3k,0}} > 0$ is to occur, it must also be the case that $S_{T'} - S_T = 2k$. By the independence of disjoint sections of the random walk, it follows that

$$\mathbf{P}\{S_{T_{3k,0}} > 0\} \leq \mathbf{P}\{S_T = k\}\mathbf{P}\{S_{T'} - S_T = 2k\}.$$

In the notation of Lemma 14, T is a stopping time of the form $T_{k,k}$, and T_2 is a stopping time of the form $T_{2k,2k}$, so we have

$$\mathbf{P}\{S_{T_{3k,0}} > 0\} \le \mathbf{P}\{S_{T_{k,k}} > 0\}\,\mathbf{P}\{S_{T_{2k,2k}} > 0\}\,.$$

Furthermore, for a general random walk S we can use Griffin and McConnell's bounds on the overshoot together with the approach of Lemma 14 to prove bounds $\mathbf{P}\{S_{T_{k,k}} > 0\}$. In general, we consider a sequence of stopping times $T_1, T_2, \ldots$, where T_{i+1} is the first time after T_i that $\left|S_{T_{i+1}} - S_{T_i}\right| \ge 2^i$, and apply Griffin and McConnell's results to bound the probability that the random walk goes positive at each step. By applying their results to such a sequence of stopping times, we are able to ensure that the error in our bounds resulting from the "overshoot" does not accumulate, and thereby prove the stronger bounds we require.

The second difficulty we must overcome is due to the fact that in order to remove the superfluous $\log n$ factor in the bound of Theorem 13, we need to replace the stopping time T_{r^*} (with $r^* = O\left(\sqrt{n/\log n}\,\right)$) by a stopping time $T_{r'}$ (with $r' = \Theta\left(\sqrt{n}\,\right)$). However, for such a value r', $\mathbf{E}T_{r'} = \Theta(n)$, and our upper tail bounds on $T_{r'}$ are not strong enough to ensure that $T_{r'} \le \lfloor n/4 \rfloor$ with sufficiently high probability.

To deal with this problem, we apply the ballot theorem inductively. Instead of stopping the walk at a stopping time $T_{r'}$, we stop the walk *deterministically* at time $t_1 = \lfloor n/4 \rfloor$. In order that $S_n > 0$ for all $0 < i < n$ occur, it must be the case that *either* $T_{r'} \le t_1$ and $S_{T_{r'}} > 0$, *or* there is $0 < k \le r' - C$ such that $k \le S_{t_1} \le k + A$ and additionally, $S_i > 0$ for all $0 < i < t_1$. We bound the probability of the former event using our strengthening of Lemma 14, and bound the probability of the latter event by inductively applying the ballot theorem. Of course, an identical analysis applies to the negative reversed random walk S', and allows us to strengthen our control of the end of the random walk correspondingly.

Finally, we give some idea of our lower bound. We fix some value r' of order $\Theta\left(\sqrt{n}\,\right)$; paralleling the proof of Theorem 13, we let T be the first time t that $S_t > r'$ or $S_t \le 0$, and let T' be the first time t that $S'_t > r' - k$ or $S'_t \le -k$. In order that $k \le S_n \le k + A$, and $S_i > 0$ for all $0 < i < n$, it suffices that the following three events occur (these events control the behavior of the beginning, end, and middle of the random walk, respectively):

E_1: $S_T > 0$, $S_T \le 2r'$, and $T < n/4$,

E_2: $S'_{T'} > 0$, $S'_{T'} \le 2r' - k$, and $T' < n/4$,

E_3: letting $\Delta = S'_{T'} - S_T$, we have $k + \Delta \leq S_{n-T'} - S_T \leq k + \Delta + A$ and $S_i - S_T > -r'$ for all $T < i < n - T'$.

Using an approach very similar to that of Theorem 13, we are able to show that in fact $\mathbf{P}\{E_1\} = \Theta(1/r')$ and that $\mathbf{P}\{E_2\} = \Theta((k+1)/r')$. The two key observations of that allow us to prove a lower bound on $\mathbf{P}\{E_3\}$ are the following:

- Given that E_1 and E_2 occur, $k + \Delta = O(\sqrt{n})$, and so it is not hard to show using Theorem 10 that $\mathbf{P}\{k + \Delta \leq S_{n-T'} - S_T \leq k + \Delta + A \mid E_1, E_2\} = \Theta(1/\sqrt{n})$.

- Since $n - T - T' = O(n)$, we expect the random walk $S_T, S_{T+1}, \ldots, S_{n-T'}$ to have a spread of order $O(\sqrt{n})$. Since $r' = \Theta(\sqrt{n})$, it easy to see (again using Theorem 10, or by the classical central limit theorem) that $S_i - S_T > -r'$ for all $T < i < n - T'$ with probability $\Omega(1)$.

Based on these two observations, we trust that the reader will find it plausible that given E_1 and E_2, the intersection of the events in E_3 occurs with probability $\Theta(1/\sqrt{n})$; in this case, combining our bounds much as in Theorem 13 yields a lower bound on $\mathbf{P}\{E_1, E_2, E_3\}$ of order $(k+1)/(r')^2\sqrt{n} = \Theta((k+1)/n^{3/2})$.

5. Conclusion

In writing this survey, we hoped to convince the reader the theory of ballots is not only rich and beautiful, in-and-of itself, but is also very much alive. Our new results are far from conclusive in terms of when ballot-style behavior can be expected of sums of independent random variables, and more generally of permutations of sets of real numbers. In the final paragraphs, we highlight some of the questions that remain unanswered.

The results of Section 3 are unsatisfactory in that they only yield "true" (conditional) ballot theorems when $S_n = O(\sqrt{n})$. Ideally, we would like such results to hold *whatever* the range of S_n. Two key weaknesses of our approach are that it (a) relies on estimates for $\mathbf{P}\{x \leq S_n \leq x + c\}$ that are based on the central limit theorem, and these estimates are not good enough when S_n is not $O(\sqrt{n})$, and (b) relies on bounds on the "overshoot" that only hold when the step size X is in the range of attraction of the normal distribution, [21] and, independently, [14], have derived necessary and sufficient conditions in order that $\mathbf{P}\{S_{T_{r,r}}\} \to 1/2$ as $r \to \infty$; in particular they show that for any $\alpha < 2$, there are distributions with $\mathbf{E}\{X^{\alpha}\} = \infty$ for which $\mathbf{P}\{S_{T_{r,r}}\} \to 1$. Therefore, we can not expect to use a doubling argument in this case, which seriously undermines our approach.

As we touched upon at various points in the paper, aspects of our technique seem as though they should work for analyzing more general random permutations of sets of real numbers. Since Andersen observed the connection between conditioned random walks and random permutations and [1, 2, 26] pointed out the full generality of Andersen's observations, just about every result on conditioned random walks has been approached from the permutation-theoretic perspective sooner or later. There is no reason our results should not benefit from such an approach.

References

[1]   Erik Sparre Andersen, Fluctuations of sums of random variables, *Mathematica Scandinavica*, **1** (1953), 263–285.

[2]   Erik Sparre Andersen, Fluctuations of sums of random variables ii, *Mathematica Scandinavica*, **2** (1954), 195–223.

[3]   Désiré André, Solution directe du probleme resolu par M. Bertrand, *Comptes Rendus de l'Academie des Sciences*, **105** (1887), 436–437.

[4]   N. Balakrishnan, *Advances in Combinatorial Methods and Applications to Probability and Statistics*, Birkhäuser, Boston, MA, first edition (1997).

[5]   Émile Barbier, Generalisation du probleme resolu par M. J. Bertrand, *Comptes Rendus de l'Academie des Sciences*, **105** (1887), 407.

[6]   J. Bertrand, Observations, *Comptes Rendus de l'Academie des Sciences*, **105** (1887), 437–439.

[7]   J. Bertrand, Sur un paradox analogue au problème de Saint-Pétersburg, *Comptes Rendus de l'Academie des Sciences*, **105** (1887), 831–834.

[8]   J. Bertrand, Solution d'un probleme, *Comptes Rendus de l'Academie des Sciences*, **105** (1887), 369.

[9]   Aryeh Dvoretzky and Theodore Motzkin, A problem of arrangements, *Duke Mathematical Journal*, **14** (1947), 305–313.

[10]   Meyer Dwass, A fluctuation theorem for cyclic random variables, *The Annals of Mathematical Statistics*, **33(4)** (December 1962), 1450–1454.

[11]   William Feller, *An Introduction to Probability Theory and Its Applications, Volume 1*, volume 1. John Wiley & Sons, Inc, third edition (1968).

[12]   I. I. Gikhman and A. V. Skorokhod, *Introduction to the Theory of Random Processes*, Saunders Mathematics Books. W. B. Saunders Company (1969).

[13] Philip S. Griffin and Terry R. McConnell, On the position of a random walk at the time of first exit from a sphere, *The Annals of Probability,* **20(2)** (April 1992), 825–854.

[14] Philip S. Griffin and Terry R. McConnell, Gambler's ruin and the first exit position of random walk from large spheres, *The Annals of Probability,* **22(3)** (July 1994), 1429–1472.

[15] Howard D. Grossman. Fun with lattice-points, *Duke Mathematical Journal,* **14** (1950), 305–313.

[16] A. Hald, *A History of Probability and Statistics and Their Applications before 1750,* John Wiley & Sons, Inc, New York, NY (1990).

[17] Olav Kallenberg, *Foundations of Modern Probability,* Probability and Its Applications, Springer Verlag, second edition (2003).

[18] Olav Kallenberg, Ballot theorems and sojourn laws for stationary processes, *The Annals of Probability,* **27(4)** (1999), 2011–2019.

[19] Harry Kesten, Sums of independent random variables – without moment conditions, *Annals of Mathematical Statistics,* **43(3)** (June 1972), 701–732.

[20] Harry Kesten, Frank spitzer's work on random walks and Brownian motion, *Annals of Probability,* **21(2)** (April 1993), 593–607.

[21] Harry Kesten and R. A. Maller, Infinite limits and infinite limit points of random walks and trimmed sums, *The Annals of Probability,* **22(3)** (1994), 1473–1513.

[22] Takis Konstantopoulos, Ballot theorems revisited, *Statistics & Probability Letters,* **24(4)** (September 1995), 331–338.

[23] Sri Gopal Mohanty, An urn problem related to the ballot problem, *The American Mathematical Monthly,* **73(5)** (1966), 526–528.

[24] Émile Rouché. Sur la durée du jeu, *Comptes Rendus de l'Academie des Sciences,* **106** (1888), 253–256.

[25] Émile Rouché, Sur un problème relatif à la durée du jeu, *Comptes Rendus de l'Academie des Sciences,* **106** (1888), 47–49.

[26] Frank Spitzer, A combinatorial lemma and its applications to probability theory, *Transactions of the American Mathematical Society,* **82(2)** (July 1956), 323–339.

[27] Charles J. Stone, On local and ratio limit theorems, in: *Proceedings of the Fifth Berkeley Symposium on Mathematical Statistics and Probability* (1965), pp. 217–224.

[28] Lajos Takács, Ballot problems, *Zeitschrift für Warscheinlichkeitstheorie und verwandte Gebeite,* **1** (1962), 154–158.

[29] Lajos Takács, A generalization of the ballot problem and its application in the theory of queues, *Journal of the American Statistical Association,* **57(298)** (1962), 327–337.

[30] Lajos Takács, The time dependence of a single-server queue with poisson input and general service times, *The Annals of Mathematical Statistics,* **33(4)** (December 1962), 1340–1348.

[31] Lajos Takács, The distribution of majority times in a ballot, *Zeitschrift für Warscheinlichkeitstheorie und verwandte Gebeite,* **2(2)** (January 1963), 118–121.

[32] Lajos Takács, Combinatorial methods in the theory of dams, *Journal of Applied Probability,* **1(1)** (1964), 69–76.

[33] Lajos Takács, Fluctuations in the ratio of scores in counting a ballot, *Journal of Applied Probability,* **1(2)** (1964), 393–396.

[34] Lajos Takács, A combinatorial theorem for stochastic processes, *Bulletin of the American Mathematical Society,* **71** (1965), 649–650.

[35] Lajos Takács, On the distribution of the supremum for stochastic processes with interchangeable increments, *Transactions of the American Mathematical Society,* **119(3)** (September 1965), 367–379.

[36] Lajos Takács, *Combinatorial Methods in the Theory of Stochastic Processes,* John Wiley & Sons, Inc, New York, NY, first edition (1967).

[37] Lajos Takács, On the distribution of the maximum of sums of mutually independent and identically distributed random variables, *Advances in Applied Probability,* **2(2)** (1970), 344–354.

[38] Lajos Takács, On the distribution of the supremum for stochastic processes, *Annales de l'Institut Henri Poincaré B,* **6(3)** (1970), 237–247.

[39] J. C. Tanner, A derivation of the borel distribution, *Biometrika,* **48(1–2)** (June 1961), 222–224.

L. Addario-Berry
Department of Statistics
University of Oxford
U.K.

B. A. Reed
School of Computer Science
McGill University
Canada
and
Projet Mascotte
I3S (CNRS/UNSA)-INRIA
Sophia Antipolis
France

BOLYAI SOCIETY
MATHEMATICAL STUDIES, 17

Horizons of Combinatorics
Balatonalmádi
pp. 37–66.

STATISTICAL INFERENCE ON RANDOM STRUCTURES

VILLŐ CSISZÁR, LÍDIA REJTŐ and GÁBOR TUSNÁDY

INTRODUCTION

Randomness for a statistician must have some structure. In traditional combinatorics the word random means uniform distribution on a set which may be the set of all graphs with n vertices, the set of all permutations of the numbers $N = (1, 2, \ldots, n)$, the set of all partitions of N, or any other set of simple structure. In practice the statistician meets a subset of the structures and she or he is interested in the question, what was the mechanism which generated the sample. Uniform distribution and independence are shapeless and they have low complexity for catching the character of samples produced by real life situations. In [12] Persi Diaconis investigated a sample consisting of the votes in an election of the American Psychological Association. The sample was investigated by others but without achieving a reasonable goodness of fit, because the present collection of distribution of permutations is not large enough. Investigating the sample we found a hidden property leading to a new class of distributions of permutations.

Classical statistics developed around the multidimensional Gaussian distribution. Even in Euclidean space the family of useful distributions is still meager. On other sample spaces the collection of distributions is much less developed. Graphs appear in applications as structured relations. In many cases rather heavy simplifications are needed for reducing the complexity of the investigated situation to a graph. One source of our interest in graphs is the system of metabolic interactions, which may have some fractal structure: the enzymatic interactions may be leveled, they may be sensitive for situations, their control might be hierarchic. Changes of the concentration of different enzymes in a cell follow their dynamical rule what is reflected imperfectly in the graph of enzymatic interactions.

In modern combinatorics the stochastic method is rapidly extending. We shall use the ideas of papers [7], [8] and [9] written by Christian Borgs, Jennifer Chayes, László Lovász, Vera T. Sós, Balázs Szegedy and Katalin Vesztergombi in defining new classes of random permutations.

SVD of real matrices. Let M be an arbitrary digital picture: a face, a tree, a hill or some other natural object which is not very complicated. Let us suppose that the colours are ordered according to their wave lengths and M is an m times n real matrix containing the codes of the colours in the individual pixels. Let α be a random permutation of the integers $1, \ldots, m$ and β of $1, \ldots, n$. Let

$$R(i, j) = M\big(\alpha(i), \beta(j)\big)$$

be the randomly reordered copy of M. How can we reconstruct M from R?

One possible method is the singular value decomposition (SVD) of R which is invariant under random permutations. The singular values of matrices M and R are identical. We refer to them as the *spectra* of the corresponding matrix. If the picture is simple, then the spectra is J-shaped: there are few large singular values and the corresponding singular vectors concentrate the majority of the relevant information in M. The coordinates of the leading singular vectors of M reflect the topology of M, while the coordinates of the singular vectors of R follow the permutations α, β. It implies that the traveling salesman problem may be easily solved in the space of leading eigenvectors independently of rows and columns.

Microarray analysis. The previous problem arises in microarray analysis where the rows are genes and the columns are the different conditions used in the experiment for controlling the expression of the genes. It is natural to postulate that the genes and conditions are embedded in Euclidean spaces and the expression level is a continuous function of the embedding. Sometime we get well defined clusters when applying SVD of microarray data: clusters in genes come from the metabolic networks of the proteins they code and the clusters of conditions come from the structure of the plan of the experiments. The phenomenon is known in the literature as the *chequerboard structure:* after appropriate reordering, gene-expression matrices become chequerboard like. Batches of genes express similarly under batches of conditions. Interestingly, rather good reorderings are supplied by simple hierarchical clusterings of rows and columns simultaneously.

GRAPHS

Graph complexity. There are natural ways to assign matrices to a graph: the off-diagonal entries reflect the connectivity and the diagonal entries may be set to zero or to the degree multiplied by -1. In the second case the sum in each row is zero and a non-zero vector with equal coordinates is an eigenvector with zero eigenvalue. All eigenvalues are non-positive in the second case. We call the matrix in first case the adjacency matrix and the second one the Laplacian ([3], [4], [5], [10], [18], [23], [35]). For regular graphs the spectra of the two matrices differ only by a constant.

An arbitrary graph is a free sequence of $\binom{n}{2}$ bits. Without fathoming the inner structure of the graph we can not catch the complexity of a graph. In the simplest case the spectra is J-shaped: there is some topology on the vertices and the edges follow that. For Albert–Barabási graphs ([1], [6]) the topology comes from preference: the degrees of the vertices control the choice of the edges. According to Wigner's semicircle law ([17], [20], [25]) for random graphs the spectra of the adjacency matrix forms a semicircle, which is definitely not J-shaped. Incidentally: we do not know what is the asymptotic for the spectra for random symmetrical matrices with i.i.d. off-diagonal entries but putting the sums (multiplied by -1) in the diagonal. If the entries of a random matrix are independent Wiener processes, the eigenvalues $\lambda_i = \lambda_i(t)$ follow the system of stochastic differential equation

$$d\lambda_i = dW_i + dt \prod_{j \neq i} \frac{1}{\lambda_i - \lambda_j}, \qquad i = 1, \ldots, n$$

showing that the eigenvalues repel each other. Do eigenvalues of random graphs repel each-other? Does this depend on which eigenvalue definition we use and what model of random graphs?

Fractals. An other intriguing question is, whether there are fractals in large graphs? To catch the fractal behavior we propose the following potential defined for connected graphs. For a given vertex x let y be the vertex closest to x of degree not smaller than that of x, and let D_x be the set of vertices different from x that are strictly closer to x than y is. This D_x is the *estate* and its size the *asset* of x. (If there is only one vertex with maximal degree then its estate is empty.) The *wealth* V_x of x is the sum of the assets of all vertices in D_x. Finally, the potential of the graph Γ is

$$Q(\Gamma) = \sum_x V_x^\alpha V_y^\alpha d^\beta(x, y),$$

where the summation runs on all pairs (x, y) of vertices, d is the distance on the graph and $\alpha, \beta > 0$ are fixed constants. What is the graph which maximizes this potential for fixed number of vertices? For $n = 254$, $\alpha = 0.75$, $\beta = 0.25$ we constructed several graphs. Revealing the structure of optimal graphs created by exhaustive stochastic search we generated the graph presented in the Appendix. For this graph $Q(\Gamma) = 14,343$. The structure of the graph is shown in Figure 1. The empty circles represent virtual vertices, which help only in building up the structure. We tend to believe that real complexity is connected with the repelling property of the eigenvalues, while the concentration of the eigenvalues comes from the equivalence of the vertices.

Equivalent vertices. Equivalence of vertices have two features:

- equivalent vertices may prefer each other: the edge-density inside equivalent clusters is larger than outside
- vertices belonging to equivalent clusters behave similarly.

The first case is reflected by the spectra of the Laplacian and the second case is Szemerédi's regularity property ([14], [21], [27], [34], [37]): we say that the bipartite graph with vertex sets A, B is ε-regular if

$$\left| \frac{E(X, Y)}{|X||Y|} - \Delta \right| \leq \varepsilon,$$

holds true for all $X \in A$, $Y \in B$ such that $|X| \geq \varepsilon|A|$, $|Y| \geq \varepsilon|B|$, where

$$\Delta = \frac{E(A, B)}{|A||B|}$$

is the edge-density in the whole graph.

Regularity lemma for a statistician. Roughly speaking, Szemerédi's regularity lemma states that the vertices of *every* graph may be clustered

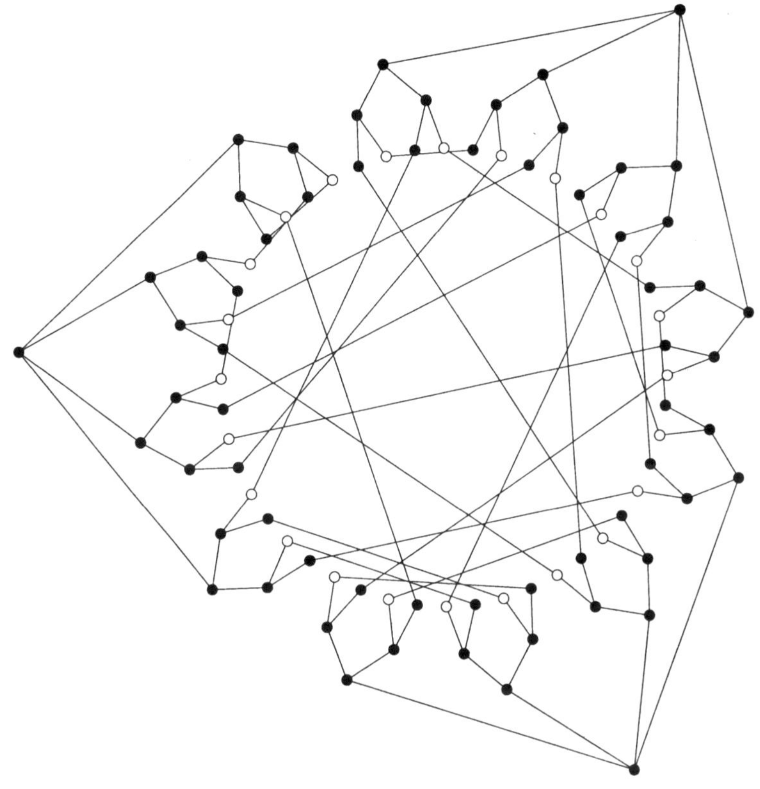

Fig. 1

in such a way that the bipartite graphs corresponding to different clusters are ε regular with a small exceptional fraction of the pairs if the number of vertices is large enough.

For a statistician the condition in the definition of ε-regularity is a statistical test resembling to Rényi's version of the Kolmogorov test. Let n be an arbitrary number, for integers i between 1 and n let $\alpha(i)$ be arbitrary integers between 1 and k, where $k < n$. Let $p_{i,j}$, $1 \le i,j \le k$ be an arbitrary symmetric matrix with $0 \le p_{i,j} \le 1$. We call the random graph *checkerboard graph* if vertices i, j, where $1 \le i \le j \le n$ are connected with probability $p_{\alpha(i),\alpha(j)}$ and the edges are independent. At first instance, the regularity lemma seems to state that the collection of checkerboard graphs is *bold enough* for having the power to generate all graphs. The striking effect of the lemma is its simplicity: the random mechanisms used in a possible rigorous formalization are quite natural and, what is more, they are not capable of catching all the possible information out of a graph.

The riddle of Szemerédi's lemma is hidden in the definition of regularity. It fixes, prescribes a test on graphs for the use of testing the hypothesis that the graph comes from the class of checkerboard distributions. Being true statisticians we propose to develop other tests, possibly with relevant power for testing the hypothesis. One natural aspirant is the spectra of the adjacency matrix: for checkerboard graphs it has to be J-shaped, *and* the eigenvectors have to show clear clusters. Any deviation from these properties may lead to rejecting the hypotheses.

A universal lemma might state that *any* maximum likelihood estimate is bold enough to have the property that it is optimal for all measures in the statistical field. You can never use the picture given by a maximum likelihood estimate for testing the hypothesis concerning the completeness of the investigated measures. Inside the world the statistical field they have to be bold enough just by definition of the maximum likelihood estimate. But we can test the hypothesis by other accordingly chosen statistics which are usually orthogonal to the logic of the likelihood. Recently one of the most interesting fields for an extension of the lemma are the hypergraphs. Accordingly we have to learn the precise use of the *stochastic method:* it is better to formulate minor sets of conditions under which a useful theorem of stochastics holds true and extend it to as wide a territory as possible but we can never forget effectiveness. In case of Szemerédi's lemma it is the blow up property.

Blow up property states that in a large enough graph all the small graphs appear with a frequency proportional with their probabilities. The statement is also called the *Counting Lemma.* Taking a large enough distance from the details of the affair investigated we think that the situation resembles quantum physics: first you choose what you are interested in, then the analytic machinery answers your question as you like it. If we want to ensure ε-regular colouring for all graphs then we have to choose the number of colours enormously large. But according to our experience, checkerboard graphs are applicable to small graphs too. The "bold enough" property appears only for large graphs, but the property being universal for large graphs may be present for a special family of small graphs.

Other models. Let $f(x, y)$ be a differentiable function for $0 \leq x, y \leq 1$ such that $0 \leq f(x, y) \leq 1$. Let $x_1, x_2, \ldots, x_n$ be arbitrary numbers in $(0, 1)$. The random graph connecting the vertices i, j independently with probability $p_{i,j} = f(x_i, x_j)$ represents the function f and numbers x_i. We can try to reconstruct the model parameters $f, x_1, \ldots, x_n$ by a maximum likelihood method. Maximizing the likelihood the following two-phase algorithm is applicable:

- for given f the gradient method applies to the x_i-s
- for given x_i-s the function $f(x, y)$ may be estimated by the edge-density for $|x_i - x| < \varepsilon$, $|x_j - y| < \varepsilon$.

We say that the function f is the *face* of the graph and the x_i-s are its *core*. For large n and $x_i = \frac{i-0.5}{n}$ the spectra of the random graph is close to the spectra of f. If $f(x, y) = \frac{x+y}{2}$ the eigengenvalues are uniformly distributed in $\left(\frac{1}{4}, \frac{3}{4}\right)$ in contrast with the chequerboard case when they are clustered around a few points. It goes without saying that the uniform distribution may be approximated by a discrete distribution concentrated on finitely many points, but we may detect the difference with appropriate statistics. What is the case with Szemerédi's statistics

$$\frac{\left(E(X, Y) - F(X, Y)\right)^2}{G(X, Y)},$$

where

- $E(X, Y)$ is the number of edges between the disjoint sets X, Y
- $F(X, Y) = \sum_{x_i \in X, x_j \in Y} p_{i,j}$ is the expected value of $E(X, Y)$
- $G(X, Y) = \max\left(1, \sum_{x_i \in X, x_j \in Y} p_{i,j}(1 - p_{i,j})\right)$ is the truncated variance of $E(X, Y)$?

Of course it has to be applicable to detect the difference, but in the regularity lemma the constants are chosen loosely for that aim. The spectra of the adjacency matrix shows more characteristic effect of the checkerboard structure than the Laplacian, but a rigid SVD of the matrix $p_{i,j}$ is usually not flexible enough for detecting real structures because it poorly approximates matrices with entries in the interval $(0, 1)$. The logistic transform $p_{i,j} = b/\left(c + \exp(a_{i,j})\right)$ offers an easy bridge between real numbers and the $(0, 1)$ interval. More generally, we can use any monotone increasing function for this role. The nonparametric maximum likelihood estimator is a step function usually with a small number of steps and a remarkable portion of the edges has probability zero or one and thus the fitted model has moderately random character only on the borderline of the two subsets of edges where we can explicitly predict their existence.

Dynamics. The most complicated matrix $p_{i,j}$ is unable to reflect fine interactions between the edges. We can build up systematically stochastic models starting with a joint distribution of two or three edges or subsets of vertices, but in case of graphs presented by real life situations the structure of stochastic interactions is mostly multifactorial. Firstly, gathering all the available information, we can try to describe with words the characteristic features of the investigated graph. Next we translate our own words to mathematical formulas and we define some potential function measuring the perfection of individual graphs and we develop algorithms to maximize the potential following a kind of Darwinian path. The algorithms may resemble to the mechanisms creating the studied graphs. But typically the optimization procedure reveals something that is rather far from our ideals formulated originally in words. In such situation the whole procedure starts again and we should recycle it until convergence.

The potential $\mathcal{Q}(\Gamma)$ defined by assets and wealths led to the following procedure. We start with one vertex. Step by step, each vertex in the graph is divided into two daughters, and in the new graph

- we join two daughter points with probability $p = 0.06$, if their mothers were joined
- otherwise we join them with probability $q = 0.005$, and
- we join them with probability $p = 0.03$ if they have the same mother.

The fractal structure is imprinted in the algorithm. The reason for the low probabilities is that the potential prefers spare graphs. One source of the potential is

$$\Psi(\Gamma) = \sum_{(x,y)\in\mathcal{E}} f_x f_y,$$

where f_x is the degree of x. In Albert–Barabási dynamics Ψ is maximal among graphs with given degrees, which is unnatural in the majority of cases: the hubs are in most cases separated, they are far from other hubs.

PERMUTATIONS

The Thurstonean. Let $F_1, F_2, \ldots, F_n$ be arbitrary continuous real distributions, for each $1 \leq i \leq n$ let X_i be a random variable with distribution

F_i and let the variables $X_1, X_2, \ldots, X_n$ be independent. Let

$$\pi = \big(\pi(1), \pi(2), \ldots, \pi(n)\big)$$

be the permutation ordering the X_i-s monotone increasingly:

$$X_{\pi(1)} < X_{\pi(2)} < \cdots < X_{\pi(n)}.$$

The model was proposed by Louis Leon Thurstone in [38] (see also in [33]) thus we call the distribution defined by the model *Thurstonean*.

It is easy to see that if the distributions F_i are exponentials with parameters λ_i then

$$P\big(\pi(a+1) = t \mid \pi(1), \pi(2), \ldots, \pi(a)\big) = \frac{\lambda_t}{\sum_{i \notin L_a} \lambda_i},$$

$$a = 0, 1, \ldots, n-1, \quad t \notin L_a,$$

where $L_a = \big\{\pi(1), \pi(2), \ldots, \pi(a)\big\}$ with $L_0 = \emptyset$.

If Y_t, $t = 1, 2, \ldots$, are i.i.d. with distribution

$$P(Y_1 = t) = p_t, \quad t = 1, 2, \ldots, n,$$

and we delete all elements from the sequence that we have seen earlier, then the remaining random numbers form a permutation in N with the same distribution as the exponential Thurstonean one, whenever

$$p_t = \frac{\lambda_t}{\sum_{i=1}^n \lambda_i}.$$

Interestingly, in these two models the EM-algorithm [24] leads to different iterations. In the general case the Baum–Welch algorithm [29] leads to the following iteration. For the sake of simplicity let us suppose that π is the identity. In this case we have to calculate the conditional distributions

$$Q_i(t) = P(X_i < t \mid X_1 < X_2 < \cdots < X_n), \quad i = 1, 2, \ldots, n.$$

In the forward phase of the algorithm we calculate recursively the conditional distributions

$$G_i(t) = P(X_i < t \mid X_1 < X_2 < \cdots < X_i)$$

by

$$g_i(t) = f_i(t)G_{i-1}(t),$$

where $f_i = F_i'$, $g_i = G_i'$. Similarly, for

$$H_i(t) = P(X_i < t \mid X_i < X_{i+1} < \cdots < X_n)$$

$$h_i(t) = f_i(t) H_{i+1}(t)$$

holds true where $h_i = H_i'$ while $G_1 = F_1$, $H_n = F_n$. Then

$$q_i(t) = f_i(t) G_{i-1}(t) H_{i+1}(t),$$

where $q_i = Q_i'$ and $G_0 = H_{n+1} = 1$.

Let us denote by $\mathcal{T}_n$ the set of all Thurstonean random permutations with n elements. A possible generalization is to drop the independence of the X_i-s. Let $\mathcal{G}_n$ be the set of n-dimensional Gaussian distributions with expectation μ and covariance Σ. $\mathcal{G}_n$ is described by

$$\binom{n+1}{2} + n - 2$$

parameters, which suggests that for $n = 2, 3$ the model is overparametrized. Indeed, for $n = 2$, Σ may be reduced to the identity matrix and $\mathcal{G}_2 = \mathcal{T}_2 = \mathcal{P}_2$ where $\mathcal{P}_n$ stands for the set of all possible distributions of permutations on N. If $n = 3$, then the distribution rendering half probability to the permutations $(1, 2, 3)$, $(3, 2, 1)$ is definitely not in $\mathcal{T}_3$ yet, it is in $\mathcal{G}_3$ for μ equals zero and with a covariance ensuring that $X_1 = 2X_2, X_3 = 0$ or $X_1 = -X_3, X_2 = 0$. For large n the set $\mathcal{T}_n$ should be larger than $\mathcal{G}_n$, for the number of degree of freedoms goes to infinity more rapidly in the first case. In the Gaussian case a possible reduction of the number of parameters is the control on the rank of Σ as it is usual in factor analysis and principal component analysis. But the covariance of π is unable to catch the rank of Σ, it is visible only in the covariance of π^{-1}. Permutations in practice mostly come from some one-to-one correspondence between two different unordered sets. The row-ordering and column-ordering of the chequerboard representing the permutations usually is lurking behind. If X has some multidimensional stochastic structure one cannot find it in π, because $\pi(i)$ gives the *coordinates* of the i-th element of the ordered sample answering the question: *who* stays on the i-th position. But the order of coordinates in X is arbitrary. In contrary, $\pi^{-1}(j)$ tells us *where* the j-th coordinate X_j is in the ordered sample which is a nearly linear function of the values of the coordinates, hence the covariances of X and π^{-1} are close to each other.

Statistics. Models and statistics on a structure are the two legs of any inference. All models have their natural statistics or sufficient statistics and for a given family we can test the goodness of fit of the whole family. For permutations the primary marginals are the positions of a subsets of the elements among the whole set: here a void mark is substituted for the elements outside the group, this means that the order of the elements of the chosen group is filled in with some void marks:

$$* * * 3 * 5\,1 * * * * * 4 * 2 *$$

means that $\pi(4) = 3$, $\pi(6) = 5$, $\pi(7) = 1$, $\pi(12) = 4$, $\pi(14) = 2$. Dropping the stars we get the permutation of the chosen elements which is another marginal. We say that the permutation 35142 is the *shrunken* version of the original one into the set $(1, 2, 3, 4, 5)$. For one element only the filled marginals contain information, one is tempted to use these one-element positions as aspirants for the unknown distributions in the Thurstonean case. Turning to the inverse, other one-element marginals appear and the distribution, having simultaneously a given row marginal and a given column marginal is of the form

$$P(\pi) = \kappa \prod_{i=1}^{n} a_{i,\pi(i)}^{\Delta},$$

where the matrix $A = (a_{i,j})$ is an arbitrary doubly stochastic matrix, Δ is a positive number, and κ is the normalizing factor. It is well known that for any doubly stochastic A there is at least one permutation with positive probability. We call the distribution *simple rook* distribution because representing the permutations on a chequerboard the probability of the permutation is proportional to the product of the numbers in the occupied pixels. For simple rook distributions the sufficient statistics are

$$\nu(i, j) = \#\{\pi(i) = j\}, \quad 1 \le i, j \le n,$$

which are simultaneously the matrices of unnormed row and column marginals. The statistics $\nu(i, j)$ are useful for distributions

$$P(\pi) = \kappa \exp\left(-d^2(\pi, \pi_0)/T\right),$$

where d is some distance function, π_0 is the centrum of the distribution, T is a positive constant, and κ is the norming factor. The family was introduced by Mallows in [22] (see also [15], [28] and [36]).

Row cuttings. Let us say that an element of $\mathcal{P}_n$ has the property of *row cutting at a* if

$$P(\pi \mid L_a) = f\big(\pi(1), \pi(2), \ldots, \pi(a)\big) \, g\big(\pi(a+1), \ldots, \pi(n)\big)$$

holds true with some a variate function f and $(n-a)$ variate function g, where $2 \le a \le n-2$. We denote by $\mathcal{R}_a$ the set of all distributions with the property row cutting at a. Row cutting at a means that the permutations $\big(\pi(1), \pi(2), \ldots, \pi(a)\big)$, $\big(\pi(a+1), \pi(a+2), \ldots, \pi(n)\big)$ are conditionally independent on the statistics L_a. We say that a random permutation is *row-free* if it has the row cutting property for all a. It is easy to see that row-free random permutations have the form

$$P(\pi) = \prod_{a=0}^{n-1} c\big(\pi(a+1), L_a\big),$$

where the conditional probabilities $c(u, V)$ are concentrated on $u \in N \setminus V$. The degree of freedom of the set $\mathcal{R}$ of row-free permutations is

$$r_n = \sum_{a=1}^{n}(a-1)\binom{n}{a} = \left(\frac{n}{2} - 1\right) 2^n + 1.$$

The set $\mathcal{C}_b$ is the set of all distributions with the property column cutting at b, and the set $\mathcal{C}$ of column-free permutations is similarly defined with substituting π^{-1} for π. A possible representation of n element permutations is putting rooks on the n by n chequerboard: here the properties of row- and column-freeness are symmetrical. The sample presented by Persi Diaconis happens to be in a certain sense inside of the intersection of the sets $\mathcal{R}$ and $\mathcal{C}$. Our main theorem states that the degree of freedom of the intersection is

$$\nu_n = \sum_{a=1}^{n-1} a^2.$$

We call the elements in the intersection *free* distributions. Exponential Thurstonean distributions are row-free and the simple rook distribution is free. A possible set of sufficient statistics for free distributions is the following:

$$\nu(a,b) = \sum_{i=1}^{a} \mathcal{I}\big(\pi(i) \le b\big), \quad 1 \le a, b \le n-1,$$

$$\mu(a,b) = \sum_{i=1}^{a} \mathcal{I}\big(\pi(i) \le b, \pi(a+1) = b+1\big), \quad 1 \le a, b \le n-2.$$

If all permutations have positive probabilities then a free distribution has
the form

$$P(\pi) = \prod_{a=1}^{n-2} g(a, \pi(a+1), \mu\big(a, \pi(a+1)\big)) \prod_{a=1}^{n-1} \prod_{b=1}^{n-1} f\big(a, b, \nu(a,b)\big)$$

In the intersection of the sets $\mathcal{R}_a$, $\mathcal{C}_b$ the distributions have the form

$$P(\pi) = \alpha(\pi_a^b)\beta\big(\pi_a^{\bar{b}}\big)\gamma\big(\pi_{\bar{a}}^b\big)\delta\big(\pi_{\bar{a}}^{\bar{b}}\big),$$

where $\alpha, \beta, \gamma, \delta$ are positive functions, and

π_a^b denotes the shrunken version of $\big(\pi(1), \pi(2), \pi(a)\big)$ to the set $(1, 2, \ldots, b)$

$\pi_a^{\bar{b}}$ denotes the shrunken version of $\big(\pi(1), \pi(2), \pi(a)\big)$ to the set $(b + 1, b + 2, \ldots, n)$

$\pi_{\bar{a}}^b$ denotes the shrunken version of $\big(\pi(a+1), \pi(a+2), \pi(n)\big)$ to the set $(1, 2, \ldots, b)$

$\pi_{\bar{a}}^{\bar{b}}$ denotes the shrunken version of $\big(\pi(a+1), \pi(a+2), \pi(n)\big)$ to the set

$$(b + 1, b + 2, \ldots, n).$$

The product of the four functions in $P(\pi)$ means that random permutations
in the intersection of $\mathcal{R}_a$ and $\mathcal{C}_b$ have the property that the events in the four
quarters of the chequerboard are conditionally independent whenever the
subsets of rows and columns occupied inside them is given and the occupied
rows and columns in the left upper quarter are conditionally independent
from the ones in the right lower quarter under the condition that the number
of rows and columns is given. (Observe that the number of rows should be
equal to the number of columns.)

Structural zeros in row-free distributions may appear independently:
any conditional probability $C(u, V)$ may be zero as long as there is at least
one permutation with positive probability. For free distributions structure
zeros may be generated by the parameters f, g, but the intersection of row-
free and column-free distributions with structural zeros is larger than this
set. For $n = 4$ the uniform distribution concentrated on permutations such
that only the permutations 1234, 2341, 2413, 2431, 3124, 3142, 3241, 4321
is free. However, for any set of the structural parameters $f(a, b, c)$, $g(a, b, c)$
such that the probabilities of the above eight permutations are positive, all
permutations have positive probabilities.

Estimation of the parameters of row-free random permutations is straightforward: the estimators of the conditional probabilities $c(u, V)$ are the corresponding conditional relative frequencies. For free random permutations there are two iterative procedures:

- we can use alternating divergence projections on the sets $\mathcal{R}, \mathcal{C}$ or

- we can apply iterative fitting procedures on the statistics $\nu(a, b), \mu(a, b)$.

An exact implementation of these algorithms consume $n!$ steps what renders them to small n-s. Metropolis algorithm and Bayes machine apply both for generating i.i.d. free samples and estimating model parameters. As an estimator of the expectations in the likelihood equations we may use the averages of i.i.d. samples generated by the iteratively changing parameters.

Let $M \subset N$ be such that $2 \leq |M| \leq n - 2$. Let us denote by L_M the set $\{\pi(i),\ i \in M\}$. We say that the random permutation has the M-cutting property if the ordered numbers $(\pi(i),\ i \in M)$ and $(\pi(i),\ i \notin M)$ are conditionally independent on L_M. Random permutations having M-cutting property for all M are the simple rook distributions. The number of model parameters may be reduced by controlling the rank of the matrix A. The rank has to be at least 2 because if it is equal to 1 then all elements of A are equal.

Partitions

Once upon a time there was a party with 14 participants labeled by integers from 1 to 14. As it is usual in parties they formed groups which were sensed and recorded by devices offered by our modern technology. The data can be found on the home page of G. Tusnády as SIRP DATA (http://www.renyi.hu/~tusnady/). The first part is given in Table 1.

Each record of the data represents one grouping (partition) formed in the course of the party. The first number means the time in hours when the actual grouping occurred and the next 14 integers denote the partition. Each set of a partition is labelled by its smallest number what we call leading member.

For example

	1	2	3	4	5	6	7	8	9	10	11	12	13	14
0.900159	1	1	1	4	5	5	5	5	5	5	11	5	11	5

Table 1. First 24 records of SIRP.DATA

TIME	1	2	3	4	5	6	7	8	9	10	11	12	13	14
0.019238	1	1	1	4	5	6	5	6	5	5	5	6	13	5
0.064438	1	1	1	4	5	6	5	6	5	5	5	6	4	5
0.107385	1	1	1	4	5	6	5	6	5	5	5	6	13	5
0.119281	1	1	1	4	5	6	7	6	5	7	7	6	13	7
0.127421	1	1	1	4	5	5	7	5	5	7	7	5	13	7
0.159595	1	1	1	4	5	6	7	6	5	7	7	5	13	7
0.244247	1	1	1	4	5	6	1	6	5	1	1	5	13	1
0.246863	1	1	1	4	5	4	1	4	5	1	1	5	13	1
0.393910	1	2	2	4	5	4	1	4	5	1	1	5	13	1
0.466802	1	1	1	4	5	4	1	4	5	1	1	5	13	1
0.506604	1	1	1	4	5	4	7	4	5	7	7	5	13	7
0.518243	1	1	1	4	5	6	7	6	5	7	7	5	13	7
0.519593	1	1	1	4	5	4	7	4	5	7	7	5	13	7
0.576503	1	1	1	4	5	4	5	4	5	5	5	5	13	5
0.707155	1	1	1	4	5	4	5	4	5	5	11	5	13	5
0.716638	1	1	1	4	5	6	5	6	5	5	11	5	13	5
0.727348	1	1	1	4	5	6	5	6	5	5	11	5	11	5
0.733247	1	1	1	4	5	4	5	4	5	5	11	5	11	5
0.834109	1	1	1	4	5	6	5	6	5	5	11	5	11	5
0.900159	1	1	1	4	5	5	5	5	5	5	11	5	11	5
0.918424	1	1	1	4	5	6	5	6	5	5	11	5	11	5
0.998953	1	1	1	4	5	6	5	6	5	5	11	5	13	5
1.155627	1	1	1	4	5	1	5	1	5	5	11	5	13	5
1.252516	1	1	1	4	5	6	5	6	5	5	11	5	13	5

means that after 0.900159 hours from the beginning of the party the following groups were sensed by our detectors: $1 + 2 + 3$, 4, $5 + 6 + 7 + 8 + 9 + 10 + 12 + 14$, $11 + 13$. Poor 4, seemingly a lonely person walked alone, the noisy central body $5 + 6 + 7 + 8 + 9 + 10 + 12 + 14$ was situated around the dinner table, while $1 + 2 + 3$ had a very important discussion in a secret corner and $11 + 13$ were playing table tennis. There is a natural way to order a graph to partitions: the vertices are the participants and they are connected whenever they belong to the same group. However, partitions are special graphs, because they contain only disjunct complete subgraphs called sometime a clique.

Visualization of partitions. In multidimensional data analysis a general idea is to compress objects whenever they have something common. The trouble is that without any constraints the population shrinks to a single point. We use multidimensional covariance standardization as a constraint: the data are centered by subtracting their average, dividing them by the standard deviation and using covariances to keep the scales finite and non-zero. The effect resembles opening an umbrella: the wires spread out what the canopy pulls together.

In a good party there are appropriate places for people willing to do something together. But to use different positions for each subset is prohibitive: there is a combinatorial explosion. We restrict our algorithm to pairs: all pairs of the participants have same special meeting point and the groups are located at the average of the positions of their pairs. The formal description of the algorithm is the following.

Let S be the number of different partitions occurring in the party and let x_k, y_k be the coordinates of the point representing the k-th partition $(k = 1, 2, .., S)$. The initial values of the coordinates are random standard normal numbers. The iteration consists of the following steps:

Step 1. Opening the umbrella (Schmidt orthogonalization):

$$\tilde{x}_k = \frac{x_k - \overline{x}}{w(x)},$$

where

$$\overline{x} = \frac{1}{S} \sum_{k=1}^{S} x_k,$$

and

$$w(x) = \sqrt{\sum_{k=1}^{S} (x_k - \overline{x})^2};$$

$$\tilde{y}_k = \frac{y_k^* - c(xy) * \tilde{x}_k}{w(y)},$$

where

$$y_k^* = y_k - \overline{y},$$

$$\overline{y} = \frac{1}{S} \sum_{k=1}^{S} y_k,$$

$$c(xy) = \sum_{k=1}^{S} y_k^* * \tilde{x}_k,$$

$$w(y) = \sqrt{\sum_{k=1}^{S} \left(y_k^* - c(xy) * \tilde{x}_k \right)^2}.$$

Step 2. Positioning pairs of persons (averaging the partitions where the given pair happens to be in the same group):

$$u_{i,j} = \frac{\sum_{k:\, p(i,k)=p(j,k)} \tilde{x}_k * t_k}{\sum_{k:\, p(i,k)=p(j,k)} t_k},$$

$$v_{i,j} = \frac{\sum_{k:\, p(i,k)=p(j,k)} \tilde{y}_k * t_k}{\sum_{k:\, p(i,k)=p(j,k)} t_k},$$

where $p(i,k)$ denotes the leading person of the group containing the i-th person in the k-th partition and t_k is the duration of the k-th partition.

Step 3. Dynamics (relocating the partitions with the gradient of the pairs they unite):

$$x_k^{\text{new}} = \tilde{x}_k - \gamma * xx_k,$$

where γ denotes a small positive constant (it controls the speed of the algorithm) and

$$xx_k = \sum_{i,j:\, p(i,k)=p(j,k)} u_{i,j},$$

$$y_k^{\text{new}} = \tilde{y}_k - \gamma * yy_k,$$

where

$$yy_k = \sum_{i,j:\, p(i,k)=p(j,k)} v_{i,j}.$$

The pair potential model. Our data are generated by the distribution

$$P_A(\pi) = \frac{1}{\Gamma(A)} \exp\big(Q(\pi, A)\big),$$

where the potential $Q(\pi, A)$ is defined by

$$Q(\pi, A) = \sum_{1 \leq i < j \leq c:\ \pi(i) = \pi(j)} a_{i,j},$$

and

$$\Gamma(A) = \sum \exp\big(Q(\pi, A)\big)$$

is the scaling factor where the summation runs over all partitions π. The matrix $A = a_{i,j}$ is symmetric and given by Table 2. The maximum of the potential $Q(\pi, A)$ is 39.64 and it is attained for the partition

$$\pi = \{1, 1, 1, 4, 5, 6, 5, 6, 5, 5, 5, 6, 13, 5\}.$$

The distribution can be sampled by the Metropolis algorithm [26], which is based on a graph where the vertices are the partitions. We say that two partitions are connected by an edge whenever one is formed by the other with uniting two of its groups. The price of the edge is the product of the numbers of persons in the united groups. The distance of two arbitrary vertices is the price of the cheapest path between them. This is

$$d(\pi_1, \pi_2) = \sum_{i=1}^{n} \sum_{j=1}^{n-1} \sum_{k=j+1}^{n} \big(\nu(i,j)\nu(i,k) + \nu(j,i)\nu(k,i)\big),$$

where

$$\nu(i,j) = \sum_{k=1}^{n} \mathcal{I}\big(\pi_1(k) = i,\ \pi_2(k) = j\big),$$

where n is the number of persons.

Concerning the partition function $\Gamma(A)$ one can prove that

$$E_A \exp\big(Q(\pi, B)\big) = \frac{\Gamma(A + B)}{\Gamma(A)},$$

and

$$\Gamma(A) \leq \prod_{i=1}^{n-1} \prod_{j=i+1}^{n} \big(1 + \exp(a_{i,j})\big),$$

Table 2. Model parameters of SIRP.DATA

0.00	−3.08	−0.40	0.23	0.75	−1.23	1.13	−0.02	0.51	−2.62	0.78	2.93	0.05	−2.66
−3.08	0.00	−4.63	1.18	1.43	1.19	4.73	−0.59	2.61	−5.01	2.08	1.94	3.91	1.94
−0.40	−4.63	0.00	4.31	1.65	2.94	−1.17	0.87	−1.37	−0.72	1.66	−0.92	0.19	0.01
0.23	1.18	4.31	0.00	2.15	−1.10	−2.77	1.63	4.13	6.79	6.29	1.51	0.41	2.92
0.75	1.43	1.65	2.15	0.00	−0.06	−2.31	−0.35	−1.52	−2.30	2.48	−2.23	2.25	−0.72
−1.23	1.19	2.94	−1.10	−0.06	0.00	2.75	−4.47	1.29	−4.41	0.38	−4.07	3.75	2.54
1.13	4.73	−1.17	−2.77	−2.31	2.75	0.00	−2.86	−2.50	−6.20	−3.35	0.30	3.98	−0.68
−0.02	−0.59	0.87	1.63	−0.35	−4.47	−2.86	0.00	−1.16	3.12	−0.27	3.61	5.84	2.25
0.51	2.61	−1.37	4.13	−1.52	1.29	−2.50	−1.16	0.00	3.29	5.46	−1.05	1.64	−7.29
−2.62	−5.01	−0.72	6.79	−2.30	−4.41	−6.20	3.12	3.29	0.00	−5.14	−2.20	−1.12	−1.77
0.78	2.08	1.66	6.29	2.48	0.38	−3.35	−0.27	5.46	−5.14	0.00	5.45	−0.32	−4.20
2.93	1.94	−0.92	1.51	−2.23	−4.07	0.30	3.61	−1.05	−2.20	5.45	0.00	3.07	1.74
0.05	3.91	0.19	0.41	2.25	3.75	3.98	5.84	1.64	−1.12	−0.32	3.07	0.00	2.82
−2.66	1.94	0.01	2.92	−0.72	2.54	−0.68	2.25	−7.29	−1.77	−4.20	1.74	2.82	0.00

but we do not have an explicit form for $\Gamma(A)$. We generated the matrix A as random Gaussian number with zero expectation and standard deviation 2.5 thus the model has the flavor of spin glass processes: there is an abundance of local maxima of the potential and the process spends the majority of the time in the potential valleys with short time jumps between them. This might be the case with real world parties where the different partitions are evaluated by the well-being of the persons inside the actual groups. Our model is the simplest possible one because it is based on pair-relations only. Generalization to higher order interactions is straightforward.

Estimation of model parameters. The pair-potentials $a_{i,j}$ can be estimated by the maximum likelihood equation (see in [2], [16])

$$\frac{1}{S} \sum_{k:\, p(i,k)=p(j,k)} 1 = P_A\big(\pi(i) = \pi(j)\big) \qquad 1 \le i < j \le n,$$

or by simple weighted linear regression. The probabilities $b_{i,j} = P_A\big(\pi(i) = \pi(j)\big)$ are given in Table 3.

Corresponding relative frequencies $\beta_{i,j}$ are given in Table 4. The pair potential model is loglinear: the logarithms of probabilities $P_A(\pi)$ are linear functions of the model parameters $a_{i,j}$. There is no direct relation between $a_{i,j}$ and $b_{i,j}$, and the $\beta_{i,j}$ relative frequencies are closer to the theoretical probabilities $b_{i,j}$ than the estimators of $a_{i,j}$ to $a_{i,j}$. The estimation of the model parameters is a typical ill-conditioned problem and to compare different data sets, the $b_{i,j}$ parameters may be more useful.

The specific feature of our data is that successive partitions can be either the union or the splitting of the previous one. For the first part of our data

Table 3. Probabilities of equivalence

.000	.750	.710	.018	.075	.087	.238	.108	.111	.413	.319	.024	.022	.294
.750	.000	.937	.006	.006	.051	.080	.085	.035	.256	.161	.034	.010	.123
.710	.937	.000	.001	.025	.029	.101	.066	.068	.249	.155	.055	.016	.149
.018	.006	.001	.000	.005	.142	.011	.087	.001	.000	.000	.076	.301	.000
.075	.006	.025	.005	.000	.279	.541	.287	.672	.415	.342	.494	.002	.530
.087	.051	.029	.142	.279	.000	.161	.663	.139	.208	.148	.548	.001	.014
.238	.080	.101	.011	.541	.161	.000	.279	.452	.792	.776	.129	.004	.682
.108	.085	.066	.087	.287	.663	.279	.000	.257	.167	.210	.244	.000	.101
.111	.035	.068	.001	.672	.139	.452	.257	.000	.278	.268	.363	.006	.678
.413	.256	.249	.000	.415	.208	.792	.167	.278	.000	.816	.123	.010	.592
.319	.161	.155	.000	.342	.148	.776	.210	.268	.816	.000	.021	.058	.588
.024	.034	.055	.076	.494	.548	.129	.244	.363	.123	.021	.000	.005	.156
.022	.010	.016	.301	.002	.001	.004	.000	.006	.010	.058	.005	.000	.002
.294	.123	.149	.000	.530	.014	.682	.101	.678	.592	.588	.156	.002	.000

Table 4. Relative frequencies of equivalence

.000	.773	.737	.012	.077	.084	.243	.100	.111	.423	.326	.023	.016	.297
.773	.000	.948	.004	.006	.054	.085	.079	.032	.266	.167	.030	.007	.130
.736	.948	.000	.001	.027	.030	.105	.058	.066	.259	.161	.055	.011	.153
.012	.004	.001	.000	.002	.118	.011	.070	.000	.000	.000	.064	.282	.000
.077	.006	.027	.002	.000	.292	.563	.314	.712	.430	.366	.510	.001	.554
.084	.054	.030	.118	.292	.000	.166	.688	.154	.208	.151	.566	.000	.015
.243	.085	.105	.011	.563	.166	.000	.287	.483	.794	.783	.138	.001	.706
.100	.079	.058	.070	.314	.688	.287	.000	.274	.173	.213	.274	.000	.108
.111	.032	.066	.000	.712	.154	.483	.274	.000	.307	.300	.384	.003	.692
.423	.266	.259	.000	.430	.208	.794	.173	.307	.000	.823	.126	.007	.610
.326	.169	.161	.000	.366	.151	.783	.213	.300	.823	.000	.029	.052	.607
.023	.030	.055	.064	.510	.566	.138	.274	.384	.126	.029	.000	.002	.165
.016	.007	.011	.282	.001	.000	.001	.000	.003	.007	.052	.002	.000	.001
.297	.130	.153	.000	.554	.015	.706	.108	.692	.610	.607	.165	.001	.000

the operations are given in Table 5. Having the information at hand that the data were generated by the Metropolis [26] algorithm, one may develop more efficient estimators. The abundance of inverted pairs of union and splitting among the operators is remarkable here.

Table 5. Operators of data in Table 1

NAME	FIRST GROUP	SIGN	SECOND GROUP
UNION	4	$+$	13
SPLITTING	4	$\|$	13
SPLITTING	5, 9	$\|$	7, 10, 11, 14
UNION	5, 9	$+$	6, 8, 12
SPLITTING	5, 9, 12	$\|$	6, 8
UNION	1, 2, 3	$+$	7, 10, 11, 14
UNION	4	$+$	6, 8
SPLITTING	1, 7, 10, 11, 14	$\|$	2, 3
UNION	1, 7, 10, 11, 14	$+$	2, 3
SPLITTING	1, 2, 3	$\|$	7, 10, 11, 14
SPLITTING	4	$\|$	6, 8
UNION	4	$+$	6, 8
UNION	5, 9, 12	$+$	7, 10, 11, 14
SPLITTING	5, 7, 9, 10, 12, 14	$\|$	11
SPLITTING	4	$\|$	6, 8
UNION	11	$+$	13
UNION	4	$+$	6, 8
SPLITTING	4	$\|$	6, 8
UNION	5, 7, 9, 10, 12, 14	$+$	6, 8
SPLITTING	5, 7, 9, 10, 12, 14	$\|$	6, 8
SPLITTING	11	$\|$	13
UNION	1, 2, 3	$+$	6, 8
SPLITTING	1, 2, 3	$\|$	6, 8

Independent participants. One may ask at this point whether any simpler stochastic model would be able to generate the same $\beta_{i,j}$ frequencies. In the above model the participants are intrinsically correlated because the whole matrix A is involved forming the probabilities of groups. The following model emerges from the idea of independence. Let us offer the possibility to the participants of the party to choose *independently* from finitely many options, like:

– to have a delicate food,

– to play hide and seek,

– to watch TV,

– to discuss Shakespeare,

– to make a small excursion.

With the program in hand, people having their preferences make their choices independently in the programs and the groups are formed in a natural way by the programs. Let us denote by $w(i, r)$ the probability that the i-th participant chooses the r-th possibility then

$$P\big(\pi(i) = \pi(j)\big) = \sum_{r=1}^{R} w(i,r)w(j,r),$$

where R denotes the number of possibilities.

Nonnegative matrix factorization was investigated in [13]. We have the constraint

$$\sum_{r=1}^{R} w(i,r) = 1, \qquad 1 \le i \le n,$$

which leads to a poor fit of our data. Interestingly, dropping the constraint, the frequencies $\beta_{i,j}$ has a good factorization with $w(i,r)$ given in the Table 6. If the number of participants goes to infinity, the size of groups is the most important feature of the distribution. It may remain bounded or slowly increasing as it is the case in politics when the groups are the political parties having the tendency to become of small number mostly because preference choice. The second possibility is the square root law: the size of groups and their number both are around the square root of n. Third possibility is represented in chemistry: the size of groups remains small and the number of groups increases with c for example for proteins. In the independent model the situation is easily controlled by R but in the case of pair-potential model we do not know the answer. Our guess is the third possibility on the argument that $Q(\pi, A)$ may achieve the size n^2 for π with small groups. A natural way to control the size of groups is to add a constant to the pair-potentials, i.e. to apply the pair-potentials $\tilde{a}_{i,j} = a_{i,j} + \Delta$. Negative Δ shrinks the groups and positive Δ increases them. When all $\tilde{a}_{i,j}$ become positive, all participants are in the same group with large probability. As a matter of fact, independence is not far from the pair-potential model: it is equivalent to the random graph model conditioned on the restriction to graphs representing only partitions. For large c we substituted all of the $a_{i,j}$-s with zero and used the parameter Δ only. According to our computer experiments $\Delta = 0.03$ seems to be the critical value for $n = 10\,000$.

Table 6. Factorization of equivalence probabilities

0.23545	0.77188	–	0.06091	0.01306	0.05126	0.01362
–	0.99511	0.00546	0.03724	0.00529	–	–
0.00937	0.94614	0.03934	0.00191	0.00865	0.02693	–
–	–	0.01632	0.10669	0.21538	–	–
0.25981	–	0.62682	0.03831	–	0.37697	0.45547
0.04035	0.01241	0.37992	1.09130	–	–	–
0.72835	0.08377	0.12272	0.08269	–	0.14629	0.51251
–	0.05948	0.16962	0.56842	–	0.00001	0.41019
0.15838	0.03633	0.34507	0.02085	0.00001	0.63947	0.46404
0.86278	0.26262	0.15723	0.09302	0.00299	–	0.21790
0.79914	0.16356	–	0.10403	0.03747	0.00339	0.34943
–	0.01224	0.65654	0.28753	0.00381	0.22826	–
–	–	–	–	1.30074	–	0.00391
0.60732	0.13486	–	–	–	0.72583	0.26880

Checkerboard model. Parties and hypergraphs are appeared as early as 1941 in the literature [11] where 18 ladies attending on 14 parties are investigated. It is a special case of partitions when only two groups are considered, whether each person is present or absent in a party. In the next table we show the application of checkerboard model to the Table 7. We reordered slightly the ladies and parties and grouped them into 8 and 6 clusters respectively. The probabilities that a lady belonging to the ith cluster takes part in the party belonging to the jth cluster are given in Table 8.

There are 22 pairs of (i, j)-s with probability zero, for example $i = 3$, $j = 6$, accordingly ladies 5, 6, 7 did not attended in parties 10, 12, 13, 14. There are 12 pairs of (i, j)-s with probability one, for example $i = 1$, $j = 2$. accordingly ladies 1, 3 attended in parties 3, 5, 6. For the remaining 14 pairs the range of probabilities are between 0.11 and 0.83. There is only one pair ($i = 7$, $j = 4$) with probability 0.5 which means the maximal uncertainty. In microarray analysis this is the so-called chequerboard structure. We characterize the uncertainty of the data with the reciprocal of the delogarithmized averaged log-likelihood which is 1.205627. In case this quantity equals 2, the uncertainty is maximal, for all (i, j) pairs the probabilities are equal to 0.5. We can test the power of the model by mixing randomly the bits in the data. In this case the uncertainty is between 1.36 and 1.42.

Table 7. Davis – Gardner – Gardner data

		1	2	4	3	5	6	7	8	9	11	10	12	13	14
		1	1	1	2	2	2	3	3	4	5	6	6	6	6
1	1	1	1	1	1	1	1	0	1	1	0	0	0	0	0
3	1	0	1	1	1	1	1	1	1	1	0	0	0	0	0
2	2	1	1	0	1	1	1	1	1	0	0	0	0	0	0
4	2	1	0	1	1	1	1	1	1	0	0	0	0	0	0
5	3	0	0	1	1	1	0	1	0	0	0	0	0	0	0
6	3	0	0	0	1	1	1	0	1	0	0	0	0	0	0
7	3	0	0	0	0	1	1	1	1	0	0	0	0	0	0
8	4	0	0	0	0	0	1	0	1	1	0	0	0	0	0
9	4	0	0	0	0	1	0	1	1	1	0	0	0	0	0
10	5	0	0	0	0	0	0	1	1	1	0	0	1	0	0
11	5	0	0	0	0	0	0	0	1	1	0	1	1	0	0
16	5	0	0	0	0	0	0	0	1	1	0	1	1	0	0
12	6	0	0	0	0	0	0	0	1	1	0	1	1	1	1
13	6	0	0	0	0	0	0	1	1	1	0	1	1	1	1
14	7	0	0	0	0	0	1	1	0	1	1	1	1	1	1
15	7	0	0	0	0	0	0	1	1	0	1	1	1	1	1
17	8	0	0	0	0	0	0	0	0	1	1	0	0	0	0
18	8	0	0	0	0	0	0	0	0	1	1	0	0	0	0

Table 8. Structural probabilities of checkerboard model

	1	2	3	4	5	6
1	0.83	1	0.75	1	0	0
2	0.67	1	1	0	0	0
3	0.11	0.78	0.67	0	0	0
4	0	0.33	0.75	1	0	0
5	0	0	0.67	1	0	0.42
6	0	0	0.75	1	0	1
7	0	0.17	0.75	0.50	1	1
8	0	0	0	1	1	0

Cluster numbers 8 and 6 seem to be large, considering the numbers of ladies and parties but with smaller cluster numbers we were unable to present satisfactory clustering. In statistical investigations, in cluster analysis partitions appear mostly in the following two different aspects:

- we may form groups from the investigated objects on the basis that any connection is possible only inside the groups
- we may form the groups of similar objects

The second possibility is used in checkerboard model. Its extension to the pair-potential model is a numbering $f(k)$, $k = 1, \ldots, n$, of the participants such that

- $1 \leq f(k) \leq g;\ k = 1, \ldots, n$
- for all $1 \leq j \leq g$ there is a $1 \leq i \leq c$ such that $f(i) = j$
- there is a $g * g$ matrix D with entries $d_{u,v}$ such that $a_{i,j} = d_{f(i),f(j)}$ for all $1 \leq i < j \leq n$

this is called blown-up of the matrix D into matrix A. We investigated partition-clustering in [30] and interactive networks in [31]. A widely investigated process on partitions is Kingman's coalescent process [19]. In the Table 9. we give the groups where the first participants spent the most time.

The number of partitions. There is a recursion for P_n which denotes the number of partitions of n elements:

$$P_{n+1} = \sum_{j=0}^{n} \binom{n}{j} P_j, \qquad n = 0, 1, \ldots,$$

where $P_0 = 1$. (Especially $P_{14} = 190,899,322$.) There is an explicit form as well for P_n,

$$P_n = \sum_{j=1}^{n} j^n \frac{\delta(n-j)}{j!},$$

where

$$\delta(k) = \sum_{s=0}^{k} \frac{(-1)^s}{s!}.$$

Acknowledgements. We thank to R. Albert, D. Bancroft, L. Barabási, I. Bárány, M. Bolla, T. Breuer, I. Csiszár, K. Friedl, P. Hussami, M. Ispány, J. Komlós, A. Krámli, L. Lovász, K. Marton, I. Miklós, M. Simonovits, G. Simonyi, V. T. Sós, E. Szemerédi and T. Vicsek the enlightening conversations.

Table 9. Most frequent partitions in SIRP.DATA

	1	2	3	4	5	6	7	8	9	10	11	12	13	14
7271.18	1	1	1	0	0	0	0	0	0	0	0	0	0	0
1370.10	1	1	1	0	0	0	0	0	0	1	0	0	0	0
1134.26	1	1	1	0	0	0	1	0	0	1	1	0	0	1
1071.91	1	0	0	0	0	0	1	0	0	1	1	0	0	1
724.33	1	1	1	0	0	0	0	0	0	1	1	0	0	1
713.87	1	0	0	0	1	0	1	0	1	1	1	0	0	1
666.55	1	1	1	0	0	0	0	0	0	1	1	0	1	0
654.82	1	1	1	0	0	0	0	0	0	1	1	0	0	0
536.01	1	1	1	0	0	0	0	1	0	0	0	0	0	0
445.13	1	0	0	0	1	0	1	0	0	1	1	0	0	1
407.04	1	0	0	0	0	0	0	0	0	0	1	0	0	1
329.61	1	1	1	0	0	1	0	0	0	1	0	1	0	0
313.40	1	1	1	0	0	0	0	0	1	0	0	0	0	1
261.53	1	0	0	0	0	0	0	0	1	0	0	0	0	1
241.93	1	1	0	0	0	1	0	1	0	1	0	0	0	0
229.54	1	0	0	0	0	0	0	0	0	0	0	0	0	0
204.26	1	0	0	0	0	1	1	1	0	1	1	0	0	0
168.22	1	0	0	0	0	1	0	1	0	0	0	0	0	0
159.01	1	1	1	0	0	0	1	0	0	1	1	0	0	0
154.32	1	0	0	0	0	0	1	0	1	1	1	0	0	1
118.99	1	1	1	0	0	1	0	0	0	1	0	0	0	0
118.45	1	1	0	0	0	0	0	0	0	0	0	0	0	0

REFERENCES

[1] A. L. Barabási and R. Albert, Emergence of scaling in random networks, *Science*, **286** (1999), 509–512.

[2] O. Barndorff-Nielsen, *Information and Exponential families in statistical theory*, Chichester (Wiley, 1978).

[3] M. Bolla, Distribution of the eigenvalues of random block-matrices, *Linear Algebra and its Applications*, **377** (2004), 219–240.

[4] M. Bolla, Recognizing linear structure in noisy matrices, *Linear Algebra and its Applications*, **402** (2005), 228–240.

[5] M. Bolla and G. Tusnády, Spectra and optimal partitions of weighted graphs, *Discrete Mathematics*, **128** (1994), 1–20.

[6] B. Bollobás, O. Riordan, J. Spencer and G. Tusnády, The degree sequence of a scale-free random graph, *Random Structures Algorithms*, **18** (2001), 279–290.

[7] C. Borgs, J. Chayes, L. Lovász, V. T. Sós, B. Szegedy and K. Vesztergombi, Counting graph homomorphisms, in: *Topics in Discrete Mathematics* (eds. M. Klazar, J. Kratochvil, M. Loebl, J. Matousek, R. Thomas, P. Valtr), 315–371 (Springer, 2006).

[8] C. Borgs, J. Chayes, L. Lovász, V. T. Sós and K. Vesztergombi, *Graph limits and parameter testing*, STOC (2006).

[9] C. Borgs, J. Chayes, L. Lovász, V. T. Sós and K. Vesztergombi, *Convergent sequences of dense graphs I: subgraph frequencies, metric properties and testing*, arXiv:math.CO/0702004v1 31Jan2007.

[10] F. Chung, *Spectral graph theory*, revised (2006).

[11] A. Davis, B. B. Gardner and M. R. Gardner, *Deep south: a social anthropological study of caste and class*, University of Chicago Press (1941).

[12] P. Diaconis, A generalization of spectral analysis with applications to ranked data, The 1987 Wald memorial lectures, *The Annals of Statistics*, **17** (1989), 949–979.

[13] L. Finesso and P. Spreij, Nonnegative matrix factorization and I-divergence alternating minimization, *Linear Algebra and its Applications*, **416** (2006), 270–287.

[14] A. Frieze and R. Kannan, *Quick approximation to matrices and applications*, manuscript (2006).

[15] J. Gupta and P. Damien, Conjugacy class prior distributions on metric-based ranking models, *Journal of the Royal Statistical Society Series B*, **64** (2002), 433–445.

[16] X. Guyon, *Random fields on a network, modelling, statistics and applications*, Springer (1995).

[17] F. Hiai and D. Petz, The semicircle law, free random variables and entropy, *AMS Mathematical Surveys and Monograph*, **77** (2000).

[18] O. Khorunzhiy, W. Kirch and P. Müller, Lifschitz tails for spectra of Erdős–Rényi random graphs, *The Annals of Applied Probability*, **16** (2006), 295–309.

[19] J. F. C. Kingman, Origins of the coalescent: 1974–1982, *Genetics*, **156** (2000), 1461–1463.

[20] W. König, Orthogonal polynomial ensembles in probability theory, *Probability Survey*, **2** (2005), 385–447.

[21] L. Lovász and B. Szegedy, *Szemerédi's regularity lemma for the analyst*, manuscript (2006).

[22] C. L. Mallows, Non null ranking models I. *Biometrika*, **44** (1957), 114–130.

[23] B. D. McKay, The expected eigenvalue distribution of a large regular graph, *Linear Algebra and its Applications*, **40** (1981), 203–216.

[24] G. J. McLachlan and T. Krishnan, *The EM algorithm and extensions*, John Wiley &Sons (New York, 1997).

[25] M. L. Mehta, *Random matrices*, 3rd edn, New York Cademic Press (2004).

[26] N. Metropolis, A. W. Rosenbluth, M. N. Rosenbluth, A. H. Teller and E. Teller, Equations of state calculations by fast computing machines, *Journal of Chemical Physics*, **21** (1953), 1087–1092.

[27] T. Nepusz, L. Négyessy, G. Tusnády and F. Bazsó, *Predicting key areas and uncharted connections in the cerebral cortex using Szemerédi's regularity lemma*, manuscript (2006).

[28] R. L. Plackett, The analysis of permutations, *Appl. Statist.*, **24** (1975), 193–202.

[29] L. R. Rabiner, A tutorial on hidden Markov models and selected applications in speech recognition, *Proceedings of the IEEE*, **77** (1989), 257–286.

[30] L. Rejtő and G. Tusnády, Clustering methods in microarrays, *Periodica Mathematica Hungarica*, **50** (2005), 199–221.

[31] L. Rejtő and G. Tusnády, Reconstruction of Kauffman networks applying trees, *Linear Algebra and Application*, **417** (2006), 220–244.

[32] C. G. Small, Multidimensional medians arising from geodesics on graphs, *Annals of Statistics*, **25** (1997), 478–494.

[33] H. Stern, Models for distributions on permutations, *Journal of the American Statistical Association*, **85** (1990), 558–564.

[34] E. Szemerédi, Regular partitions of graphs, *Colloquies Internulionule C.N.R.S.*, **260**, Proalèkes Combinatorics et Théories des Graphes, Oracy (1976), pp. 399–401.

[35] A. D. Szlam, M. Maggioni, R. R. Coifman and J. C. Bremer, Jr.: Diffusion-driven multiscale analysis on manifolds and graphs: top-down and bottom-up constructions, *SPIE Wavelets XI*, **5914** (2005).

[36] G. M. Tallis and B. R. Dansie, An alternative approach to the analysis of permutations, *Appl. Statist.*, **32** (1983), 110–114.

[37] T. Tao, *Szemerédi's regularity lemma revisited*, arXiv:math.CO/0504472v2 16Nov2005.

[38] L. L. Thurstone, A law for comparative judgement, *Psychological Review*, **34** (1927), 278–286.

Villő Csiszár, Lídia Rejtő & Gábor Tusnády
Rényi Institute
Budapest, Hungary

e-mail: `villo@renyi.hu`, `rejto@renyi.hu`, `tusnady@renyi.hu`

APPENDIX

sign: vertex label d: degree, a: asset, w: wealth, n_i: i-th neighbor

sign	d	a	w	n_1	n_2	n_3	n_4	n_5	n_6
A	6	109	321	A1	A2	A3	A4	A5	A6
B	6	134	403	B1	B2	B3	B4	B5	B6
C	6	152	431	C1	C2	C3	C4	C5	C6
D	6	149	421	D1	D2	D3	D4	D5	D6
E	6	167	459	E1	E2	E3	E4	E5	E6
F	6	159	451	F1	F2	F3	F4	F5	F6
G	6	158	440	G1	G2	G3	G4	G5	G6
H	6	143	428	H1	H2	H3	H4	H5	H6
I	6	151	397	I1	I2	I3	I4	I5	I6
J	6	155	467	J1	J2	J3	J4	J5	J6
a	5	16	20	a1	b1	c1	d1	e1	
b	5	15	20	k1	l1	m1	n1	o1	
c	5	17	8	a2	b2	c2	d2	e2	
d	5	16	16	k2	l2	m2	n2	o2	
e	5	17	12	a3	b3	c3	d3	e3	
f	5	15	24	k3	l3	m3	n3	o3	
g	5	15	20	a4	b4	c4	d4	e4	
h	5	15	24	k4	l4	m4	n4	o4	
i	5	15	24	a5	b5	c5	d5	e5	
j	5	12	16	k5	l5	m5	n5	o5	
k	5	16	16	a6	b6	c6	d6	e6	
l	5	15	24	k6	l6	m6	n6	o6	
m	5	17	12	f1	g1	h1	i1	j1	
n	5	15	12	p1	q1	r1	s1	t1	
o	5	14	16	f2	g2	h2	i2	j2	
p	5	18	12	p2	q2	r2	s2	t2	
q	5	16	16	f3	g3	h3	i3	j3	
r	5	13	16	p3	q3	r3	s3	t3	
s	5	16	20	f4	g4	h4	i4	j4	
t	5	16	16	p4	q4	r4	s4	t4	

sign	d	a	w	n_1	n_2	n_3	n_4	n_5
u	5	16	16	f5	g5	h5	i5	j5
v	5	17	12	p5	q5	r5	s5	t5
w	5	16	12	f6	g6	h6	i6	j6
x	5	17	16	p6	q6	r6	s6	t6
a6	4	0	0	05	A2	A6		
B1	4	0	0	b2	l1	c4		
B2	4	0	0	b2	l2	d4		
b2	4	0	0	B3				
c1	4	0	0	25	35	C1		
c2	4	0	0	08	11	C2		
D1	4	0	0	d3	d1	n1		
d2	4	0	0	11	D2	02		
n2	4	0	0	11	26	D2		
D3	4	0	0	d3	a1	n3		
d3	4	0	0	26				
e3	4	0	0	03	14	E3		
f1	4	0	0	25	28	F1		
F2	4	0	0	f3	f2	s3		
f3	4	0	0	08	F3			
q2	4	0	0	05	07	11		
g4	4	0	0	04	28	G4		
g6	4	0	0	07	14	G6		
q6	4	0	0	08	G6	02		
r2	4	0	0	H2	H3	40		
H4	4	0	0	r4	g2	h4		
r4	4	0	0	04	28			
r5	4	0	0	03	13	H5		
r6	4	0	0	13	35	H6		
s2	4	0	0	08	13	I2		
i5	4	0	0	07	25	I5		

Appendix (continued)

sign	d	a	w	n_1	n_2	n_3	n_4
s5	4	0	0	35	I1	I5	
j1	4	0	0	03	07	J1	
t1	4	0	0	26	J1	40	
J5	4	0	0	i1	j5	t5	
j6	4	0	0	05	14	J6	
01	4	4	0	m6	p2	t4	j5
03	4	0	0	a2			
04	4	0	0	a4	g5		
05	4	0	0	p6			
06	4	4	0	o3	h4	h5	i1
09	4	4	0	a4	l1	c6	f5
10	4	4	0	l4	b5	e1	i6
12	4	4	0	e2	p3	g3	t2
13	4	0	0	j3			
14	4	0	0	q1			
15	4	4	0	a3	k4	n1	d6
16	4	4	0	o1	p6	j2	j3
17	4	4	0	n3	e1	e6	q4
18	4	4	0	b3	n4	s1	t5
19	4	4	0	k1	b4	h1	n5
20	4	4	0	a2	l5	c5	o5
21	4	4	0	m1	p5	i3	s6
22	4	4	0	k5	k6	l2	c4
23	4	4	0	a1	a5	q3	t3
24	4	4	0	m6	p2	s4	s6
25	4	0	0	b6			
26	4	0	0	d6			
27	4	4	0	m4	e4	p4	i4
28	4	0	0	k4			
29	4	4	0	b6	e5	f6	j4
30	4	4	0	k3	l6	c3	f2
31	4	4	0	l3	d5	h2	r3

sign	d	a	w	n_1	n_2	n_3	n_4
32	4	4	0	a5	l3	d1	f4
33	4	4	0	q5	h3	s3	t6
34	4	4	0	k2	b1	o4	g5
35	4	0	0	h3			
36	4	4	0	m2	p1	g1	r1
37	4	4	0	l4	m3	o6	h6
38	4	4	0	d4	n6	o2	m5
39	4	4	0	f4	q1	g2	i2
A1	3	0	0	k1	l5		
A2	3	0	0	k2			
a3	3	0	0	A3			
k3	3	0	0	E5			
k5	3	0	0	A5			
A6	3	0	0	k6			
b1	3	0	0	D4			
B3	3	0	0	b3			
B4	3	0	0	b4	o2		
b5	3	0	0	B5			
B6	3	0	0	l6	e4		
C1	3	0	0	m1			
C2	3	0	0	m2			
C3	3	0	0	c3	m3		
m4	3	0	0	C4			
c5	3	0	0	D6			
c6	3	0	0	E1			
n4	3	0	0	i4			
d5	3	0	0	D5			
D6	3	0	0	n6			
E1	3	0	0	o1			
e2	3	0	0	E2			
E3	3	0	0	o3			
o4	3	0	0	E4			

sign	d	a	w	n_1	n_2
E5	3	0	0	e5	
o5	3	0	0	A4	
E6	3	0	0	e6	o6
F1	3	0	0	p1	
F3	3	0	0	p3	
p4	3	0	0	F4	
f5	3	0	0	F5	
p5	3	0	0	H2	
f6	3	0	0	F6	
g1	3	0	0	G1	
g3	3	0	0	40	
q3	3	0	0	G5	
G4	3	0	0	q4	
G5	3	0	0	q5	
II1	3	0	0	r1	j4
h1	3	0	0	J2	
h2	3	0	0	G2	
H3	3	0	0	t6	
r3	3	0	0	J3	
h5	3	0	0	I6	
h6	3	0	0	I4	
I1	3	0	0	s1	
I2	3	0	0	i2	
i3	3	0	0	02	
I4	3	0	0	s4	
I6	3	0	0	i6	
j2	3	0	0	G3	
t2	3	0	0	I3	
J3	3	0	0	t3	
t4	3	0	0	J4	

BOLYAI SOCIETY
MATHEMATICAL STUDIES, 17

Horizons of Combinatorics
Balatonalmádi
pp. 67–78.

Proof Techniques for Factor Theorems

YOSHIMI EGAWA

1. Introduction

In this paper, we consider only finite, undirected, simple graphs with no loops and no multiple edges. The purpose of this paper is to illustrate three different types of proof techniques for theorems concerning the existence of a 2-factor.

We start with definitions. Let G be a graph. We let $V(G)$ and $E(G)$ denote the set of vertices and the set of edges of G, respectively. For $x \in V(G)$, we let $N_G(x)$ denote the set of vertices of G adjacent to x, and set $d_G(x) = |N_G(x)|$. We let $\delta(G)$ denote the minimum of $d_G(x)$ as x ranges over $V(G)$. For $S \subseteq V(G)$, we let $N_G(S) = \bigcup_{x \in S} N_G(x)$. For S, $T \subseteq V(G)$ with $S \cap T = \emptyset$, $E_G(S,T)$ denotes the set of those edges of G which join a vertex in S and a vertex in T. For $S \subseteq V(G)$, $G[S]$ denotes the subgraph induced by S in G, and $G - S$ denotes the subgraph obtained from G by deleting S; thus $G - S = G[V(G) - S]$. We say that S is independent in G if $E(G[S]) = \emptyset$. We let $\alpha(G)$ denote the maximum cardinality of an independent subset of $V(G)$. Note that $\alpha(G) + \delta(G) \leq |V(G)|$. A subgraph of G is often identified with its vertex set; for example, if H is a subgraph of G, we write $N_G(H)$ for $N_G(V(H))$. A subset S of $V(G)$ is called a cutset if $G - S$ is disconnected. For a nonnegative integer k, G is said to be k-connected if $|V(G)| \geq k + 1$ and G has no cutset with cardinality strictly less than k. We let $\kappa(G)$ denote the maximum integer k such that G is k-connected. Clearly $\kappa(G) \leq \delta(G)$.

A 2-factor of G is a subgraph F of G such that $V(F) = V(G)$ and $d_F(x) = 2$ for all $x \in V(G)$. Thus a subgraph F of G with $V(F) = V(G)$ is a 2-factor if and only if each component of F is a cycle. For a graph Q, we say that G is Q-free if G does not contain a subgraph isomorphic to Q as an induced subgraph. For a family $\mathscr{Q}$ of graphs, G is said to be $\mathscr{Q}$-free if G is Q-free for each $Q \in \mathscr{Q}$.

For an integer $t \geq 2$, $K_{1,t}$ denotes the complete bipartite graph with partite sets of cardinalities 1 and t. Note that for a connected graph G, G is $K_{1,2}$-free if and only if G is a complete graph. Finally we let P_4 denote the path with four vertices (i.e., with length 3).

2. $K_{1,t}$-FREE GRAPHS

In this section and the next section, we consider $K_{1,t}$-free graphs. The following theorem was proved by Ota and Tokuda in [10].

Theorem 2.1. *Let $t \geq 3$ be an integer. Let G be a $K_{1,t}$-free graph, and suppose that $\delta(G) \geq 2t - 2$. Then G has a 2-factor.*

In Theorem 2.1, the bound $2t - 2$ on $\delta(G)$ is best possible. However, if we assume that $\kappa(G) \geq 2$, then, as was shown in [1], we can lower the bound as follows.

Theorem 2.2. *Let $t \geq 3$ be an integer. Let G be a $K_{1,t}$-free graph, and suppose that $\kappa(G) \geq 2$ and $\delta(G) \geq t$. Then G has a 2-factor.*

In Theorem 2.2, the bound t on $\delta(G)$ is sharp. For $t \geq 4$, this can be seen from the following example (the case where $t = 3$ is somewhat exceptional, and thus we focus our attention on the case where $t \geq 4$). Let $n \geq t - 1$ be an integer, and define a graph $G = G(t, n)$ of order $3n - 1$ by

$$V(G) = \{x_i \mid 2 \leq i \leq n\} \cup \{y_j, z_j \mid 1 \leq j \leq n\},$$

$$E(G) = \{y_j z_j \mid 1 \leq j \leq n\}$$

$$\cup \big\{ x_i y_j, x_i z_j \mid 2 \leq i \leq n, \quad 1 \leq j \leq n,$$

$$j - i \in \{0, 1, \ldots, t - 1\} \pmod{n} \big\}.$$

Then G is $K_{1,t}$-free and $\kappa(G) \geq 2$ and $\delta(G) = t - 1$, but G has no 2-factor. Actually $\kappa\big(G(t, n)\big) = t - 2$. Thus the bound t remains sharp even if we assume that $\kappa(G) \geq t - 2$ in Theorem 2.2. However, the situation is different if $\kappa(G) \geq t - 1$; in fact, the following result was also proved in [1].

Theorem 2.3. *Let $t \geq 4$ be an integer. Let G be a $K_{1,t}$-free graph, and suppose that $\kappa(G) \geq t - 1$. Then G has a 2-factor.*

We now describe the proof of Theorem 2.2. For that purpose, we need some preparations. Let S, T be disjoint subsets of the vertex set of a graph G. A component H of $G - S - T$ is called an odd component if $\left| E_G(H, T) \right|$ is odd. We let $h_G(S, T)$ denote the number of odd components, and set $f_G(S, T) = 2|S| - \sum_{y \in T} \left(2 - d_{G-S}(y) \right) - h_G(S, T)$. The following theorem is a special case of the g-Factor Theorem of Tutte [11].

Theorem 2.4. *A graph G has a 2-factor if and only if $f_G(S, T) \geq 0$ for all subsets S, T of $V(G)$ with $S \cap T = \emptyset$.*

The following lemma was also proved in [11].

Lemma 2.5. *For any two disjoint subsets S, T of the vertex set of a graph G, $f_G(S, T)$ is even.*

The following lemma follows from Lemma 2.5.

Lemma 2.6. *Let G be a graph with no 2-factor, and choose disjoint subsets S, T of $V(G)$ with $f_G(S, T) < 0$ so that $|S \cup T|$ is as large as possible. Then $|V(H)| \geq 3$ for every component H of $G - S - T$.*

Proof of Theorem 2.2. for $t \geq 4$. Suppose that G has no 2-factor. By Theorem 2.4, there exist S, $T \subseteq V(G)$ with $S \cap T = \emptyset$ such that $f_G(S, T) < 0$. We choose S and T so that $|S \cup T|$ is as large as possible. Set

$$A = \left\{ y \in T \mid d_{G[T]}(y) = 0 \right\}, \qquad B = \left\{ y \in T \mid d_{G[T]}(y) = 1 \right\}.$$

Then

$$f_G(S, T) = 2|S| - \sum_{y \in T} \left(2 - d_{G[T]}(y) \right) + \left| E_G\big(T, V(G) - S - T\big) \right| - h_G(S, T)$$

$$\geq 2|S| - 2|A| - |B| + \left| E_G\big(T, V(G) - S - T\big) \right| - h_G(S, T).$$

Let $H_1, \ldots, H_r$ be the components of $G - S - T$. We may assume that there exist integers r_0, r_1, r_2, r_3 with $r_0 + r_1 + r_2 + r_3 = r$ such that $E(H_i, T) = \emptyset$ for each $1 \leq i \leq r_0$, $\left| E(H_i, T) \right| = 1$ for each $r_0 + 1 \leq i \leq r_0 + r_1$, $\left| E(H_i, T) \right| = 2$ for each $r_0 + r_1 + 1 \leq i \leq r_0 + r_1 + r_2$, and $\left| E(H_i, T) \right| \geq 3$ for each $r_0 + r_1 + r_2 + 1 \leq i \leq r_0 + r_1 + r_2 + r_3$. Then $h_G(S, T) \leq r_1 + r_3$. Set

$$U_1 = \bigcup_{r_0 + 1 \leq i \leq r_0 + r_1} V(H_i), \qquad U_2 = \bigcup_{r_0 + r_1 + 1 \leq i \leq r_0 + r_1 + r_2 + r_3} V(H_i).$$

Then $\left|E_G(U_1, T)\right| = r_1$ and $\left|E_G(U_2, T)\right| \geq 2r_2 + 3r_3$. Consequently $f_G(S, T) \geq 2|S| - 2|A| - |B| + 2\left|E_G(U_2, T)\right|/3$. Since $t \geq 4$, it follows that

$$(1) \qquad f_G(S, T) \geq 2|S| - 2|A| - |B| + 2\left|E_G(U_2, T)\right|/(t-1).$$

For each i with $r_0 + 1 \leq i \leq r_0 + r_1$, if we write $E_G(H_i, T) = \{v_i y_i\}$ ($v_i \in V(H_i), y_i \in T$), then since $\left|V(H_i)\right| \geq 2$ by Lemma 2.6, there exists $x_i u_i \in E_G(S, H_i)$ ($x_i \in S, u_i \in V(H_i)$) with $u_i \neq v_i$ by the assumption that $\kappa(G) \geq 2$ (note that if $r_1 \neq 0$, then $S \neq \emptyset$). For each $x \in S$, set $M(x) = \{u_i \mid x_i = x\}$. Then $\sum_{x \in S}\left|M(x)\right| = r_1$. Since G is $K_{1,t}$-free, $\left|M(x)\right| + \left|E_G(\{x\}, A)\right| + \left|E_G(\{x\}, B)\right|/2 \leq t - 1$ for all $x \in S$. Hence

$$(2) \qquad r_1 + \left|E_G(S, A)\right| + \left|E_G(S, B)\right|/2 \leq (t-1)|S|.$$

On the other hand, since $\delta(G) \geq t$, we have

$$(3) \qquad \left|E_G(S, A)\right| + \left|E_G(U_1, A)\right| + \left|E_G(U_2, A)\right| \geq t|A| \geq (t-1)|A|,$$

$$(4) \qquad \left|E_G(S, B)\right| + \left|E_G(U_1, B)\right| + \left|E_G(U_2, B)\right| \geq (t-1)|B|.$$

Since $r_1 \geq \left|E_G(U_1, A)\right| + \left|E_G(U_1, B)\right| \geq \left|E_G(U_1, A)\right| + \left|E_G(U_1, B)\right|/2$, it follows from (2) through (4) that $-\left|E_G(U_2, A)\right| - \left|E_G(U_2, B)\right|/2 \leq (t-1)|S| - (t-1)|A| - (t-1)|B|/2$, which clearly implies $(t-1)|S| - (t-1)|A| - (t-1)|B|/2 + \left|E_G(U_2, T)\right| \geq 0$. Therefore it follows from (1) that $f_G(S, T) \geq 0$, which contradicts the assumption that $f_G(S, T) < 0$. This completes the proof of Theorem 2.2 for $t \geq 4$. $\blacksquare$

3. $\{P_4, K_{1,t}\}$-FREE GRAPHS

In this section, in addition to the assumption that G is $K_{1,t}$-free, we assume that G is P_4-free. Then in Theorems 2.1 and 2.2, we can replace the lower bound on $\delta(G)$ virtually by $t - 1$.

To state the result, we introduce the following notation. Let $t \geq 2$ be an integer. Define $G(t, t-1)$ as in Section 2. Thus

$$V\big(G(t, t-1)\big) = \{x_i \mid 2 \leq i \leq t-1\} \cup \{y_j, z_j \mid 1 \leq j \leq t-1\},$$

$$E\big(G(t, t-1)\big) = \{y_j z_j \mid 1 \leq j \leq t-1\}$$

$$\cup \{x_i y_j, x_i z_j \mid 2 \leq i \leq t-1, \ 2 \leq j \leq t-1\}.$$

We let $\mathscr{G}(t)$ denote the family of graphs G with $V(G) = V\big(G(t, t-1)\big)$ and $E\big(G(t, t-1)\big) \subseteq E(G) \subseteq E\big(G(t, t-1)\big) \cup \{x_i x_j \mid 2 \leq i, j \leq t-1,\ i \neq j\}$ (we remark that not all members of $\mathscr{G}(t)$ are P_4-free). Note that for each member G of $\mathscr{G}(t)$, G is $K_{1,t}$-free and $\kappa(G) = \max\{t-2, 1\}$ and $\delta(G) = t-1$, and G has no 2-factor. The following result was proved in [5].

Theorem 3.1. *Let t, k be integers with $t \geq 2$ and $1 \leq k \leq t-1$. Let G be a $\{P_4, K_{1,t}\}$-free graph, and suppose that $\kappa(G) \geq k$ and $\delta(G) \geq t-1$. Then G has a 2-factor with at most $t-k$ components, unless $(1 \leq k \leq \max\{t-2, 1\}$ and) G is isomorphic to a member of $\mathscr{G}(t)$.*

Although there are some similarities between the statements of Theorems 2.2 and 3.1, the proofs are entirely different. The proof of Theorem 3.1 is based on the following simple observation, which is implicitly stated in Faudree–Faudree–Ryjacek [6].

Lemma 3.2. *Let G be a connected noncomplete P_4-free graph, and let S be a cutset with $|S| = \kappa(G)$. Then each vertex in S is adjacent to all vertices in $V(G) - S$.*

The following lemma is an immediate consequence of Lemma 3.2.

Lemma 3.3. *Let $t \geq 2$ be an integer, and let G be a connected P_4-free graph. Then G is $K_{1,t}$-free if and only if $\alpha(G) \leq t-1$.*

In view of Lemma 3.3, we can restate Theorem 3.1 in the following form.

Theorem 3.4. *Let t, k be integers with $t \geq 2$ and $1 \leq k \leq t-1$. Let G be a P_4-free graph, and suppose that $\kappa(G) \geq k$, $\alpha(G) \leq t-1$ and $\delta(G) \geq t-1$. Then G has a 2-factor with at most $t-k$ components, unless G is isomorphic to a member of $\mathscr{G}(t)$.*

To make in relief the point of argument, we here prove the following weaker statement.

Proposition 3.5. *Let $t \geq 2$ be an integer. Let G be a connected P_4-free graph, and suppose that $\alpha(G) \leq t-1$ and $\delta(G) \geq t-1$. Then G has a 2-factor unless G is isomorphic to a member of $\mathscr{G}(t)$.*

Naturally, in the proof of Proposition 3.5, we make use of the following theorem, which is due to Chvátal and Erdős [4].

Theorem 3.6. *Let G be a graph of order at least 3, and suppose that $\alpha(G) \leq \kappa(G)$. Then G contains a connected 2-factor, i.e., a hamiltonian cycle.*

Proof of Proposition 3.5. We proceed by induction on t. If $t = 2$, then G is a complete graph, and hence the proposition clearly holds. Thus let $t \geq 3$, and assume that the theorem is proved for smaller values of t. In view of Theorem 3.6, we may assume $\alpha(G) > \kappa(G)$. Let S be a cutset with $|S| = \kappa(G)$. Let $H_1, \ldots, H_r$ be the components of $G - S$. If there exists i such that $|V(H_i)| = 1$, then $|S| \geq \delta(G)$, and hence $\kappa(G) = \delta(G) \geq t - 1 \geq \alpha(G)$, a contradiction. Thus $|V(H_i)| \geq 2$ for all $1 \leq i \leq r$. Also since $\alpha\big(G[S]\big) \leq |S| = \kappa(G) < \alpha(G)$, it follows from Lemma 3.2 that $\alpha(G) = \max\big\{\alpha(G - S), \alpha\big(G[S]\big)\big\} = \alpha(G - S) = \sum_{1 \leq i \leq r} \alpha(H_i)$. Note that this in particular implies $|V(G) - S| \geq \alpha(G) > |S|$.

Claim 1. *Let $B \subseteq S$, and suppose that $G - B$ has a 2-factor. Then G has a 2-factor.*

Proof. Let F be a 2-factor of $G - B$. Then $\big|E(F) \cap E(G - S)\big| \geq |V(G) - B| - 2|S - B| = |V(G) - S| - |S - B| > |B|$. Hence by Lemma 3.2, we can "insert" all vertices of B into F, to obtain a 2-factor of G. $\blacksquare$

We return to the proof of the proposition. For each i, set $t_i = \alpha(H_i)$ and $d_i = \delta(H_i)$. Note that $|V(H_i)| \geq t_i + d_i$.

Claim 2. *Let $1 \leq i \leq r$, and let A be a subset of S with $|A| = \max\{t_i - d_i + 1, 0\}$. Then $G\big[V(H_i) \cup A\big]$ has a 2-factor.*

Proof. Set $H = G\big[V(H_i) \cup A\big]$. By Lemma 2.2, $\alpha(H) = \max\big\{t_i, |A|\big\} = t_i$ and $\delta(H) \geq \min\big\{d_i + |A|, |V(H_i)|\big\} = d_i + |A| \geq t_i + 1$. Hence H has a 2-factor by the induction assumption. $\blacksquare$

We may assume that there exists $q \geq 0$ such that $d_i \leq t_i$ for all $1 \leq i \leq q$, and $d_i \geq t_i + 1$ for all $q + 1 \leq i \leq r$.

Case 1. $q = 0$.

It follows from Claim 2 that for each i, H_i contains a 2-factor F_i. Then $\bigcup_{1 \leq i \leq r} F_i$ is a 2-factor of $G - S$. Hence, applying Claim 1 with $B = S$, we see that G has a 2-factor.

Case 2. $q \geq 1$.

Set $p = \max\{q, r - 1\}$.

Claim 3.

$$|S| \geq \sum_{1 \leq i \leq p} (t_i - d_i + 1) + \left(\sum_{p+1 \leq i \leq r} t_i\right) - 1.$$

Proof. Since $\delta(G) \geq t - 1$,

$$|S| \geq t - 1 - d_1 \geq \sum_{1 \leq i \leq r} t_i - d_1 = (t_1 - d_1 + 1) + \sum_{2 \leq i \leq r} t_i - 1$$

$$\geq \sum_{1 \leq i \leq p} (t_i - d_i + 1) + \sum_{p+1 \leq i \leq r} t_i - 1. \qquad \blacksquare$$

By Claim 3, there exist disjoint subsets $A_1, A_2, \ldots, A_p, A_r$ of S such that $|A_i| = t_i - d_i + 1$ for all $1 \leq i \leq p$ and $|A_r| = t_r - 1$. It follows from Claim 2 that for each $1 \leq i \leq p$, $G[V(H_i) \cup A_i]$ has a 2-factor. If $1 \leq q \leq r - 1$, then by Claim 2, H_i has a 2-factor for each $q + 1 \leq i \leq r$, and hence we see that G has a 2-factor by applying Claim 1 with $B = S - \bigcup_{1 \leq i \leq q} A_i$. Thus we may assume $q = r$. At the cost of relabelling, we may assume $|V(H_1)| \leq |V(H_2)| \leq \cdots \leq |V(H_r)|$. Set $H = G[V(H_r) \cup A_r]$. We have $\alpha(H) = t_r$ and $\delta(H) = t_r + d_r - 1$. If H has a 2-factor, we can argue as in the case where $1 \leq q \leq r - 1$. Thus in view of the induction assumption, we may assume $H \in \mathscr{G}(t_r + 1)$. Then $\delta(H) = t_r$, and hence $d_r = 1$. If $|V(H_r)| \geq 3$, then $\{v \in V(H) \mid d_H(v) = t_r\} = \{v \in V(H_r) \mid d_{H_r}(v) = 1\}$ is independent, which contradicts the structure of graphs in $\mathscr{G}(t_r + 1)$. Thus $|V(H_r)| = 2$, and hence $|V(H_i)| = 2$ for all i. This implies $r = \alpha(G)$ and $|S| \geq \delta(G) - 1 \geq t - 2$. Since $|S| < \alpha(G) \leq t - 1$, this forces $r = \alpha(G) = t - 1$ and $|S| = t - 2$, and we therefore obtain $G \in \mathscr{G}(t)$. This completes the proof of Proposition 3.5. $\blacksquare$

4. INDEPENDENCE NUMBER

In this section, we take up a conjecture concerning the existence of a 2-factor with a specified number of components. Modifying the proof of Proposition 3.5, we can prove the following proposition.

Proposition 4.1. *Let p be a positive integer. Let G be a P_4-free graph of sufficiently large order, and suppose that $\alpha(G) \leq \kappa(G)$. Then G has a 2-factor with precisely p components.*

Note that Theorem 3.6 shows that for $p = 1$, Proposition 4.1 holds even if we drop the condition that G is P_4-free. Thus it is natural to make the following conjecture, which was formulated by Kaneko and Yoshimoto in [8].

Conjecture 4.2. *Let p be a positive integer. Let G be a graph of sufficiently large order, and suppose that $\alpha(G) \leq \kappa(G)$. Then G has a 2-factor with precisely p components.*

In connection with the case $p = 2$ of Conjecture 4.2, the following theorem was proved in [8].

Theorem 4.3. *Let G be a 4-connected graph of order at least 6, and suppose that $\alpha(G) \leq \kappa(G)$. Then G has a 2-factor with precisely two components.*

Before describing the proof of Theorem 4.3, we state a related result. Let $r(m, n)$ denote the Ramsey number, i.e., the minimum of those integers r for which the proposition that every graph of order at least r contains either a complete subgraph of order m or an independent subset of cardinality n holds. The following theorem was proved by Chen et al. in [3].

Theorem 4.4. *Let p, t be positive integers. Let G be a graph with $\alpha(G) = t$ and $|V(G)| \geq p \cdot r(t + 4, t + 1)$, and suppose that $\alpha(G) \leq \kappa(G)$. Then G has a 2-factor with precisely p components.*

Note that Theorems 4.3 and 4.4 settle Conjecture 4.2 for $r = 2$; i.e., the following corollary holds.

Corollary 4.5. *Let p be a positive integer. Let G be a graph with $|V(G)| \geq p \cdot r(7, 4)$, and suppose that $\alpha(G) \leq \kappa(G)$. Then G has a 2-factor with precisely two components.*

In the proof of Theorem 4.3, we use arguments which are often used in proving the existence of a hamiltonian cycle. To outline the proof, we make some preparations. Let $D = x_1 x_2 \ldots x_m x_1$ be a cycle of a graph G with length $m \geq 6$. Let i, j be indices with $i + 3 \leq j \leq i + m - 3$, and suppose that $x_i x_{j+1}, x_{i+1} x_j \in E(G)$. In this situation, following [8], we call the cycle $x_{i+1} x_j x_{j+1} x_i x_{i+1}$ of length 4 a collar of D. In establishing the existence of a cycle with a collar, we make use of the following theorem concerning the existence of a cycle of length 4, which was proved by Amar, Fournier and Germa in [2] (see also [7] and [9]).

Theorem 4.6. *Let G be a graph with $|V(G)| \geq 6$, and suppose that $\alpha(G) \leq \kappa(G)$. Then G contains a cycle of length 4.*

Sketch of Proof of Theorem 4.3. It is easy to verify the theorem in the case where $|V(G)| \leq 7$. Thus assume $|V(G)| \geq 8$. By Theorem 4.6, G contains a cycle C of length 4. From the assumption that $\kappa(G) \geq 4$, it follows that G contains a cycle D such that C is a collar of D. We assume that we have chosen C and D so that $|V(D)|$ is as large as possible. If $V(D) = V(G)$, G has a desired 2-factor. Thus we may assume $V(D) \neq V(G)$. Write $D = x_1 x_2 \ldots x_m x_1$ so that $C = x_1 x_a x_{a+1} x_m x_1$. For each $x_i \in V(D)$, we define $f(x_i)$ by

$$
f(x_i) = \begin{cases} x_{i+1} & (\text{if } 1 \leq i \leq a-1) \\ x_1 & (\text{if } i = a) \\ x_m & (\text{if } i = a+1) \\ x_{i-1} & (\text{if } a+2 \leq i \leq m). \end{cases}
$$

It follows from the maximality of $|V(D)|$ that if H is a component of $G - V(D)$ and if $x_i \in \big(N_G(H) \cap V(D)\big) - \{x_a, x_{a+1}\}$, then $f(x_i) \notin N_G(H)$. Set $k = \kappa(G)$. Then $|V(D)| \geq k$, and $|N_G(H) \cap V(D)| \geq k$ for each component H of $G - V(D)$.

Claim 1. *Let H be a component of $G - V(D)$. Then $f\big(\big(N_G(H) \cap V(D)\big) - \{x_a, x_{a+1}\}\big)$ is independent unless $x_{a-1}, x_{a+2} \in N_G(H)$, and $f^{-1}\big(\big(N_G(H) \cap V(D)\big) - \{x_1, x_m\}\big)$ is independent unless $x_2, x_{m-1} \in N_G(H)$.*

Proof. By symmetry, it suffices to show that $f\big(\big(N_G(H) \cap V(D)\big) - \{x_a, x_{a+1}\}\big)$ is independent unless $x_{a-1}, x_{a+2} \in N_G(H)$. Suppose that there exist $x_i, x_j \in \big(N_G(H) \cap V(D)\big) - \{x_a, x_{a+1}\}$ such that $f(x_i)f(x_j) \in E(G)$. If $1 \leq i, j \leq a-1$ or $a+2 \leq i, j \leq m$, we easily get a contradiction. Thus we may assume $1 \leq i \leq a-1$, $a+2 \leq j \leq m$ and $\{i, j\} \neq \{a-1, a+2\}$. Let P be a path in H connecting a vertex in $N_G(x_i) \cap V(H)$ to a vertex in $N_G(x_j) \cap V(H)$. Then C is a collar of the cycle

$$
x_1 x_2 \ldots x_i P x_j x_{j+1} \ldots x_m x_{a+1} x_{a+2} \ldots x_{j-1} x_{i+1} x_{i+2} \ldots x_a x_1,
$$

which contradicts the maximality of $|V(D)|$.   ■

Since $\alpha(G) \leq k$, the following claim follows from Claim 1 (here and in what follows, we omit the discussion concerning technical difficulties arising from the possible existence of an edge in $G\big[f\big(\big(N_G(H) \cap V(D)\big) -$

$\{x_a, x_{a+1}\})\big]$ and $G\big[f^{-1}\big(\big(N_G(H) \cap V(D)\big) - \{x_1, x_m\}\big)\big]$; in fact, the possibility of existence of such an edge is ignored in [8], and errata to [8] will shortly be written by the authors of [8]).

Claim 2. *Let H be a component of $G - V(D)$. Then $\alpha(H) \leq 2$. Further if $\alpha(H) = 2$, then $N_G(H) \supseteq V(C)$ and $\big|N_G(H) \cap V(D)\big| = k$.* ∎

For simplicity, we henceforth assume that $G - V(D)$ is connected. Thus set $H = G - V(D)$.

Case 1. $\big|V(H)\big| \geq 3$.

If H contains a hamiltonian cycle D_0, then D_0 and D form a desired 2-factor. Thus in view of Claim 2 and Theorem 3.6, we may assume that there exists $v \in V(H)$ such that $H - \{v\}$ is disconnected. Since $\alpha(H) \leq 2$, $H - \{v\}$ consists of two components H_1 and H_2, and each H_i is a complete graph. Since $\kappa(G) = k$, it follows from Claim 2 that $\big|N_G(H_i) \cap V(C)\big| \geq 3$ for each $i = 1, 2$. Hence by symmetry, we may assume $x_1 \in N_G(H_1)$ and $x_a \in N_G(H_2)$. We may also assume that H contains a path P with $V(P) = V(H)$ connecting a vertex in $N_G(x_a) \cap V(H_2)$ to a vertex in $N_G(x_1) \cap V(H_1)$. Then the cycles $x_1 x_2 \ldots x_a P x_1$ and $x_{a+1} x_{a+2} \ldots x_m x_{a+1}$ form a desired 2-factor.

Case 2. $\big|V(H)\big| \leq 2$.

We here consider only the case where $\big|V(H)\big| = 1$ (the case where $\big|V(H)\big| = 2$ is quite similar). Thus write $V(H) = \{v\}$. We may assume $\{x_1, x_a\} \not\subseteq N_G(v)$ and $\{x_{a+1}, x_m\} \not\subseteq N_G(v)$. If $N_G(v) \cap \{x_a, x_{a+1}\} = \emptyset$, then by Claim 1, $f\big(N_G(v) \cap V(D)\big) \cup \{v\}$ is independent, which contradicts the assumption that $\alpha(G) \leq k$ because $\big|N_G(v) \cap V(D)\big| \geq k$. Thus $N_G(v) \cap \{x_a, x_{a+1}\} \neq \emptyset$. Similarly $N_G(v) \cap \{x_1, x_m\} \neq \emptyset$. Consequently $N_G(v) \cap V(C) = \{x_a, x_m\}$ or $\{x_1, x_{a+1}\}$. We may assume $N_G(v) \cap V(C) = \{x_a, x_m\}$. Since $\alpha(G) \leq k$, it follows from Claim 1 that $\big|N_G(v) \cap V(D)\big| = k$. We consider x_1, which is equal to $f(x_a)$. If there exists $x_i \in N_G(v) \cap V(D)$ with $2 \leq i \leq a - 1$ such that $x_1 f(x_i) \in E(G)$, then $x_1 x_2 \ldots x_i v x_a x_{a-1} \ldots x_{i+1} x_1$ and $x_{a+1} x_{a+2} \ldots x_m x_{a+1}$ form a 2-factor; if there exists $x_i \in N_G(v) \cap V(D)$ with $a + 2 \leq i \leq m - 1$ such that $x_1 f(x_i) \in E(G)$, then $x_1 x_2 \ldots x_{i-1} x_1$ and $x_i x_{i+1} \ldots x_m v x_i$ form a 2-factor. Thus we may assume that $f\big(\big(N_G(v) \cap V(D)\big) - \{x_m\}\big)$ is independent. Then $f\big(\big(N_G(v) \cap V(D)\big) - \{x_m\}\big) \cup \{v\}$ is independent. On the other hand, since $\alpha(G) \leq k$, $f\big(N_G(v) \cap V(D)\big) \cup \{v\}$ is not independent. In view of Claim 1, this implies $x_1 x_{m-1} \in E(G)$. Similarly $x_{a+1} x_{a-1} \in E(G)$. We now argue as above with x_1 replaced by x_2. If $x_2 v \in E(G)$, then $x_2 x_3 \ldots x_a v x_2$ and $x_{a+1} x_{a+2} \ldots x_{m-1} x_1 x_m x_{a+1}$ form a 2-factor.

Thus we may assume $x_2v \notin E(G)$. If there exists $x_i \in N_G(v) \cap V(D)$ with $3 \leq i \leq a-1$ such that $x_2 f(x_i) \in E(G)$, then $x_2 x_3 \ldots x_i v x_a x_{a-1} \ldots x_{i+1} x_2$ and $x_{a+1} x_{a+2} \ldots x_{m-1} x_1 x_m x_{a+1}$ form a 2-factor; if there exists $x_i \in N_G(v) \cap V(D)$ with $a+2 \leq i \leq m-1$ such that $x_2 f(x_i) \in E(G)$, then $x_2 x_3 \ldots x_{i-1} x_2$ and $x_i x_{i+1} \ldots x_{m-1} x_1 x_m v x_i$ form a 2-factor. Thus we may assume that $f\big(\big(N_G(v) \cap V(D)\big) - \{x_a, x_m\}\big) \cup \{x_2\}$ is independent. As above, this implies $x_2 x_{m-1} \in E(G)$. Now $x_2 x_3 \ldots x_{a-1} x_{a+1} x_{a+2} \ldots x_{m-1} x_2$ and $x_1 x_a v x_m x_1$ form a 2-factor with the desired properties. This completes the proof of Theorem 4.3. ∎

References

[1] R. E. L. Aldred, Y. Egawa, J. Fujisawa, K. Ota and A. Saito, *The existence of a 2-factor in $K_{1,n}$-free graphs with large connectivity and large edge-connectivity*, preprint.

[2] D. Amar, I. Fournier and A. Germa, Pancyclism in Chvátal–Erdős's graphs, *Graphs Combin.*, **7** (1991), 101–112.

[3] G. Chen, R. J. Gould, K. Kawarabayashi, K. Ota, A. Saito and I. Schiermeyer, *The Chvátal–Erdős condition and 2-factors with a specified number of components*, preprint.

[4] V. Chvátal and P. Erdős, A note on Hamilton circuits, *Discrete Math.*, **2** (1972), 111–113.

[5] Y. Egawa, J. Fujisawa, S. Fujita and K. Ota, *2-factors in r-connected $\{K_{1,k}, P_4\}$-free graphs*, preprint.

[6] J. R. Faudree, R. J. Faudree and Z. Ryjacek, Forbidden subgraphs that imply 2-factors, *Discrete Math.*, to appear.

[7] B. Jackson and O. Ordaz, Chvátal–Erdős conditions for paths and cycles in graphs and digraphs. A survey, *Discrete Math.*, **84** (1990), 241–254.

[8] A. Kaneko and K. Yoshimoto, A 2-factor with two components of a graph satisfying the Chvátal–Erdős condition, *J. Graph Theory*, **43** (2003), 269–279.

[9] D. Lou, The Chvátal–Erdős condition for cycles in triangle-free graphs, *Discrete Math.*, **152** (1996), 253–257.

[10] K. Ota and T. Tokuda, A degree condition for the existence of regular factors in $K_{1,n}$-free graphs, *J. Graph Theory*, **22** (1996), 59–64.

[11] W. T. Tutte, The factors of graphs, *Canadian J. Math.*, **4** (1952), 314–328.

Yoshimi Egawa

*Department of Mathematical Information
Science
Tokyo University of Science
Shinjuku-ku, Tokyo
162-8601 Japan*

BOLYAI SOCIETY
MATHEMATICAL STUDIES, 17

Horizons of Combinatorics
Balatonalmádi
pp. 79–103.

ERDŐS–HAJNAL-TYPE RESULTS ON INTERSECTION PATTERNS OF GEOMETRIC OBJECTS

JACOB FOX[*] and JÁNOS PACH[†]

In their seminal paper [21], Erdős and Hajnal raised the following question. Is it true that for any graph G there exists a constant $c = c(G) > 0$ with the property that every graph of n vertices that contains no induced subgraph isomorphic to G has a complete or an empty induced subgraph of size n^c? We answer this question in the affirmative for some special classes of graphs defined by geometric methods.

1. Introduction, Definitions

A classic result of Erdős and Szekeres [23] in Ramsey theory states that every graph on n vertices contains a clique (that is, a complete subgraph) or an independent set of size[1] at least $\frac{1}{2}\log n$. This bound, which has been slightly improved by Conlon [15], is tight up to a constant factor: Erdős [20] showed that there exists a graph on n vertices, for every integer $n > 1$, with no clique or independent of more than $2\log n$ vertices.

Here we consider the same problem for intersection graphs of geometric objects. Given a system $\mathcal{S}$ of n sets, their *intersection graph* is a graph $G_{\mathcal{S}}$ whose vertices are the elements of $\mathcal{S}$, two vertices $S, T \in \mathcal{S}$ being connected by an edge if and only if $S \cap T \neq \emptyset$. Applying the above theorem to $G_{\mathcal{S}}$,

[*]Supported by an NSF Graduate Research Fellowship and a Princeton Centennial Fellowship.

[†]Supported by NSF Grant CCF-05-14079, and by grants from NSA, PSC-CUNY, the Hungarian Research Foundation OTKA, and BSF.

[1]All logarithms in this paper are of base two.

we obtain that $\mathcal{S}$ always contains at least $\frac{1}{2}\log n$ members that are either pairwise intersecting or pairwise disjoint. Can we say more than this, if we assume that the elements of $\mathcal{S}$ are "nice" geometric objects in some, say, Euclidean space? At first glance it seems that the answer is no. It was shown by Tietze [52] that every finite graph can be realized as the intersection graph of convex closed polytopes in $\mathbb{R}^3$, and we can even assume no two of these polytopes have an interior point in common. However, this statement is certainly not true in the plane. It is not hard to see that, for instance, the bipartite graph on 15 vertices formed by replacing each edge of the clique K_5 by a path of length 2 has no such realization [19]. (See Fig. 1.) This immediately implies that *most* graphs with n vertices are not realizable in this way, as $n \to \infty$, since they almost surely contain a 15-vertex induced subgraph isomorphic to the one depicted in Fig. 1.

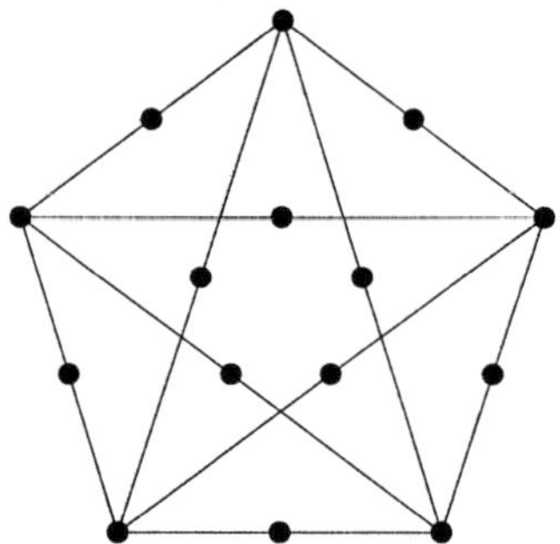

Fig. 1. The fifteen vertex graph formed by replacing each edge of K_5 by a path of length two is not an intersection graph of connected sets in the plane.

In a seminal paper written in 1989, Erdős and Hajnal [21] showed that, given any graph G, the family $\mathcal{F}(G)$ of all graphs that do not contain G as an induced subgraph have much stronger Ramsey-type properties than the family of all graphs. More precisely, they proved that there exists a constant $c = c(G) > 0$, depending only on G, such that every graph of n vertices that belongs to $\mathcal{F}(G)$ has a clique or an independent set of size at least $e^{c\sqrt{\log n}}$. They raised question whether one can always find a complete or an empty induced subgraph of size n^c. This remains one of the most challenging open problems in Ramsey theory.

A complete bipartite graph is said to be *balanced* if its vertex classes differ in size by at most one. A balanced complete bipartite graph with n vertices is called a *bi-clique of size n*.

Erdős, Hajnal, and Pach [22] proved a bipartite variant: There is a constant $c = c(G) > 0$ such that every graph on n vertices that belongs to

$\mathcal{F}(G)$ or its complement contains a bi-clique of size n^c. Recently, Fox and Sudakov [30] strengthened this result: there is a constant $c = c(G) > 0$ such that every graph on n vertices that belongs to $\mathcal{F}(G)$ contains a bi-clique or an independent set of size n^c.

It is easy to see that the Erdős–Hajnal theorem generalizes to *hereditary* families of graphs, that is, to any family $\mathcal{F}$ (other than the family of all finite graphs) that is closed under taking induced subgraphs. The families of all graphs that can be realized as intersection graphs of connected sets, convex sets, disks, segments, etc. in the plane obviously belong to this category.

For convenience, we use the following terminology.

Definition. A family $\mathcal{F}$ of graphs has the

1. *Erdős–Hajnal property* if there is a constant $c(\mathcal{F}) > 0$ such that every graph in $\mathcal{F}$ on n vertices contains a clique or an independent set of size $n^{c(\mathcal{F})}$;

2. *strong Erdős–Hajnal property* if there is a constant $b(\mathcal{F}) > 0$ such that for every graph G in $\mathcal{F}$ on n vertices, G or its complement $\overline{G}$ contains a bi-clique of size $b(\mathcal{F})n$.

The above terminology is justified by the following observation of Alon *et al.* [7]: If a hereditary family of graphs has the strong Erdős–Hajnal property, then it also has the Erdős–Hajnal property.

To see this, we need the notion of *cographs* (or *complement reducible graphs*), also used by Erdős and Hajnal [16, 21]. The trivial graph with one vertex is a cograph, and so are the disjoint union and the join of two cographs. (The join can be obtained from the disjoint union by adding all edges between the two parts.)

Suppose now that $\mathcal{F}$ has the strong Erdős–Hajnal property with a constant $b > 0$. That is, every $G \in \mathcal{F}$ with $n \geq 2$ vertices has two disjoint sets of vertices V_1, V_2, each of size at least bn, such that G contains either *all* edges between V_1 and V_2 or *no* edges running between V_1 and V_2. Let $s(n)$ denote the largest number s such that every $G \in \mathcal{F}$ with n vertices contains a cograph with s vertices. Applying the condition to the subgraphs of G induced by V_1 and V_2, we obtain that $s(n) \geq 2s(bn)$. Solving this recurrence, we conclude that $s(n) \geq n^c$, where $c = \frac{1}{\log 1/b}$. It remains to notice that every cograph of s vertices is a perfect graph, therefore it contains a clique or an independent set of size $\sqrt{s}$. Thus, G or its complement has a clique of size at least $n^{c/2}$, showing that $\mathcal{F}$ has the (weak) Erdős–Hajnal property.

It is certainly not true that all hereditary families of graphs have the strong Erdős–Hajnal property. For instance, the family of all triangle-free graphs does not have the strong Erdős–Hajnal property.

2. Convex Sets and Dilworth's Theorem

It was shown by Larman *et al.* that the family of intersection graphs of plane convex sets has the Erdős–Hajnal property. In fact, a somewhat stronger statement is true. We call a connected set *vertically convex* if any vertical line intersects it in an interval.

Theorem 2.1 [41]. *Any family of n vertically convex sets in the plane contains at least $n^{1/5}$ members that are either pairwise disjoint or pairwise intersecting.*

For the proof of Theorem 2.1, we need Dilworth's theorem [17], according to which any partially ordered set of more than pq elements contains a chain whose length is larger than p or an antichain that has more than q elements. Larman *et al.* [41] and Pach and Törőcsik [50] introduced *four* partial orders $<_1, <_2, <_3, <_4$ on the family of all vertically convex sets in the plane such that any two disjoint sets are comparable with respect to at least one of these partial orders, but no two intersecting elements are. Applying Dilworth's theorem four times, we obtain that any family of n plane convex sets has at least $n^{1/5}$ members that form a chain with respect to *some* $<_i$ or an antichain with respect to *all* $<_i$ ($1 \leq i \leq 4$). In the first case, these sets are pairwise disjoint, in the second case pairwise intersecting.

The best possible exponent in Theorem 2.1 is not known. Károlyi *et al.* [34] constructed a family of n segments in the plane, which has no more than $n^{\log 4/\log 27}$ members that are either pairwise disjoint or pairwise crossing. Recently, Jan Kynčl [40] has found a slightly better construction, for which the exponent is $\log 8/\log 169$.

Dilworth's theorem implies that every partially ordered set of n elements contains a chain or an antichain of size at least $\sqrt{n}$. If we have r partial orders on the same n-element ground set, then, by repeated application of Dilworth's theorem, there are at least $n^{\frac{1}{r+1}}$ elements such that no two are comparable by any of the r orders, or we can choose one of the r orders so that any two elements are comparable by it. Dumitrescu and G. Tóth [18] proved that for large values of r this statement is not very far from being optimal.

Theorem 2.2. [18] *For any $r \in \mathbb{N}$, there are n-element sets P and r partial orders defined on P such that connecting two elements of X if and only if they are comparable by at least one of these partial orders, the resulting graph contains neither a clique nor an independent set of size larger than $n^{\frac{1+\log r}{r+1}}$.*

The first named author established the existence of a much larger "homogeneous" *bipartite* pattern in partially ordered sets. For any partially ordered set $(P, >)$, we write $a \perp b$ if a and b are incomparable. For any pair of subsets A and B of P, we write $A > B$ if $a > b$ for all $a \in A$ and $b \in B$. Likewise, we write $A \perp B$ if $a \perp b$ for all $a \in A$ and $b \in B$.

Theorem 2.3 [24]. *Every n-element partially ordered set $(P, >)$ has two subsets $A, B \subset P$ with $|A| = |B| \geq \frac{n}{4 \log_2 n}$ such that $A > B$ or $A \perp B$, provided that n is sufficiently large. This result is tight up to a constant factor.*

In another paper we generalized the last result to multiple partial orders.

Theorem 2.4 [25]. *Let r be a fixed positive integer, and let $>_1, \ldots, >_r$ be partial orders on an n-element set P. Then there are two disjoint subsets $A, B \subset P$, each with at least $\frac{n}{2^{(1+o(1))(\log \log n)^r}}$ elements, such that either $A >_i B$ for at least one i, or $A \perp_i B$ for all $1 \leq i \leq r$.*

Applying Theorem 2.4 to the $r = 4$ partial orders defined on the family of vertically convex sets, which were mentioned above, we can conclude that any collection of n vertically convex sets in the plane has two disjoint subcollections, A and B, each with at least $\frac{n}{2^{(1+o(1))(\log \log n)^4}}$ members, such that either every member of A intersects all members of B or every member of A is disjoint from all members of B. This is only slightly weaker than saying that the family of intersection graphs of vertically convex sets in the plane has the strong Erdős–Hajnal property. In fact, it was shown in [49] that this is not the case.

A continuous curve that intersects every vertical line in at most one point is called x-*monotone*. Obviously, every x-monotone curve is vertically convex. It is easy to see that the *incomparability graph* of every partially ordered set $(P, >)$, defined by connecting two elements of P if and only if they are *not* comparable by $>$, can be realized as the intersection graph of a collection of x-monotone curves [49, 51]. Therefore, it follows from the tightness of Theorem 2.3 that the family of intersection graphs of x-monotone curves does not have the strong Erdős–Hajnal property.

In [28], Theorem 2.3 was slightly strengthened as follows. There is a positive constant c such that every n-element partially ordered set $(P, >)$ has two subsets $A, B \subset P$ such that either $|A| = |B| \geq \frac{cn}{\log n}$ and $A \perp B$, or $|A| = |B| \geq cn$ and $A > B$. For x-monotone curves, we have a similar result.

Theorem 2.5 [28]. *There exists a constant $c > 0$ with the property that the intersection graph G of any collection of n x-monotone curves in the plane satisfies at least one of the following two conditions:*

1. *G contains a bi-clique of size $\frac{cn}{\log n}$; or*

2. *$\overline{G}$, the complement of G, contains a bi-clique of size cn.*

For convex sets, we have a stronger result. In fact, it is as strong as it can get.

Theorem 2.6 [28]. *The family of intersection graphs of finite collections of convex sets in the plane has the strong Erdős–Hajnal property.*

The proof relies on Turán-type results for incomparability graphs.

3. Semialgebraic Sets and Ramsey-graphs

Pach and Solymosi [48] proved that the family of intersection graphs of line segments in the plane has the strong Erdős–Hajnal property. Later, Alon *et al.* [7] generalized this result to semialgebraic sets. To formulate this result more precisely, we need to agree about the definitions.

A *semialgebraic set* in $\mathbb{R}^d$ is the locus of all points that satisfy a given finite Boolean combination of polynomial equations and inequalities in the d coordinates. We say that the *description complexity* of such a set S is at most κ if in some representation of S the dimension d is at most κ, the number of equations and inequalities is at most κ, and each of them has degree at most κ. (See [13].)

Every element S of a family $\mathcal{F}$ of semialgebraic sets of constant description complexity κ can be represented by a point S^* of a κ^*-dimensional Euclidean space (in which the coordinates are, say, the coefficients of the monomials in the polynomials that define S). We say that a binary *relation* R on $\mathcal{F} \times \mathcal{F}$ is *semialgebraic,* if the corresponding set

$$\left\{ (S^*, T^*) \in \mathbb{R}^{2\kappa^*} \mid S, T \in \mathcal{F}, \ (S, T) \in R \right\}$$

is semialgebraic.

Theorem 3.1 [7]. *Let $\mathcal{F}$ be a family of semialgebraic sets of constant description complexity, and let $R \subseteq \mathcal{F} \times \mathcal{F}$ be a fixed semialgebraic relation on $\mathcal{F}$. Then there exists a constant $c > 0$, which depends only on the maximum description complexity of the sets in $\mathcal{F}$ and of R, with the following property. Any collection of n elements of $\mathcal{F}$ has two subcollections $\mathcal{F}_1$ and $\mathcal{F}_2$, each containing at least cn elements, such that either $\mathcal{F}_1 \times \mathcal{F}_2 \subseteq R$, or $(\mathcal{F}_1 \times \mathcal{F}_2) \cap R = \emptyset$.*

It is easy to verify that the relation that two sets $S, T \in \mathcal{F}$ has non-empty intersection is semialgebraic. Thus, Theorem 3.1 indeed implies that any family of intersection graphs of (real) semialgebraic sets of constant description complexity has the strong (and, therefore, the weak) Erdős–Hajnal property.

Corollary 3.2. *For any family $\mathcal{F}$ of semialgebraic sets of constant description complexity, there is a constant $c = c(\mathcal{F}) > 0$ with the following property. Any collection of n elements of $\mathcal{F}$ has two subcollections $\mathcal{F}_1$ and $\mathcal{F}_2$, each containing at least cn elements, such that either every element of $\mathcal{F}_1$ intersects all elements of $\mathcal{F}_2$, or no element of $\mathcal{F}_1$ intersects any element of $\mathcal{F}_2$.*

Recently, Basu [12] has further extended this result for a broader class of algebraically defined sets.

Let us call an n-vertex graph *t-Ramsey* if it contains no clique and no independent set of size at least t. According to the results of Erdős [20] and Erdős and Szekeres [23] quoted in the introduction, there are n-vertex graphs that are $2\log n$-Ramsey, but no n-vertex graph is $\frac{1}{2}\log n$-Ramsey. Erdős's construction is probabilistic, and it appears to be a formidable task to find comparably good efficient constructions. More precisely, there is no known polynomial time deterministic algorithm for the construction of $O(\log n)$-Ramsey graphs on n vertices. Theorem 3.1 above shows that no such construction can be given by defining graphs using semialgebraic relations on a family of semialgebraic sets of constant description complexity. In fact, any n-vertex graph constructed in such a way will necessarily have a clique or an independent set of size at least n^c for some $c > 0$. This settles a conjecture of Babai [10], and improves a previous result [6] that showed that such graphs cannot be t-Ramsey for $t = e^{o(\sqrt{\log n})}$. The best known polynomial time construction is due to Barak *et al.* [11] and it produces $2^{(\log n)^{o(1)}}$-Ramsey graphs. The previous record was held by Frankl and Wilson [31].

As we remarked in the Introduction, in three and higher dimensions the family of intersection graphs of convex sets does not have the Erdős–Hajnal property. Somewhat surprisingly, Corollary 3.2 holds in any fixed dimension. There is another special class of intersection graphs of higher dimensional convex bodies that has the Erdős–Hajnal property. A set $S \subset \mathbb{R}^d$ is called K-fat if there are two d-dimensional balls B_1 and B_2 such that $B_1 \subseteq S \subseteq B_2$ and the ratio of the radius of B_2 to the radius of B_1 is at most K.

Theorem 3.3. *For any constant $K \geq 1$ and for any positive integer d, the family of intersection graphs of K-fat convex bodies in $\mathbb{R}^d$ has the strong Erdős–Hajnal property.*

Theorem 3.3 can be strengthened to say that there is a constant $c = c(K, d) > 0$ such that every intersection graph of K-fat convex bodies in $\mathbb{R}^d$ contains a clique or its complement contains a bi-clique of size cn. One proof of this result uses the separator theorem of Miller, Teng, Thurston, and Vavasis [45] discussed in the next section. Theorem 3.3 does not remain true for arbitrary (not necessarily convex) K-fat bodies even in the case $d = 2$.

Theorem 3.1 can be proved by a standard linearization process (see e.g. [3]) to transform the elements of $\mathcal{F}$ into vectors in a higher dimensional space, and the relation R to the set of all pairs of vectors (u, v) with a nonnegative scalar product $\langle u, v \rangle$. Thus, Theorem 3.1 can be reduced to the following

Lemma 3.4 [7]. *Let U and V be finite multisets of vectors in $\mathbb{R}^d$. Then there are subsets $U' \subset U$ and $V' \subset V$ such that $|U'| \geq \frac{1}{2^{d+1}}|U|$, $|V'| \geq \frac{1}{2^{d+1}}|V|$, and either $\langle u, v \rangle \geq 0$ for all $u \in U'$, $v \in V'$, or $\langle u, v \rangle < 0$ for all $u \in U'$, $v \in V'$.*

For $d > 2$, in spite of the apparent simplicity of Lemma 3.4, we do not have any "elementary" proof that would avoid using the probabilistic method or some form of the Borsuk–Ulam theorem (specifically, a partition theorem of Yao and Yao [53]). We challenge the reader to come up with such a proof.

4. String Graphs and Separator Theorems

A graph that can be realized as the intersection graph of finitely many continuous curves (strings) in the plane is called a *string graph*. As it was pointed out toward the end of Section 2, the family of intersection graphs of x-monotone curves, and thus the family of string graphs, do not have the strong Erdős–Hajnal property. However, it was conjectured in [49], that the situation is different if we restrict our attention to collections of curves with a bounded number of intersections per pair.

A collection of curves in the plane is called *t-intersecting* if any two of them intersect in at most t points. The elements of a 1-intersecting collection of curves are called *pseudo-segments*. A collection of portions of algebraic curves of maximum degree d in general position is d^2-intersecting. Clearly, the intersection graphs of t-intersecting collections of curves form a hereditary family.

Theorem 4.1 [29]. *For every $t \in \mathbb{N}$, the family of intersection graphs of t-intersecting collections of curves in the plane has the strong Erdős–Hajnal property. That is, for every $t \in \mathbb{N}$, there is a constant $c_t > 0$ such that the intersection graph G of every t-intersecting collection of n curves in the plane contains a bi-clique of size $c_t n$ or its complement $\overline{G}$ contains a bi-clique of size $c_t n$.*

For the proof, we need to extend the Lipton–Tarjan separator theorem [42].

A *separator* for a graph $G = (V, E)$ is a subset $V_0 \subset V$ such that there is a partition $V = V_0 \cup V_1 \cup V_2$ with $|V_1|, |V_2| \le \frac{2}{3}|V|$ and no vertex in V_1 is adjacent to any vertex in V_2. The Lipton–Tarjan separator theorem states that every planar graph with n vertices has a separator of size $O(\sqrt{n})$. By an important theorem of Koebe [37], every planar graph can be represented as the intersection (incidence) graph of nonoverlapping closed disks in the plane. Miller, Teng, Thurston, and Vavasis [45] proved that for every $d \ge 2$, the intersection graph of any collection of n balls in $\mathbb{R}^d$ such that no point belongs to more than k of them has a separator of size $O(dk^{1/d}n^{1-1/d})$.

A *Jordan region* is a subset of the plane that is homeomorphic to a closed disk. We say that a Jordan region R *contains* another Jordan region S if S lies in the interior of R. A *crossing* between R and S is either a crossing between their boundaries or a containment between them. The following result is a generalization of the separator theorems of Lipton and Tarjan and of Miller, Teng, Thurston, and Vavasis [45] in two dimensions.

Theorem 4.2 [26]. *If C is a finite collection of Jordan regions with a total of m crossings, then the intersection graph of C has a separator of size $O(\sqrt{m})$.*

By slightly fattening curves in the plane, Theorem 4.2 implies that it is also true for curves in the plane instead of Jordan regions. The proof of Theorem 4.1 has another interesting feature. The statement guarantees the existence of a large bi-clique in the graph G or in its complement $\overline{G}$. As it turns out, in many cases it can be proved that both G and $\overline{G}$ contain large bi-cliques. If G is a "dense" graph, then it must contain a bi-clique. Otherwise, $\overline{G}$ contains a bi-clique.

Theorem 4.3 [29]. *Let C be a t-intersecting collection of n curves in the plane such that at least εn^2 pairs of them intersect. Then the intersection graph of C contains a bi-clique of size at least $c_t \varepsilon^{64} n$, where $c_t > 0$ depends only on t.*

Fox *et al.* [29] have also generalized Theorem 4.1 in another direction. We first need a few definitions. Define an *r-region* to be a subset of the plane that is union of at most r Jordan regions. Call these (at most r) Jordan regions of an r-region the *components* of the r-region. A family of Jordan regions is *t-intersecting* if the boundaries of any two of them intersect in at most t points. A collection of r-regions is t-intersecting if the collection of all of its components is t-intersecting. They showed that for all $r, t \in \mathbb{N}$, the family of intersection graphs of finite collections of t-intersecting families of r-regions has the strong Erdős–Hajnal property.

Note that the last result can also be regarded as a generalization of Corollary 3.2 in the planar case, which states that the family of intersection graphs of collections of semialgebraic sets of constant description complexity has the strong Erdős–Hajnal property. Indeed, the boundary of a semialgebraic set of bounded description complexity in the plane is the union of a bounded number of algebraic curves of bounded degree, any two of which either intersect in a bounded number of points or overlap. By slightly perturbing semialgebraic sets, while maintaining their intersection pattern and their description complexity, we can assume that the boundaries of no two semialgebraic sets overlap. We can further assume, by slightly fattening the sets, if necessary, that each of them is the union of a constant number of Jordan regions, so that the above result applies.

In Table 1, we summarize the discussed results concerning Erdős–Hajnal properties for various families of intersection graphs.

Table 1: Erdős–Hajnal properties

family of intersection graphs of	Erdős–Hajnal Property	Strong Erdős–Hajnal Property
convex sets in $\mathbb{R}^3$	no	no
fat convex sets in $\mathbb{R}^d$	yes	yes
convex sets in $\mathbb{R}^2$	yes	yes
x-monotone curves in $\mathbb{R}^2$	yes	no
curves in $\mathbb{R}^2$	?	no
t-intersecting collections of curves in $\mathbb{R}^2$	yes	yes
t-intersecting collections of r-regions in $\mathbb{R}^2$	yes	yes
semialgebraic sets of constant description complexity	yes	yes
fat connected sets in $\mathbb{R}^2$	?	no

5. ASYMMETRIC RAMSEY-TYPE QUESTIONS

So far we discussed a variety of results that guarantee the existence of unexpectedly large homogeneous (sometimes bipartite) subgraphs in intersection graphs of various geometric objects. These results were *symmetric*, in the sense that in most of them *empty* and *complete* subgraphs played symmetric roles. In the spirit of so-called *"off-diagonal"* Ramsey theory, we can consider asymmetric variants of these questions.

A classical asymmetric result is the following theorem of Ajtai, Komlós, and Szemerédi [4]: Every triangle-free graph on n vertices contains an independent set of size $\Omega(\sqrt{n \log n})$. Kim [35] proved that this bound is tight up to a constant factor. If we restrict our attention to certain types of planar intersection graphs, this bound can be substantially improved.

Theorem 5.1 [27]. *If G is a K_k-free intersection graph of a t-intersecting family of $n \geq k$ curves in the plane, then G contains an independent set of size at least $n\left(c_t \frac{\log k}{\log n}\right)^{c \log k}$, where c is an absolute constant and $c_t > 0$ only depends on t.*

Taking δ such that $\varepsilon = c\delta \log \frac{1}{c_t \delta}$, we have the following corollary.

Corollary 5.2 [27]. *For each $\varepsilon > 0$ and positive integer t, there is $\delta = \delta(\varepsilon, t) > 0$ such that if G is an intersection graph of a t-intersecting family of n curves in the plane, then G has a clique of size at least n^δ or an independent set of size at least $n^{1-\varepsilon}$.*

Note that Corollary 5.2 is stronger than saying that the family of intersection graphs of t-intersecting families of curves in the plane has the Erdős–Hajnal property.

By slightly fattening curves in the plane, it is easy to see that if G is an intersection graph of a t-intersecting collection of curves, then G is also an intersection graph of a $4t$-intersecting collection of Jordan regions.

As usual, let $\chi(G)$ and $\alpha(G)$ denote the *chromatic number* and the size of the largest *independent set* of a graph G. Clearly, we have $\alpha(G) \geq \frac{n}{\chi(G)}$. It is not hard to generalize Theorem 5.1, as follows.

Theorem 5.3 [27]. *If G is a K_k-free intersection graph of a t-intersecting family of n r-regions, then*

$$\chi(G) \leq \left(c_{t,r}\frac{\log n}{\log k}\right)^{cr \log k},$$

where $c_{t,r}$ only depends on t and r and c is an absolute constant.

In the plane, every semialgebraic set of constant description complexity is the intersection graph of a t-intersecting collection of r-regions, where r and t depend only on the description complexity. Therefore, we have the following corollary of Theorem 5.3.

Corollary 5.4 [27]. *If G is a K_k-free intersection graph of a collection of semialgebraic sets in the plane of description complexity d, then*

$$\chi(G) \leq \left(c_d\frac{\log n}{\log k}\right)^{c_d \log k},$$

where c_d only depends on d.

A pair of *convex sets* or a pair of *x-monotone curves* can have arbitrarily many intersection points between their boundaries. Thus, Theorem 5.3 is not directly applicable to their intersection graphs. Nevertheless, we can show the following result.

Theorem 5.5 [27]. *If G is a K_k-free intersection graph of n convex sets in the plane, then*

$$\chi(G) \leq \left(c\frac{\log n}{\log k} \right)^{13 \log k},$$

where c is an absolute constant.

Taking δ such that $\varepsilon = 13\delta \log \frac{c}{\delta}$, and noting that $\alpha(G) \geq \frac{n}{\chi(G)}$, for every graph G with n vertices, we obtain the following corollary of Theorem 5.5.

Corollary 5.6 [27]. *For each $\varepsilon > 0$ there is $\delta = \delta(\varepsilon) > 0$ such that every intersection graph of n convex sets in the plane has a clique of size at least n^δ or an independent set of size at least $n^{1-\varepsilon}$.*

A more general form of Theorem 2.1 states that, for every positive integer k, every family of n convex sets in the plane has an independent set of size k or a clique of size at least n/k^4 [41]. Notice that Corollary 5.6 only applies in the case that the clique number is not too large while the result of Larman et al. [41] only applies when the independence number is not too large.

We can also prove the following theorem.

Theorem 5.7 [27]. *If G is a K_k-free intersection graph of n x-monotone curves in the plane, then*

$$\chi(G) \leq (c \log n)^{15 \log k},$$

where c is an absolute constant.

A collection C of curves in the plane is *grounded* if there is a closed (Jordan) curve γ such that every curve in C has one endpoint on γ and the rest of the curve lies in the exterior of γ. The intersection graph of a collection of grounded curves is called an *outerstring graph*.

McGuinness [43] proved that there is a constant C such that if G is a triangle-free intersection graph of a grounded 1-intersecting collection of curves, then G has chromatic number at most C. In Section 7, we prove an upper bound for the chromatic number of K_k-free outerstring graphs.

A survey by Kostochka [38] discusses results on coloring intersection graphs of certain geometric figures. Some of the known bounds are summarized in Table 5.

Table 2: Chromatic numbers of K_k-free intersection graphs

K_k-free intersection graphs of	upper bound on chromatic number	reference
intervals in $\mathbb{R}$	$k-1$	Gallai, Hajós
arcs along a circle	$\left\lfloor \frac{3(k-1)}{2} \right\rfloor$	Karapetian [32]
segments in $\mathbb{R}^2$	?	Erdős
half-lines in $\mathbb{R}^2$, $k = 3$	$< \infty$	McGuinness [43, 44]
chords of a circle, $k = 3$	8	Karapetian [33]
chords of a circle	$50 \cdot 2^k$	Kostochka-Kratochvíl [39]
axis-parallel rectangles in $\mathbb{R}^2$	$3k^2 - 8k + 4$	Asplund, Grünbaum [9] and C. Hendler
unit squares in $\mathbb{R}^2$, $k = 3$	$= 3$	Akiyama et al. [5]
translates of convex body in $\mathbb{R}^2$	$3k - 6$	Kim et al. [36]
homothetic copies of convex body in $\mathbb{R}^d$	$(O(d))^d k$	Pach [46] and Kostochka [38]
axis-parallel boxes in $\mathbb{R}^d$, $d \geq 3$	∞	Burling [14]

In the next two sections, we illustrate some of the ideas used for establishing the above results from [27] by giving simple proofs of some weaker versions of Theorems 5.1 and 5.7.

6. Independent Sets in String Graphs

Let $I_t(n, k)$ denote the maximum I such that every K_k-free intersection graph of n curves in the plane with no pair of curves intersecting in more than t points has an independent set of size I. The aim of this section is to establish some weaker versions of Theorem 5.1.

We first prove a very simple lower bound for $I_t(n, k)$, and then we show how to improve it with a little extra care.

Proposition 6.1. *There is an absolute constant c such that for all positive integers n, k, t with $n, k \geq 2$, we have $I_t(n, k) \geq \dfrac{n}{(ct^{1/2} \log n)^{2(k-2)}}$.* *That*

is, every K_k-free intersection graph of n curves in the plane with no pair intersecting in more than t points has an independent set of size at least

$$\frac{n}{(ct^{1/2}\log n)^{2(k-2)}}.$$

Proof. The proof is by induction on n and k. The base cases $n = 2$ and $k = 2$ are trivial. Let G be a K_k-free intersection graph of a t-intersecting collection C of n curves in the plane such that the largest independent set in G has size $I_t(n,k)$. If there is a vertex v adjacent to at least $n(ct^{1/2}\log n)^{-2}$ other vertices, then the intersection graph of the neighborhood of v has no clique of size $k-1$, and by induction, we are done in this case. So we may assume that the maximum degree of G is at most $n(ct^{1/2}\log n)^{-2}$. The number of crossings between elements of C is at most $\frac{1}{2}tn^2(ct^{1/2}\log n)^{-2} < n^2(c\log n)^{-2}$. Applying the separator theorem for curves, which is a corollary of Theorem 4.2, there is a partition $C = C_0 \cup C_1 \cup C_2$, with $|C_0| < c'n/\log n$, $|C_1|,|C_2| \le 2n/3$, and no edges between C_1 and C_2 in G, where c' is $\frac{1}{c}$ times the implied constant in the separator theorem. Letting $a_1 = |C_1|$ and $a_2 = |C_2|$, we have $a_1 + a_2 \ge n - c'n/\log n$, $a_1, a_2 \le 2n/3$, and

$$(1) \qquad\qquad I_t(n,k) \ge I_t(a_1,k) + I_t(a_2,k).$$

Using the induction hypothesis, we have

$$(2) \qquad\qquad I_t(a_i,k) \ge a_i^2\left(ct^{1/2}\log a_i\right)^{-2(k-2)}$$

for $i = 1,2$. It is straightforward to check that we can pick c large enough so that c' is small enough so that combining (1) and (2) gives the desired lower bound on $I_t(n,k)$. ∎

By also keeping track of the number of edges of the intersection graph G, we can improve the exponent of the $\log n$ factor in the lower bound in Proposition 6.1 from $2(k-2)$ to $k-2$.

Proposition 6.2. *There is an absolute constant c such that for all positive integers n, k, t with $n,k \ge 2$, we have $I_t(n,k) \ge \dfrac{n}{(ct\log n)^{(k-2)}}$.*

Proposition 6.2 follows from the next statement, by induction on k; the base case $k = 2$ is trivial.

Proposition 6.3. *There is a constant c such that if G is a nonempty intersection graph of a t-intersecting collection of $n \geq 2$ curves in the plane, then G contains an induced subgraph with at least $\frac{cn}{t \log n}$ vertices whose clique number is strictly less than the clique number of G.*

Proof. Let $D_t(m, n)$ denote the maximum D such that every graph G with n vertices and $m \geq 1$ edges, which is an intersection graph of a t-intersecting collection of curves in the plane, has an induced subgraph with D vertices such that its clique number is *strictly* smaller than the clique number of G.

It is sufficient to show that there is a constant c such that

$$D_t(m, n) \geq \frac{cn}{t \log n} + \frac{m}{n},$$

for all m and n with $n \geq 2$.

The proof is by induction on n, noting that $D_t(0, 1) = 1$. Let G be an intersection graph of a t-intersecting collection C of curves in the plane with n vertices, m edges, and every induced subgraph of G of size larger than $D_t(m, n)$ has the same clique number as G. Let Δ be the maximum degree of G. Notice that $\Delta \leq D_t(m, n)$ since the induced subgraph by the neighborhood of a vertex of maximum degree has clique number less than the clique number of G. Also $\Delta \geq 2m/n$ since $2m/n$ is the average degree of G. Hence, if $\Delta \geq 2\frac{cn}{t \log n}$, then the desired inequality holds. Therefore, we may assume $\Delta < 2\frac{cn}{t \log n}$.

By Theorem 4.2, the separator theorem, there is a partition $C = C_0 \cup C_1 \cup C_2$ with $|C_0| < c't\sqrt{m}$, $|C_1|, |C_2| \leq 2n/3$ and no curve in C_1 intersects a curve in C_2, where c' is the implied constant for the separator theorem. For $i \in \{1, 2\}$, let n_i and m_i denote the number of vertices and edges, respectively, of the subgraph of G induced by C_i. So

$$D_t(m, n) \geq D_t(m_1, n_1) + D_t(m_2, n_2),$$

with

$$n_1, n_2 \leq 2n/3,$$

$$n_1 + n_2 \geq n - |C_0| \geq n - c't\sqrt{m},$$

and

$$m_1 + m_2 \geq m - \Delta|C_0| \geq m - 2\frac{cn}{t \log n}c't\sqrt{m} = m - 2cc'\frac{n}{\log n}\sqrt{m}.$$

Notice that, by the induction hypothesis,

$$(3) \quad D_t(m,n) \geq D_t(m_1,n_1) + D_t(m_2,n_2) \geq \frac{cn_1}{t\log n_1} + \frac{m_1}{n_1} + \frac{cn_2}{t\log n_2} + \frac{m_2}{n_2}$$

$$\geq \frac{c(n_1+n_2)}{t\log(2n/3)} + \frac{m_1+m_2}{2n/3} \geq \frac{c(n_1+n_2)}{t(-1/2+\log n)} + \frac{m_1+m_2}{2n/3}.$$

Case 1: $m \geq \left(\frac{12cc'}{t}\frac{n}{\log n}\right)^2$. In this case,

$$m_1 + m_2 \geq m - 2cc'\frac{n}{\log n}\sqrt{m} \geq 5m/6.$$

Also, using (3),

$$D_t(m,n) \geq \frac{c(n_1+n_2)}{t\log n} + \frac{5}{4}m/n = \frac{cn}{t\log n} + \frac{m}{n} + \frac{m}{4n} - \frac{c\left(n-(n_1+n_2)\right)}{t\log n}$$

$$\geq \frac{cn}{t\log n} + \frac{m}{n} + \frac{m}{4n} - \frac{cc'\sqrt{m}}{t\log n} \geq \frac{cn}{t\log n} + \frac{m}{n},$$

completing the analysis in this case.

Case 2: $m < \left(\frac{12cc'}{t}\frac{n}{\log n}\right)^2$. Using (3), we have

$$D_t(m,n) \geq \frac{c(n_1+n_2)}{t(-1/2+\log n)} \geq \frac{c}{t}\left(n - c't\sqrt{m}\right)\left(\frac{1}{\log n} + \frac{1}{2\log^2 n}\right)$$

$$\geq \frac{c}{t}\left(\frac{n}{\log n} + \frac{n}{2\log^2 n} - \frac{2c't\sqrt{m}}{\log n}\right) \geq \frac{cn}{t\log n} + \frac{m}{n}.$$

as long as c is chosen originally to be at most $\frac{1}{576c'^2}$. This completes the proof. $\blacksquare$

7. Independent Sets of x-monotone Curves

In this section, we prove a weaker version of Theorem 5.7.

If a graph has small chromatic number, then it has a large independent set. The following result is an analogue of Proposition 6.1 for intersection graphs of x-monotone curves.

Proposition 7.1. *There is a constant c such that every intersection graph of n x-monotone curves with no $k > 2$ pairwise crossing has chromatic number at most $(c \log n)^{2k-3}$.*

Let $X(n,k)$ denote the maximum chromatic number over all K_k-free intersection graphs of n x-monotone curves. Let $V(n,k)$ denote the maximum chromatic number over all K_k-free intersection graphs of n x-monotone curves that each intersect a fixed vertical line L. We start the proof of Proposition 7.1 with the following lemma relating $V(n,k)$ and $X(n,k)$.

Lemma 7.2. *For all positive integers n and k, we have*

$$X(n,k) \leq X\left(\left\lfloor \frac{n}{2} \right\rfloor, k\right) + V(n,k).$$

Proof. Let C be a family of n x-monotone curves, and let $x_1 \leq \cdots \leq x_n$ be the x-coordinates of the left endpoints of the n x-monotone curves. Let L be the vertical line $x = x_{\lceil \frac{n}{2} \rceil}$. Notice that every x-monotone curve whose right endpoint has x-coordinate less than $x_{\lceil \frac{n}{2} \rceil}$ is disjoint from every x-monotone curve whose left endpoint has x-coordinate more than $x_{\lceil \frac{n}{2} \rceil}$. There are at most $\left\lfloor \frac{n}{2} \right\rfloor$ curves whose right endpoint has x-coordinate less than $x_{\lceil \frac{n}{2} \rceil}$ and at most $\left\lfloor \frac{n}{2} \right\rfloor$ curves whose left endpoint has x-coordinate greater than $x_{\lceil \frac{n}{2} \rceil}$. Hence, we can properly color the x-monotone curves in C that do not intersect L with $X\left(\left\lfloor \frac{n}{2} \right\rfloor, k\right)$ colors. We can color the remaining x-monotone curves in C, which all intersect L, with $V(n,k)$ colors. Hence, $X(n,k) \leq X\left(\left\lfloor \frac{n}{2} \right\rfloor, k\right) + V(n,k)$. ∎

By iterating Lemma 7.2, we obtain

$$X(n,k) \leq \sum_{i=0}^{\lfloor \log n \rfloor} V\left(\left\lfloor \frac{n}{2^i} \right\rfloor, k\right) \leq (1 + \log_2 n)V(n,k).$$

Recall that a collection C of curves in the plane is grounded if there is a closed (Jordan) curve γ such that every curve in C has one endpoint on γ and the rest of the curve lies in the exterior of γ. Also recall that the intersection graph of a collection of grounded curves is called an outerstring graph. Let $G(n, k)$ denote the maximum chromatic number over all K_k-free outerstring graphs with n vertices. The following lemma relates $G(n, k)$ and $V(n, k)$.

Lemma 7.3. *For all positive integers n and k, we have*

$$V(n, k) \leq G(n, k)^2.$$

Proof. Let $C = \{C_1, \ldots, C_n\}$ be a family of n x-monotone curves that intersect a vertical line $L\colon x = x_0$. Let L_i denote the intersection of C_i with the left half-plane $\{(x, y)\colon x \leq x_0\}$. Let R_i denote the intersection of C_i with the right half-plane $\{(x, y)\colon x \geq x_0\}$. Let $\mathcal{L} = \{L_1, \ldots, L_n\}$ and $\mathcal{R} = \{R_1, \ldots, R_n\}$. Notice that the intersection graph of $\mathcal{L}$ can be properly colored with $G(n, k)$ colors, and the intersection graph of $\mathcal{R}$ can be properly colored with $G(n, k)$ colors. Consider proper colorings $c_1\colon \mathcal{L} \to \{1, \ldots, G(n, k)\}$ and $c_2\colon \mathcal{R} \to \{1, \ldots, G(n, k)\}$ of the intersection graphs of $\mathcal{L}$ and $\mathcal{R}$, respectively. Assign to each x-monotone curve C_i the color $\big(c_1(L_i), c_2(R_i)\big)$. The family C is properly colored with $G(n, k)^2$ colors. Hence, $V(n, k) \leq G(n, k)^2$. $\blacksquare$

A family C of n grounded curves naturally comes with a cyclic labelling by their endpoints along the ground. Start by assigning any grounded curve the label 0 and then proceed to label the grounded curves clockwise, breaking ties arbitrarily, so the $(i + 1)^{th}$ grounded curve has label i. The labels are elements of $\mathbb{Z}_n$. Define the distance between a pair of grounded curves in C as the cyclic distance between their labels, that is, the distance $d(i, j)$ between the arc with label i and the arc with label j is $\min\big(|i - j|, n - |i - j|\big)$. We let $[i, j]$ denote the cyclic interval of elements $\{i, i + 1, \ldots, j\}$.

The following is the main lemma of this section.

Lemma 7.4. *For all integers $n \geq 2$ and $k \geq 3$, we have $G(n, k) \leq G\big(\lfloor \frac{2n}{3} \rfloor, k\big) + 4G(n, k - 1)$.*

Proof. Let $C = \{C_1, \ldots, C_n\}$ be a family of $n \geq 2$ grounded curves with curve C_i having cyclic label i. If no pair of arcs in C intersect, then the chromatic number of the intersection graph of C is 1. Therefore, we may suppose that there are a pair of curves in C that intersect.

Let (C_a, C_b) be a pair of arcs that intersect such that the distance $d(a, b)$ is the maximum distance over all pairs of curves in C that intersect.

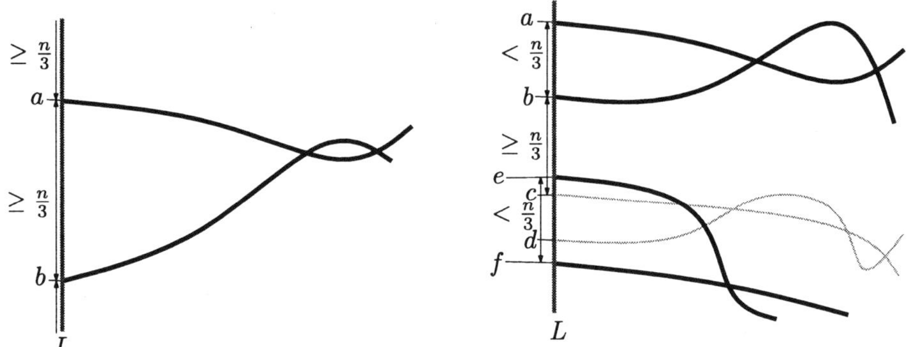

Fig. 2. On the left: there are two curves, C_a and C_b, that intersect and whose cyclic distance along L is at least $n/3$. On the right: the maximum distance between any two curves that intersect is less than $n/3$.

Case 1: $d(a, b) \geq \frac{n}{3}$ (which is depicted in the left-hand side of Figure 2). If C_i with i in the cyclic interval $[a+1, b-1]$ is disjoint from C_a and C_b, and C_j with j in the cyclic interval $[b+1, a-1]$ is disjoint from C_a and C_b, then C_i and C_j are disjoint. Hence, the curves that are disjoint from C_a and C_b can be properly colored with $G(n - d(a, b) - 1, k) \leq G\left(\left\lfloor \frac{2n}{3} \right\rfloor - 1, k\right)$ colors. We can properly color the curves in C that intersect C_a with $G(n, k-1)$ colors, and properly color the curves in C that intersect C_b with $G(n, k-1)$ colors. Therefore, C can be properly colored with $G\left(\left\lfloor \frac{2n}{3} \right\rfloor - 1, k\right) + 2G(n, k-1)$ colors.

Case 2: $d(a, b) < \frac{n}{3}$ (which is depicted in the right-hand side of Figure 2). Let $c \in \mathbb{Z}_n$ be given by $c \equiv b + \left\lceil \frac{n}{3} \right\rceil \pmod{n}$. If the curve C_c is disjoint from the other arcs in C, then the chromatic number of the intersection graph of C is the same as the chromatic number as the intersection graph of $C \setminus \{C_c\}$. If the curve C_c intersects at least one other curve in C, then let d be a label such that C_d intersects C_c and $d(c, d)$ is as large as possible. Finally, let e in the cyclic interval $[b, c]$ and f in the cyclic interval $[d, a]$ be such that C_e intersects C_f and $d(e, f)$ is as large as possible. Properly color the curves that intersect C_a with $G(n, k-1)$ colors, the remaining curves that intersect C_b with $G(n, k-1)$ colors, the remaining curves that intersect C_e with $G(n, k-1)$ colors, and the remaining curves that intersect C_f with $G(n, k-1)$ colors. Each of the remaining curves have labels in the cyclic intervals $I_1 := [a, b]$, $I_2 := [b, e]$, $I_3 := [e, f]$, or $I_4 := [f, a]$, and no remaining curve with label in I_i intersects a remaining curve with label in the interval I_j for $1 \leq i < j \leq 4$. Notice that each of the four intervals I_1, I_2, I_3, I_4 has at most $\lfloor \frac{2n}{3} \rfloor$ elements, so C can be properly colored with $G\left(\lfloor \frac{2n}{3} \rfloor, k\right) + 4G(n, k - 1)$ colors, which completes the proof. $\blacksquare$

Iterating Lemma 7.4, we have

$$G(n, k) \leq 4 \sum_{i \geq 0} G\left(\left\lfloor \left(\frac{2}{3}\right)^i n \right\rfloor, k - 1\right) \leq \frac{4}{\log 2/3}(1 + \log n)G(n, k - 1).$$

Trivially, $G(n, 2) = 1$. Therefore, there is an absolute constant c such that for $n > 1$ and $k > 2$, we have

$$G(n, k) \leq \left(\frac{4}{\log 2/3}(1 + \log n)\right)^{k-2} G(n, 2) \leq \left(c(\log n)^{k-2}\right),$$

which, with Lemmas 7.2 and Lemma 7.3, completes the proof of Theorem 7.1.

8. Open Problems

A few outstanding unsolved problems related to our subject are listed below.

Problem 8.1. Does the family of intersection graphs of continuous curves in the plane have the Erdős–Hajnal property?

Problem 8.2. Is it true that every perfect graph with n vertices or its complement contains a bi-clique of size $n^{1-o(1)}$?

Problem 8.3. Does there exist for every integer $k > 2$ a natural number C_k with the property that the intersection graph of any finite collection of convex sets in the plane with no k pairwise intersecting members is C_k-colorable?

We do not even know if every such intersection with n vertices contains an independent set of size at least $c_k n$, for a suitable constant $c_k > 0$.

A *geometric graph* is a graph whose vertices are points in the plane in general position and whose edges are straight-line segments connecting certain pairs of points. An affirmative answer to (even the weaker form of) Problem 8.3 would yield that any geometric graph with n vertices and no k pairwise crossing edges has at most $D_k n$ edges, where D_k is a constant depending only on k. This is known to be true for $k \leq 4$; see [2, 47, 1].

Problem 8.4 [8]**.** Does there exist a positive constant c such that every complete geometric graph with n vertices has cn pairwise crossing edges?

Problem 8.5. [26] Is it true that any $K_{k,k}$-free intersection graph of n segments in $\mathbb{R}^3$ has at most $D_k n$ edges, for some $D_k > 0$ depending only on k?

REFERENCES

[1]   E. Ackerman, On the maximum number of edges in topological graphs with no four pairwise crossing edges, *Proc. 22nd ACM Sympos. on Comput. Geom.* ACM Press (2006), pp. 259–263.

[2]   P. K. Agarwal, B. Aronov, J. Pach, R. Pollack and M. Sharir, Quasi-planar graphs have a linear number of edges, *Combinatorica,* **17** (1997), 1–9.

[3]   P.K. Agarwal and J. Matoušek, Range searching with semialgebraic sets, *Discrete Comput. Geom.,* **11** (1994), 393–418.

[4]   M. Ajtai, J. Komlós and E. Szemerédi, A note on Ramsey numbers, *J. Combin. Theory Ser. A,* **29** (1980), 354–360.

[5]   J. Akiyama, K. Hosono and M. Urabe, Some combinatorial problems, *Discrete Math.,* **116** (1963), 291–298.

[6]   N. Alon, Ramsey graphs cannot be defined by real polynomials, *J. Graph Theory,* **14** (6) (1990), 651–661.

[7]   N. Alon, J. Pach, R. Pinchasi, R. Radoičić and M. Sharir, Crossing patterns of semi-algebraic sets, *J. Combin. Theory Ser. A,* **111** (2) (2005), 310–326.

[8]   B. Aronov, P. Erdős, W. Goddard, D. Kleitman, M. Klugerman, J. Pach and L. J. Schulman, Crossing families, *Combinatorica,* **14** (2) (1994), 127–134.

[9]   E. Asplund and B. Grünbaum, On a coloring problem, *Math. Scand.,* **8** (1960), 181–188.

[10] L. Babai, Open problem, in: *Proc. 5th Hungar. Conf. Combin.* (A. Hajnal and V. T. Sós, eds.), Keszthely, Hungary, 1976, Vol. 2, North Holland (1978), 1189.

[11] B. Barak, A. Rao, R. Shaltiel and A. Wigderson, 2-source dispersers for sub-polynomial entropy and Ramsey graphs beating the Frankl–Wilson construction, in: *Proceedings of STOC 06* (2006), 671–680.

[12] S. Basu, Combinatorial complexity in o-minimal geometry, manuscript, 2006. `http://www.arxiv.org/abs/math.CO/0612050`

[13] S. Basu, R. Pollack and M.-F. Roy, *Algorithms in real algebraic geometry. Second edition. Algorithms and Computation in Mathematics,* **10**, Springer-Verlag (Berlin, 2006).

[14] J. P. Burling, On coloring problems of families of prototypes, Ph.D. Thesis, Univ. of Colorado (1965).

[15] D. Conlon, A new upper bound for diagonal Ramsey numbers, to appear in *Annals of Math.* (2007).

[16] D. G. Corneil, Y. Perl and L. K. Stewart, A linear recognition algorithm for cographs, *SIAM J. Comput.,* **14** (1985), 926–934.

[17] R. P. Dilworth, A decomposition theorem for partially ordered sets, *Annals of Math.,* **51** (2) (1950), 161–166.

[18] A. Dumitrescu and G. Tóth, Ramsey-type results for unions of comparability graphs, *Graphs Combin.,* **18** (2002), 245–251.

[19] G. Ehrlich, S. Even and R. E. Tarjan, Intersection graphs of curves in the plane, *J. Combin. Theory Ser. B,* **21** (1) (1976), 8–20.

[20] P. Erdős, Some remarks on the theory of graphs, *Bulletin of the Amer. Math. Soc.,* **53** (1947), 292–294.

[21] P. Erdős and A. Hajnal, Ramsey-type theorems, *Discrete Appl. Math.,* **25** (1989), 37–52.

[22] P. Erdős, A. Hajnal and J. Pach, Ramsey-type theorem for bipartite graphs, *Geombinatorics,* **10** (2000), 64–68.

[23] P. Erdős and G. Szekeres, A combinatorial problem in geometry, *Compositio Mathematica,* **2** (1935), 463–470.

[24] J. Fox, A bipartite analogue of Dilworth's theorem, *Order,* **23** (2–3) (2006), 197–209.

[25] J. Fox and J. Pach, A bipartite analogue of Dilworth's theorem for multiple partial orders, *European J. Combin.,* to appear.
http://math.nyu.edu/~pach/publications/multi060406.pdf

[26] J. Fox and J. Pach, Separator theorems and Turán-type results for planar intersection graphs, submitted (2007).

[27] J. Fox and J. Pach, Coloring planar intersection graphs, manuscript (2007).

[28] J. Fox, J. Pach and Cs. D. Tóth, Turán-type results for partial orders and intersection graphs of convex sets, *Israel J. Mathematics,* to appear.

[29] J. Fox, J. Pach and Cs. D. Tóth, A bipartite strengthening of the Crossing Lemma, *Graph Drawing 2007,* LNCS, Springer-Verlag, Berlin, to appear (2007).

[30] J. Fox and B. Sudakov, Density theorems for bipartite graphs and related Ramsey-type results, submitted.

[31] P. Frankl and R. M. Wilson, Intersection theorems with geometric consequences, *Combinatorica,* **1** (4) (1981), 357–368.

[32] I. Karapetian, On coloring of circular arc graphs, *Doklady AN ArmSSR (Notes of Armenian Acad. Sci.,* **70** (1980) (5), 306–311 (in Russian).

[33] I. Karapetian, Chordal Graphs, *Matematicheskie voprosi kibernetiki i vichislitelnoi tehniki,* **14** (1985), 6–10, Erevan (in Russian).

[34] G. Károlyi, J. Pach and G. Tóth, G. Ramsey-type results for geometric graphs. I, *Discrete Comput. Geom.*, **18** (3) (1997), 247–255.

[35] J. H. Kim, The Ramsey number $R(3, t)$ has order of magnitude $t^2 / \log t$, *Random Structures Algorithms*, **7** (1995), 173–207.

[36] S. J. Kim, A. Kostochka and K. Nakprasit, On the chromatic number of intersection graphs of convex sets in the plane, *Electron. J. Combin.*, **11** (2004), #R52.

[37] P. Koebe, Kontaktprobleme der konformen Abbildung, *Berichte über die Verhandlungen der Sachsischen Akademie der Wissenschaften, Leipzig, Mathematische-Physische Klasse*, **88** (1936), 141–164.

[38] A. Kostochka, Coloring intersection graphs of geometric graphs, in: *Towards a Theory of Geometric Graphs* (J. Pach Ed.), AMS, Providence, Rhode Island (2003), pp. 127–138.

[39] A. Kostochka and J. Kratochvíl, Covering and coloring polygon-circle graphs, *Discrete Math.*, **163** (1997), 299–305.

[40] J. Kynčl, unpublished.

[41] D. Larman, J. Matoušek, J. Pach and J. Törőcsik, A Ramsey-type result for convex sets, *Bull. London Math. Soc.*, **26** (2) (1994), 132–136.

[42] R. J. Lipton and R. E. Tarjan, A separator theorem for planar graphs, *SIAM J. Appl. Math.*, **36** (2) (1979), 177–189.

[43] S. McGuinness, Colouring arcwise connected sets in the plane. I, *Graphs Combin.*, **16** (2000), 429–439.

[44] S. McGuinness, Colouring arcwise connected sets in the plane. II *Graphs Combin.*, **17** (2001), 135–148.

[45] G. L. Miller, S.-H. Teng, W. Thurston and S. A. Vavasis, Separators for sphere-packings and nearest neighbor graphs, *J. ACM*, **44** (1) (1997), 1–29.

[46] J. Pach, Decomposition of multiple packing and covering, 2. Kolloquium über Diskrete Geometrie, Salzburg (1980), 169–178.

[47] J. Pach, R. Radoičić and G. Tóth, Relaxing planarity for topological graphs, *Discrete and Computational Geometry* (J. Akiyama, M. Kano, eds.), vol. 2866 of LNCS, Springer-Verlag (Berlin, 2003), pp. 221–232.

[48] J. Pach and J. Solymosi, Crossing patterns of segments, *J. Combin. Theory Ser. A*, **96** (2001), 316–325.

[49] J. Pach and G. Tóth, Comment on Fox News, *Geombinatorics*, **15** (2006), 150–154.

[50] J. Pach and J. Törőcsik, Some geometric applications of Dilworth's theorem, *Discrete Comput. Geom.*, **12** (1) (1994), 1–7.

[51] J. B. Sidney, S. J. Sidney and J. Urrutia, Circle orders, n-gon orders and the crossing number, *Order*, **5** (1) (1988), 1–10.

[52] H. Tietze, Über das Problem der Nachbargebiete im Raum, *Monatshefte Math.*, **16** (1905), 211–216.

[53] A. C. Yao and F. F. Yao, A general approach to d-dimensional geometric queries, *Proc. 17th Annu. ACM Sympos. Theory Comput.*, (1983), 163–168.

Jacob Fox

Department of Mathematics
Princeton University
Fine Hall, Washington Road
Princeton NJ 08544-1000
U.S.A.

e-mail:
jacobfox@math.princeton.edu

János Pach

City College, CUNY and
Courant Institute, NYU
251 Mercer Street
New York, NY 10012
U.S.A.

e-mail: pach@cims.nyu.edu

BOLYAI SOCIETY
MATHEMATICAL STUDIES, 17

Horizons of Combinatorics
Balatonalmádi
pp. 105–118.

Old and New Problems and Results in Ramsey Theory

RON GRAHAM*

In this note, I will describe a variety of problems from Ramsey theory on which I would like to see progress made. I will also discuss several recent results which do indeed make progress on some of these problems.

1. Introduction

Ramsey theory has sometimes been described as the study of unavoidable regularity in large structures. That is, one would like to know when it is the case that whenever the elements of some (sufficiently large) object are partitioned into a finite number of classes (i.e., colored with a finite number of colors), there is always at least one (color) class which contains all the elements of some regular structure. When this is the case, one additionally would like to have quantitative estimates of what "sufficiently large" means. In this sense, the guiding philosophy of Ramsey theory can be described by the phrase: *"Complete disorder is impossible"*. The roots of Ramsey theory go back to the work of Ramsey [37], Schur [40], van der Waerden [53], Rado [35, 36], Erdős [9], Szekeres [14, 15], Turán [16] and even Hilbert [27] (and even further, if you count the Pigeon-hole principle). A fuller account of this field can be found in the books [24], [23] and [31]. In this paper, I will focus on a number of both classical and new problems in this subject, and on some of the recent progress which has been made. Following the tradition popularized by Erdős, I am offering small monetary rewards for some of

*Research supported in part by NSF Grant CCR-0310991.

these problems, in the hope that these might stimulate further progress on them.

2. THE GROWTH OF $R(n)$

Define $R(n)$ to be the least integer such that any graph on $R(n)$ vertices contains **either** a clique of size n **or** an independent set of size n. It was shown in the classical paper of Ramsey [37] that $R(n)$ always exists. One of the oldest open problems in Ramsey theory is to determine or at least estimate, the rate of growth of $R(n)$. Some of the earliest estimates (due to Erdős [9]) are:

$$\frac{1}{e\sqrt{2}} n 2^{n/2} \big(1 + o(1)\big) < R(n) \leq \binom{2n - 2}{n - 1}.$$

Unfortunately, very little progress has been made in the past half century on these bounds (and this is not for lack of trying!). The best known improvements of the bounds are usually given as:

$$\frac{2}{e\sqrt{2}} n 2^{n/2} \big(1 + o(1)\big) < R(n) < n^{-1/2 + c\sqrt{\log n}} \binom{2n - 2}{n - 1}$$

for a suitable $c > 0$. The lower bound is due to Spencer [45] while the upper bound is due to Thomason [49].

However, very recently a significant improvement on the upper bound was established by Conlon [7]. He showed that for a suitable $c > 0$,

$$R(n + 1) < n^{c \frac{\log n}{\log \log n}} \binom{2n}{n}.$$

This implies that

$$R(n + 1) = O\left(\frac{1}{n^s} \binom{2n}{n} \right)$$

for every $s > 0$.

Problem 1 ($100). Prove that $\lim_{n \to \infty} R(n)^{1/n}$ exists.

Problem 2 ($250). Assuming this limit exists, what is it?

Of course, the limit would have to lie between $\sqrt{2}$ and 4. One popular guess is that it is 2. (Well, why not!)

Both of these problems (and the associated prizes) were frequently mentioned by Erdős in his uncountably many talks and problems papers. More complete descriptions of these and many other related problems in this vein can be found in the monograph **Erdős on Graphs: His Legacy of Unsolved Problems** by Fan Chung and the author [5].

Problem 3 ($100).   Give a **constructive** proof that for some $c > 0$, $R(n) > (1 + c)^n$.

As is well known, the best lower bounds currently available for $R(n)$ have been obtained using the probabilistic method [2].

One generalization of the Ramsey number $R(n)$ is the so-called off-diagonal Ramsey number $R(k, l)$, which is defined to be the least integer R such that in any red/blue coloring of the edges of the complete graph on R vertices, either there is a red complete graph on k vertices, or a blue complete graph on l vertices formed.

Problem 4 (Erdős – $250).   Prove or disprove that

$$R(4, n) > \frac{n^3}{\log^c n}$$

for some absolute constant c, provided that n is sufficiently large.

More generally, is it true that for k fixed,

$$R(k, n) > \frac{n^{k-1}}{\log^c n}$$

for some absolute constant c, provided that n is sufficiently large?

We close this section with three nice questions concerning off-diagonal Ramsey numbers. Specific references to the literature for these problems can be found in [5].

Problem 5 (Erdős/Burr).   Prove that

$$R(n + 1, n) > (1 + c)R(n, n)$$

for some fixed $c > 0$.

Conjecture 1 (Erdős/Sós).

$$R(3, n + 1) - R(3, n) \to \infty.$$

Conjecture 2 (Erdős/Sós).

$$R(3, n + 1) - R(3, n) = o(n).$$

3. The Erdős/Szekeres Problem

For each positive integer n, let $f(n)$ denote the least integer such that any set of $f(n)$ points in the plane in general position always contains a subset of size n which forms a **convex** n-gon. The general problem here is to determine or at least estimate the function $f(n)$.

This problem has a long and very interesting history which can be found, for example, in [25]. In particular, the original paper of Erdős and Szekeres [14] treating this problem contained an independent proof of Ramsey's theorem, among with other now classical results. Their original 1935 estimates

$$2^{n-2} + 1 \le f(n) \le \binom{2n - 4}{n - 2} + 1$$

remained unchanged until 1997, at which time the upper bound was improved by Fan Chung and the author [6] to:

$$f(n) \le \binom{2n - 4}{n - 2},$$

a modest improvement, to be sure! However, this was followed by a rapid series of further improvements [50], the current best being that of Tóth and Valtr [51]

$$f(n) \le \binom{2n - 5}{n - 2} + 1$$

which is about half as large as the original upper bound. It is suspected by many people that the lower bound is the truth. It has been known for some time that $f(3) = 3$, $f(4) = 5$, and $f(5) = 9$, all agreeing with the lower bound above. Very recently, it has been shown by Peters and Szekeres [47] that the lower bound is also the truth for $n = 6$, namely that $f(6) = 17$.

Their proof required the use of some 1500 hours of computing on a computer with a 2 GHz processor. The lower bound of $2^{n-2}+1$ results from an explicit construction in [14] and [15], where it is shown that there are sets of 2^{n-2} points in the plane in general position which contain no convex subset of size n. (It is a nice exercise to construct such sets if you haven't already seen the construction). This prompts the following:

Problem 6 ($1000). Prove or disprove that

$$f(n) = 2^{n-2} + 1.$$

for all $n \geq 2$. As a warm-up, one might like to first tackle

Problem 7 ($100). Show that

$$f(n) = o\left(\frac{4^n}{\sqrt{n}}\right).$$

A related question also raised by Erdős is whether the analogous results hold for **empty** convex n-gons. That is, if we define $g(n)$ to be the least integer g such any set of g points in the plane in general position contains the vertices of a convex n-gon which contains none of the other points in its interior, then does $g(n)$ always exist, and if so, what is its value? It is known that $g(3) = 3$, $g(4) = 5$ and $g(5) = 10 > 9 = f(5)$. Somewhat unexpectedly, it was shown by Horton [29] in 1983 that $g(n)$ does not exist for $n \geq 7$. The remaining open problem since then was the existence of an empty convex hexagon in a sufficiently large planar set in general position. This has just now been resolved in a very nice paper of Gerken [20]. He shows that $g(6)$ exists and in fact satisfies $30 \leq g(6) \leq 1717$. A simpler presentation (with a weaker upper bound) can be found in the paper of Valtr [52]. The reader is referred to the survey paper of Morris and Soltan [32] for a complete collection of results on the Erdős–Szekeres problem and its many variants.

4. Partition Regular Equations

We say that an equation $f(x, y, z, \ldots) = 0$ is **partition regular** if for any partition of the set of nonnegative numbers $\mathbb{N}$ into finitely many classes

$C_1, C_2, \ldots, C_r$, some C_i contains a nontrivial solution to the equation. (Nontrivial means not all the variables are equal). Often we think of the C_i as **colors**, and the solution in a single class as **monochromatic.** A rather complete theory of partition regularity for (systems of) linear equations was developed by Rado [35]. For example, $x + y = z$ and $x + y = 2z$ are partition regular, but $x + y = 3z$ is not. In fact, a single homogenous linear equation over $\mathbb{N}$ is partition regular if and only if it has a nontrivial solution in 0's and 1's, (i.e., not all 0). However, for **nonlinear** equations, the situation is much less clear. For example, it was shown by Rödl [38] that the equation $1/x + 1/y = 1/z$ is partition regular. Another recent result with a partition regularity flavor for nonlinear equations is the striking result of Croot [8] who showed that for any r-coloring of the integers greater than 1, the equation $\sum_{i \in I} 1/x_i = 1$, has a monochromatic solution for some finite set I. In fact, he proved that an appropriate I can always be found in the interval $[2, e^{167000r}]$. Very recently, Hippler [28] proved in his doctoral thesis that for $r = 2$, the exact bound for this problem is 208. That is, any 2-coloring of the set $\{2, 3, \ldots, 208\}$ must contain a monochromatic subset whose reciprocals sum to 1, and this is not true if 208 is replaced by 207.

The following problem of Erdős and the author has been open for over 30 years [10]:

Problem 7 ($250). Determine whether the equation

$$x^2 + y^2 = z^2$$

is partition regular.

There is actually very little data (in either direction) to know which way to guess.

Let us say that an equation $f(x_1, x_2, \ldots, x_n) = 0$ is r-**partition regular** if for any partition of the integers in r color classes, there is a nontrivial solution to this equation in a single color.

Conjecture (Rado). For each n, there is a least integer $M = M(n)$ so that if the linear homogeneous equation $f(x_1, x_2, \ldots, x_n) = 0$ is M-partition regular, then in fact the equation is partition regular.

Rado showed that this conjecture holds for $n = 1$ and $n = 2$. Very recently, Fox and Kleitman [17] have now shown that this conjecture also holds for $n = 3$, and in fact, that $M(3) \leq 36$.

Challenge. Prove that $M(n)$ always exists and determine (or estimate) its value.

An interesting phenomenon has been recently observed by Fox, Radoičić, Alexeev and the author [1, 18] which shows how the axioms of set theory can affect the outcome of some of these questions. For example, consider the linear equation $E : x + y + z - 4w = 0$. This is certainly not partition regular, and in fact, there is a 4-coloring of the integers which prevents E from having any (nontrivial) monochromatic solution. However, suppose we change the question and asked whether E has monochromatic solutions in **reals** for every 4-coloring of the reals. It can be shown that in ZFC, there exist 4-colorings of the reals for which E has no monochromatic solution. However, if we replace the Axiom of Choice (the "C" in ZFC) by LM (which is the axiom asserting that every set of reals is Lebesgue measurable), then in the system ZF + LM (which is consistent if ZFC is), the answer is yes. In other words, in this system every 4-coloring of the reals always contains a nontrivial monochromatic solution to E. On the other hand, this distinction does not occur for the equation $x + y - z = 0$, for example.

(Wide-open) Question. For which (systems of) equations does this distinction occur?

5. THE CHROMATIC NUMBER OF THE PLANE

In this problem, which goes back to Nelson in 1950 (and perhaps even to Hadwiger in 1944; see [4, 44]), we are asked for the minimum number r of colors needed so that the points of the Euclidean plane $\mathbb{E}^2$ can be r-colored in such a way that any two points separated by distance 1 have **different** colors. (Strictly speaking, we might think of this as an "anti-Ramsey" question, since in this case we are looking for the smallest number of colors needed to prevent the occurrence of a particular monochromatic configuration). This number, called the chromatic number of the plane, and denoted by $\chi(\mathbb{E}^2)$, is known to satisfy

$$4 \le \chi(\mathbb{E}^2) \le 7.$$

Both of these inequalities are quite easy to see, and have been known since the time the problem was proposed. If we operate in Zermelo-Fraenkel set theory with the Axiom of Choice (ZFC), then it follows by compactness that if $\chi(\mathbb{E}^2) = r$, then in fact there is a **finite** set which also requires r colors to legally color it. However, as pointed out recently by Shelah and Soifer, if we replace the Axiom of Choice by the two axioms Dependent Choice (= DC) and the assertion that every set of reals is Lebesgue measurable (= LM), then ZF + DC + LM is just as consistent as ZFC but now we no longer have compactness and the answer can change (see [42] and [43] for details).

An interesting related result of O'Donnell [33, 34] shows that for every integer g, there is a unit distance graph in $\mathbb{E}^2$ with girth greater than g which has chromatic number 4. Perhaps, this is evidence that $\chi(\mathbb{E}^2)$ is at least 5?

Problem 8 ($100). Show that $\chi(\mathbb{E}^2) \geq 5$.

Problem 9 ($250). Show that $\chi(\mathbb{E}^2) \leq 6$.

In higher dimensions, it is known that (see [4]):

$$6 \leq \chi(\mathbb{E}^3) \leq 15, \quad \left(1.239 + o(1)\right)^n < \chi(\mathbb{E}^n) < \left(3 + o(1)\right)^n.$$

6. Euclidean Ramsey Sets

Let us say that a finite subset X of Euclidean space $\mathbb{E}^n$ is **Ramsey** if for any number of colors r, there is an integer $N = N(X, r)$ such that in any r-coloring of the points of $\mathbb{E}^N$, there is always a monochromatic "copy" X' of X. In other words, X' can be obtained from X by some Euclidean motion (rotation and translation). This subject had its genesis in a series of papers by Erdős et al. [11, 12, 13]. In particular, it was shown that the Cartesian product of Ramsey sets is Ramsey, so that since a 2-point set is obviously Ramsey, then so is any subset of the vertices of a rectangular parallelepiped. On the other hand, it was also shown that any Ramsey set X must lie on the surface of some **sphere.** In such a case, we say that X is **spherical.**

Problem 10 ($1000). Prove that all spherical sets are Ramsey.

As a warm-up to this problem, one might work on the simpler:

Problem 11 ($100). Prove that any 4-point subset of a circle is Ramsey.

In order not to be too discouraging, we mention several more (presumably easier) Euclidean Ramsey problems.

Conjecture 1 ($25). For any 3-point set T, there is a 3-coloring of $\mathbb{E}^2$ which has no monochromatic copy of T.

Conjecture 2 ($50). In any 2-coloring of $\mathbb{E}^2$, a monochromatic copy of **every** 3-point set occurs, except possibly for a single equilateral triangle.

Very recently, Jelínek, Kynčl, Stolař and Valla [30] have shown that Conjecture 2 is true if one of the color classes is a closed set and the other color class is an open set. In fact, in this case, *every* 3-point set T occurs monochromatically.

Fact [46]. For any set L of 3 collinear points, there is a 16-coloring of $\mathbb{E}^n$ which contains no monochromatic copy of L.

Question. Is 16 the best possible constant here?

Perhaps the right answer is 4 (or even 3!).

7. Van der Waerden's Theorem

The classical theorem of van der Waerden [53] on arithmetic progressions asserts that for any finite coloring of $\mathbb{N}$, there always exists arbitrarily long monochromatic arithmetic progressions. The finite version (for two colors) guarantees the existence of a least number $W(n)$ such that if the integers $\{1, 2, \ldots, W(n)\}$ are 2-colored, then a monochromatic n-term arithmetic progression (n-AP) must always be formed. The estimation of the function $W(n)$ has challenged mathematicians ever since van der Waerden proved this result in 1927. The first upper bound, due to van der Waerden, grew like the Ackermann function, and was not even primitive recursive (his proof was a double induction on n and the number of colors). This was finally remedied by a new proof by Shelah [41] who reduced it to a function residing in the 5th level of the Gregorchik hierarchy (basically towers of towers). The current champion is based on a striking result of Gowers [21] concerning the upper density of subsets of $[1, N]$ which contain no n-AP. From this result, one can deduce the estimate that for all n, we have:

$$W(n) < 2^{2^{2^{2^{2^{2^{2n+9}}}}}}.$$

In particular, this settled a long-standing conjecture I had made on the size of $W(n)$ (which asserted that $W(n)$ was upper-bounded by an exponential tower of 2's of height n), and as a result, left me \$1000 poorer (but much happier). Undaunted, I now propose the following:

Conjecture (\$1000). For all n,

$$W(n) < 2^{n^2}.$$

I might point out that the best lower bound (due to Berlekamp [3]) has been around for almost 40 years:

$$W(n+1) \geq n2^n$$

for n prime (the proof uses finite fields).

Isn't it about time for some improvement here?

8. COMBINATORIAL LINES

For a finite set $A = \{a_1, a_2, \ldots, a_t\}$, let A^N denote the set of N-tuples from A. A **combinatorial line** in A^N is a set of t N-tuples $X_1, X_2, \ldots, X_t$ where

$$X_k = \big(X_k(1), X_k(2), \ldots, X_k(N) \big)$$

and for $1 \leq j \leq N$, either all $X_k(j)$ are equal, or $X_k(j) = a_k$, $1 \leq k \leq t$. This concept was first introduced in the seminal paper of Hales and Jewett [26].

In 1990, Furstenberg and Katznelson [19] proved the following beautiful theorem, generalizing Szemerédi's great density theorem for arithmetic progressions [48]:

Theorem. *For every $\varepsilon > 0$, there exists a least $N = N(\varepsilon, t)$ such if $R \subseteq A^N$ with $|R| > \varepsilon t^N$ then R must contain a combinatorial line.*

Unfortunately, the ergodic theory tools used by Furstenberg and Katznelson do not allow us to conclude anything about the growth rate of $N(\varepsilon, t)$ as $\varepsilon \to 0$.

Problem. Establish **any** upper bound on $N(\varepsilon, t)$.

The case of $t = 2$ is instructive (and is the only case we can handle!). In this case, using $A = \{0, 1\}$, we see that a combinatorial line in A^N is equivalent to having two subsets $X, Y \subseteq \{1, 2, \ldots, N\}$ with $X \subset Y$. By the well known result of Sperner, this must happen as soon as $|R| > \binom{N}{\lfloor N/2 \rfloor}$. This implies that $N(\varepsilon, 2) < c\varepsilon^{-2}$ for a suitable $c > 0$.

Warm-up Problem. Establish an upper bound on $N(\varepsilon, 3)$. In particular, it would be of great interest to obtain Gowers' type bounds on these quantities, that is, bounded towers of exponents (which might be called "Gowers towers"!).

9. CONCLUDING REMARKS

Of course, in this brief note I have only been able to touch on a few of the problems in this area that are most attractive to me, and for which I feel the time is ripe for making further progress. Much richer collections of problems and results in this subject can be found in a variety of sources, such as [4], [5], [22], [23], and [24].

REFERENCES

[1]   B. Alexeev, J. Fox and R. L. Graham, On minimal colorings without monochromatic solutions to a linear equation, (2006) (preprint), 15 pp.

[2]   N. Alon and J. Spencer, *The Probabilistic Method,* Second Edition, Wiley-Interscience, John Wiley & Sons, New York (2000). xviii+301 pp.

[3]   E. R. Berlekamp, A construction for partitions which avoid long arithmetic progressions, *Canad. Math. Bull.,* **11** (1968), 409–414.

[4]   P. Brass, W. Moser and J. Pach, *Research Problems in Discrete Geometry,* Springer, New York (2005). xii+499 pp.

[5]   F. Chung and R. Graham, *Erdős on Graphs: His Legacy of Unsolved Problems,* A K Peters, Ltd., Wellesley, MA (1998). xiv+142 pp.

[6]   F. R. K. Chung and R. L. Graham, Forced convex n-gons in the plane, *Discrete Comput. Geom.,* **19** (1998), 367–371.

[7]   D. Conlon, New upper bounds for some Ramsey numbers (2006), http://arxiv.org/abs/math/0607788.

[8]   E. Croot, On a coloring conjecture about unit fractions, *Ann. of Math. (2),* **157** (2003), 545–556.

[9]   P. Erdős, Some remarks on the theory of graphs, *Bull. Amer. Math. Soc.,* **53** (1947), 292–294.

[10]   P. Erdős and R. L. Graham, *Old and New Problems and Results in Combinatorial Number Theory,* Monographies de L'Enseignement Mathématique, **28**, Université de Genève, Geneva (1980) 128 pp.

[11] P. Erdős, R. L. Graham, P. Montgomery, B. L. Rothschild, J. H. Spencer and E. G. Straus, Euclidean Ramsey Theorems. I, *J. Combin. Theory Ser. A*, **14** (1973), 341–363.

[12] P. Erdős, R. L. Graham, P. Montgomery, B. L. Rothschild, J. H. Spencer and E. G. Straus, Euclidean Ramsey Theorems. II, *Infinite and finite sets (Colloq., Keszthely, 1973, Vol. I*, pp. 529–557. Colloq. Math. Soc. János Bolyai, Vol. **10**, North-Holland, Amsterdam (1975).

[13] P. Erdős, R. L. Graham, P. Montgomery, B. L. Rothschild, J. H. Spencer and E. G. Straus, Euclidean Ramsey Theorems. III, *Infinite and finite sets (Colloq., Keszthely, 1973, Vol. I*, pp. 559–583. Colloq. Math. Soc. János Bolyai, Vol. **10**, North-Holland, Amsterdam (1975).

[14] P. Erdős and G. Szekeres, A combinatorial problem in geometry, *Composito Math.*, **2** (1935), 464–470.

[15] P. Erdős and G. Szekeres, On some extremum problems in elementary geometry. *Ann. Univ. Sci. Budapest. Eőtvös Sect. Math.*, **3–4** (1960–1961), 53–62.

[16] P. Erdős and P. Turán, On some sequences of integers, *J. London Math. Soc.*, **11** (1936), 261–264.

[17] J. Fox and D. Kleitman, On Rado's boundedness conjecture, *J. Combin. Theory*, Ser. A, **113** (2006), 84–100.

[18] J. Fox and R. Radoičić, The axiom of choice and the degree of regularity of equations over the reals (2005) (preprint).

[19] H. Fürstenberg and Y. Katznelson, An ergodic Szemerédi theorem for commuting transformations, *J. Analyse Math.*, **34** (1979), 275–291.

[20] T. Gerken, On empty convex hexagons in planar point sets (2006) (submitted).

[21] W. T. Gowers, A new proof of Szemerédi's theorem, *Geom. Funct. Anal.*, **11** (2001), 465–588.

[22] W. T. Gowers, Some unsolved problems in additive/combinatorial number theory, (on Gowers Website at www.dpmms.cam.ac.uk/~wtg10/papers.html)

[23] R. L. Graham, *Rudiments of Ramsey Theory*, CBMS Regional Conference Series in Mathematics, **45**. American Mathematical Society, Providence, R.I., (1981). v+65pp.

[24] R. L. Graham, B. L. Rothschild and J. H. Spencer, *Ramsey Theory*, Second edition, John Wiley & Sons, Inc., New York (1990). xii+196 pp.

[25] R. L. Graham and J. Nešetřil, Ramsey Theory and Paul Erdős (recent results from a historical perspective), *Paul Erdős and his mathematics, II (Budapest, 1999)*, 339–365, Bolyai Soc. Math. Stud., **11**, János Bolyai Math. Soc., Budapest (2002)

[26] A. W. Hales and R. I Jewett, Regularity and positional games, *Trans. Amer. Math. Soc.*, **106** (1963), 222–229.

[27] D. Hilbert, Uber die Irreducibilität Ganzer Rationaler Functionen mit Ganzzahligen, *J. Reine Angew. Math.*, **110** (1892), 104–129.

[28] A. Hippler, Dissertation for "Doktor der Naturwissenshaften am Fachbereich Mathematik" der Johannes Gutenberg-Univertität in Mainz, (2002), 103 pp.

[29] J. D. Horton, Sets with no empty convex 7-gons, *Canadian Math. Bull.*, **26** (1983), 482–484.

[30] V. Jelínek, J. Kynčl, R. Stolař and T. Valla, Monochromatic triangles in two-colored plane (2006) (preprint), 25 pp.

[31] B. Landman and A. Robertson, *Ramsey Theory on the Integers*, Student Mathematical Library, **24**. American Mathematical Society, Providence, RI (2004), xvi+317.

[32] W. Morris and V. Soltan, The Erdős–Szekeres problem on points in convex position – a survey, *Bull. Amer. Math. Soc.*, **37** (2000), 437–458.

[33] P. O'Donnell, Arbitrary girth, 4-chromatic unit distance graphs in the plane. I. Graph embedding, *Geombinatorics*, **9** (2000), 180–193.

[34] P. O'Donnell, Arbitrary girth, 4-chromatic unit distance graphs in the plane. II. Graph description, *Geombinatorics*, **9** (2000), 145–152.

[35] R. Rado, Studien zur Kombinatorik, *Math. Zeit.*, **36** (1933), 242–280.

[36] R. Rado, Verallgemeinerung Eines Satzes von van der Waerden mit Anwendungen auf ein Problem der Zahlentheorie, *Sonderausg. Sitzungsber. Preuss. Akad. Wiss. Phys-Math. Klasse*, **17** (1933), 1–10.

[37] F. P. Ramsey, On a problem in formal logic, *Proc. London Math. Soc.*, **30** (1930), 264–286.

[38] V. Rődl, (personal communication).

[39] K. F. Roth, On certain sets of integers, *J. London Math. Soc.*, **28** (1953), 104–109.

[40] I. Schur, Uber die Kongruenz $x^m + y^m \equiv z^m \pmod{p}$, *Jber. Deutsch. Math.-Verein.*, **25**, (1916), 114–117.

[41] S. Shelah, Primitive recursive bounds for van der Waerden numbers, *J. Amer. Math. Soc.*, **1** (1988), 683–697.

[42] S. Shelah and A. Soifer, Axiom of choice and chromatic number of the plane, *J. Combin. Theory Ser. A*, **103** (2003), 387–391.

[43] S. Shelah and A. Soifer, Axiom of choice and chromatic number: examples on the plane, *J. Combin. Theory Ser. A*, **105** (2004), 359-364.

[44] A. Soifer, Chromatic number of the plane: its past and future. Proceedings of the Thirty-Fourth Southeastern International Conference on Combinatorics, Graph Theory and Computing, *Congr. Numer.*, **160**, (2003), 69–82.

[45] J. H. Spencer, Ramsey's theorem – a new lower bound, *J. Comb. Th. (A)* **18** (1975), 108–115.

[46] E. G. Straus, A combinatorial theorem in group theory, *Math. Computation* **29** (1975), 303–309.

[47] G. Szekeres and L. Peters, Computer solution to the 17 point Erdős–Szekeres problem, (2006) (preprint), 14 pp.

[48] E. Szemerédi, On sets of integers containing no k elements in arithmetic progression, *Acta Arith.*, **27** (1975), 199–245.

[49] A. Thomason, An upper bound for some Ramsey numbers, *J. Graph Theory*, **12** (1988), 509–517.

[50] G. Tóth and P. Valtr, Note on the Erdős–Szekeres theorem, *Discrete Comput. Geom.*, **19** (1998), 457–459.

[51] G. Tóth and P. Valtr, The Erdős–Szekeres theorem: upper bounds and related results, *Combinatorial and computational geometry*, 557–568, Math. Sci. Res. Inst. Publ., 52, Cambridge Univ. Press, Cambridge (2005).

[52] P. Valtr, On the empty hexagon theorem (2006) (preprint), 9 pp.

[53] B. L. van der Waerden, Beweis einer Baudetschen Vermutung, *Nieuw Arch. Wiskunde,* **15** (1927), 212–216.

Ron Graham

Department of Mathematics and
Computer Science
University of California
San Diego

e-mail: graham@ucsd.edu

BOLYAI SOCIETY
MATHEMATICAL STUDIES, 17

Horizons of Combinatorics
Balatonalmádi
pp. 119–140.

Forbidden Intersection Patterns in the Families of Subsets (Introducing a Method)

GYULA O.H. KATONA*

1. Introduction

Let $[n] = \{1, 2, \ldots, n\}$ be a finite set, $\mathcal{F} \subset 2^{[n]}$ a family of its subsets. In the present paper $\max |\mathcal{F}|$ will be investigated under certain conditions on the family $\mathcal{F}$. The well-known Sperner theorem ([14]) was the first such result.

Theorem 1.1. *If $\mathcal{F}$ is a family of subsets of $[n]$ without inclusion ($F, G \in \mathcal{F}$ implies $F \not\subset G$) then*

$$|\mathcal{F}| \leq \binom{n}{\lfloor \frac{n}{2} \rfloor}$$

holds, and this estimate is sharp as the family of all $\lfloor \frac{n}{2} \rfloor$-element subsets shows.

There is a very large number of generalizations and analogues of this theorem. (See e.g. [7]). Here we will consider only results when the condition on $\mathcal{F}$ excludes certain configurations what can be expressed by inclusion, only. That is, no intersections, unions, etc. are involved. The first such generalization was obtained by Erdős [8]. The family of k distinct sets with mutual inclusions, $F_1 \subset F_2 \subset \ldots \subset F_k$ is called a *chain of length k*. It will be simply denoted by P_k. Let $\mathrm{La}(n, P_k)$ denote the largest family $\mathcal{F}$ without a chain of length k.

*The work was supported by the Hungarian National Foundation for Scientific Research grant numbers NK0621321, AT048826, the Bulgarian National Science Fund under Grant IO-03/2005 and the projects of the European Community: INTAS 04-77-7171, COMBSTRU–HPRN-CT-2002-000278, FIST–MTKD-CT-2004-003006.

Theorem 1.2 [8]. $\mathrm{La}(n, P_{k+1})$ *is equal to the sum of the k largest binomial coefficients of order n.*

Let V_r denote the *r-fork*, that is the following family of distinct sets: $F \subset G_1, F \subset G_2, \ldots, F \subset G_r$. The quantity $\mathrm{La}(n, V_r)$, that is, the largest family on n elements containing no V_r was first (asymptotically) determined for $r = 2$. We use the well-known notation $\Omega(n)$ where $f(n) = \Omega(n)$ means that there is a constant $0 < c$ such that $cn \leq f(n)$ holds for all n.

Theorem 1.3 [11].

$$\binom{n}{\lfloor \frac{n}{2} \rfloor}\left(1 + \frac{1}{n} + \Omega\left(\frac{1}{n^2}\right)\right) \leq \mathrm{La}(n, V_2) \leq \binom{n}{\lfloor \frac{n}{2} \rfloor}\left(1 + \frac{2}{n}\right).$$

The first result for general r is contained in the following theorem.

Theorem 1.4 [15].

$$\binom{n}{\lfloor \frac{n}{2} \rfloor}\left(1 + \frac{r}{n} + \Omega\left(\frac{1}{n^2}\right)\right) \leq \mathrm{La}(n, V_{r+1}) \leq \binom{n}{\lfloor \frac{n}{2} \rfloor}\left(1 + 2\frac{r^2}{n} + o\left(\frac{1}{n}\right)\right).$$

The constant in the second term in the upper estimate was recently improved.

Theorem 1.5 [3].

$$\binom{n}{\lfloor \frac{n}{2} \rfloor}\left(1 + \frac{r}{n} + \Omega\left(\frac{1}{n^2}\right)\right) \leq \mathrm{La}(n, V_{r+1})$$

$$\leq \binom{n}{\lfloor \frac{n}{2} \rfloor}\left(1 + 2\frac{r}{n} + O\left(\frac{1}{n^2}\right)\right).$$

See some remarks in Section 7 explaining why this second term is difficult to improve any more.

The aim of the present paper is to introduce some recent results and show a method, proving good upper estimates, developed recently.

2. NOTATIONS, DEFINITIONS

A *partially ordered set,* shortly *poset* P is a pair $P = (X, \leq)$ where X is a (in our case always finite) set and $\leq$ is a relation on X which is *reflexive* ($x \leq x$ holds for every $x \in X$), *antisymmetric* (if both $x \leq y$ and $x \geq y$ hold for $x, y \in X$ then $x = y$) and transitive ($x \leq y$ and $y \leq z$ always implies $x \leq z$). We say that y *covers* x if $x < y$ and there is no $z \in X$ such that $x < z < y$ holds. It is easy to see that if $X = 2^{[n]}$ and the $\leq$ is defined as $\subseteq$, then these conditions are satisfied, that is the family of all subsets of an n-element set ordered by inclusion form a poset. We will call this poset the *Boolean lattice* and denote it by B_n. Covering in this poset means "inclusion with difference 1".

The definition of a *subposet* is obvious: $R = (Y, \leq_2)$ is a subposet of $P = (X, \leq_1)$ iff there is an injection α of Y into X is such a way that $y_1, y_2 \in Y, y_1 \leq_2 y_2$ implies $\alpha(y_1) \leq_1 \alpha(y_2)$. On the other hand R is an *induced subposet* of P when $\alpha(y_1) \leq_1 \alpha(y_2)$ holds iff when $y_1 \leq_2 y_2$. If $P = (X, \leq)$ is a poset and $Y \subset X$ then the poset *spanned by Y in P* is defined as $(Y, \leq^*)$ where $\leq^*$ is the same as $\leq$, for all the pairs taken from Y. Given a "small" poset R, $\mathrm{La}(n, R)$ denotes the maximum number of elements of $Y \subset 2^{[n]}$ (that is, the maximum number of subsets of $[n]$) such that R is not a subposet of the poset spanned by Y in B_n.

Redefine our "small" configurations in terms of posets. The chain P_k contains k elements: $a_1, \ldots, a_k$ where $a_1 < \ldots < a_k$. The *r-fork* contains $r + 1$ elements: $a, b_1, \ldots, b_r$ where $a < b_1, \ldots a < b_r$. It is easy to see that the definitions of $\mathrm{La}(n, P_k)$, $\mathrm{La}(n, V_r)$, in Sections 1 and 2 agree. In the rest of the paper we will use the two different terminology alternately. In the definition of $\mathrm{La}(n, R)$ we mean non-induced subposets, that is, if $R = V_2$ then P_3 is also excluded as a subposet.

A poset is *connected* if for any pair (z_0, z_k) of its elements there is a sequence $z_1, \ldots, z_{k-1}$ such that either $z_i < z_{i+1}$ or $z_i > z_{i+1}$ holds for $0 \leq i < k$. If the poset is not connected, maximal connected subposets are called its *connected components.* Given a family $\mathcal{F}$ of subsets of $[n]$, it spans a poset in B_n. We will consider its connected components Q in two different ways. First as posets themself, secondly as they are represented in B_n. In the latter case the sizes of in the sets are also indicated. This is called a *realization* of Q. A *full chain* in B_n is a family of sets $A_0 \subset A_1 \subset \ldots \subset A_n$ where $|A_i| = i$. We say that a (full) chain *goes through* a family (subposet)

$\mathcal{F}$ if their intersection is non-empty, that is if it "goes through" at least one member of the family.

3. Lubell's Proof of the Sperner Theorem

The number of full chains in $[n]$ is $n!$ since the choice of a full chain is equivalent to the choice of a permutation of the elements of $[n]$. On the other hand, the number of full chains going through a given set F of f elements is $f!(n-f)!$ since the chain "must grow" within F until it "hits" F and outside after that. Suppose that the family $\mathcal{F}$ of subsets of $[n]$ is without inclusion ($F, G \in \mathcal{F}$ implies $F \not\subset G$). Then a full chain cannot go through two members of $\mathcal{F}$. Therefore the set of full chains going through distinct members of $\mathcal{F}$ must be disjoint. Hence we have

$$\sum_{F \in \mathcal{F}} |F|!\,(n - |F|)! \leq n!\,.$$

Dividing the inequality by $n!$

$$(3.1) \qquad \sum_{F \in \mathcal{F}} \frac{1}{\binom{n}{|F|}} \leq 1$$

is obtained. Replace $\binom{n}{|F|}$ by $\binom{n}{\lfloor \frac{n}{2} \rfloor}$. Then

$$\frac{|\mathcal{F}|}{\binom{n}{\lfloor \frac{n}{2} \rfloor}} = \sum_{F \in \mathcal{F}} \frac{1}{\binom{n}{\lfloor \frac{n}{2} \rfloor}} \leq 1$$

follows, the theorem is proved.

Let us remark that inequality (3.1) is important on its own right and is called the YBLM-inequality (earlier LYM, see [17], [2], [12], [13]). ∎

4. The Method, Illustrated with an Old Result

Lubell's proof easily applies for Theorem 1.2, however, surprisingly it was not exploited for proving theorems of the present type. The reason might be that not the "excluded" configurations should be considered when using the idea, but the "allowed induced posets". (See later.)

Following the definition of the r-fork, let us define the r-*brush* (in a poset) which contains $r + 1$ elements: $a, b_1, \ldots, b_r$ where $a > b_1, \ldots, a > b_r$ and is the "dual" of the r-fork. (Here and in what follows we use the expression "dual" when the complements of the sets involved are considered.) Theorem 1.3 gives the best expected asymptotic upper bound up to the second term for V_2 in the Boolean lattice. It is easy to see that it implies the same solution for Λ_2. However the result is very different when both of them are excluded. Our notation $\mathrm{La}(n, R)$ is extended in an obvious way for the case when two subposets R_1 and R_2 are excluded: $\mathrm{La}(n, R_1, R_2)$.

Theorem 4.1 [11].

$$\mathrm{La}(n, V_2, \Lambda_2) = 2 \binom{n-1}{\lfloor \frac{n-1}{2} \rfloor}.$$

Proof. The construction giving the equality is the following:

$$\left\{ F \subset [n] : 1 \notin F, \ |F| = \left\lfloor \frac{n-1}{2} \right\rfloor \right\} \cup \left\{ F \subset [n] : 1 \in F, \ |F| = \left\lfloor \frac{n+1}{2} \right\rfloor \right\}.$$

The non-trivial part of the proof is the verification of the upper bound.

Let $\mathcal{F}$ be a family of subsets of $[n]$ which contains neither a V_2 nor a Λ_2 as subposet. Therefore it cannot contain a P_3 either. Consider the connected components of the poset spanned by $\mathcal{F}$. It is obvious that a connected component can be either a one element poset P_1 or a P_2. Let α_1 and α_2 be their respective numbers. Then

$$(4.1) \qquad\qquad |\mathcal{F}| = \alpha_1 + 2\alpha_2.$$

We will now determine the minimum number of full chains going through a one or two-element component. Let $P(1; a)$ be a one-element component which is an a-element set. The number $c\big(P(1; a)\big)$ of full chains going through $P(1; a)$ is $a!(n - a)!$. Therefore

$$\frac{c\big(P(1; a)\big)}{n!} = \frac{1}{\binom{n}{a}}.$$

It takes on its minimum at the value $a = \left\lfloor \frac{n}{2} \right\rfloor$. Hence we obtained

$$(4.2) \qquad\qquad \left\lfloor \frac{n}{2} \right\rfloor! \left\lceil \frac{n}{2} \right\rceil! \leq c\big(P(1; a)\big).$$

The two-element component consisting of an a-element subset A and a b-element subset B $(A \subset B)$ $(a < b)$ is denoted by $P(2; a, b)$. The number $c\big(P(2; a, b)\big)$ of full chains going through (at least one element of) $P(2; a, b)$ is

$$(4.3) \qquad c\big(P(2; a, b)\big) = a!(n-a)! + b!(n-b)! - a!(b-a)!(n-b)! \,.$$

Divide it by $n!$.

$$(4.4) \qquad \frac{c\big(P(2; a, b)\big)}{n!} = \frac{1}{\binom{n}{a}} + \frac{1}{\binom{n}{b}} - \frac{1}{\binom{n}{b}\binom{b}{a}} = \frac{1}{\binom{n}{a}} + \frac{1}{\binom{n}{b}}\left(1 - \frac{1}{\binom{b}{a}}\right).$$

By symmetry, let $b \geq \frac{n}{2}$. Suppose first that a is fixed and is $\leq \left\lfloor \frac{n-1}{2} \right\rfloor$. Then (4.4) takes on its minimum for $b = \left\lfloor \frac{n+1}{2} \right\rfloor$. Fix b here and consider the following variant of (4.4):

$$(4.5) \qquad \frac{c\big(P(2; a, b)\big)}{n!} = \frac{1}{\binom{n}{a}} + \frac{1}{\binom{n}{b}} - \frac{1}{\binom{n}{a}\binom{n-a}{n-b}} = \frac{1}{\binom{n}{b}} + \frac{1}{\binom{n}{a}}\left(1 - \frac{1}{\binom{n-a}{n-b}}\right).$$

This is a monotone decreasing function of a in the interval $0 \leq a \leq \left\lceil \frac{n}{2} \right\rceil$. Therefore the pair giving the minimum in this case is $a = \left\lfloor \frac{n-1}{2} \right\rfloor, b = \left\lfloor \frac{n+1}{2} \right\rfloor$.

Suppose now that $a \geq \left\lfloor \frac{n}{2} \right\rfloor$. Then b can be chosen to be $a+1$ by (4.4), and (4.3) becomes $na!(n-a-1)!$ It achieves its minimum at $a = \left\lfloor \frac{n-1}{2} \right\rfloor$, again. We obtained

$$(4.6) \qquad n\left\lfloor \frac{n-1}{2} \right\rfloor! \left\lceil \frac{n-1}{2} \right\rceil! \leq c\big(P(2; a, b)\big) \,.$$

Observe that a full chain cannot go through two distinct components, therefore

$$\sum_{P_1 \text{ is a component}} c(P_1) + \sum_{P_2 \text{ is a component}} c(P_2) \leq n!$$

holds. The left hand side can be lower estimated by (4.2) and (4.6):

$$(4.7) \qquad \alpha_1 \left\lfloor \frac{n}{2} \right\rfloor! \left\lceil \frac{n}{2} \right\rceil! + \alpha_2 n \left\lfloor \frac{n-1}{2} \right\rfloor! \left\lceil \frac{n-1}{2} \right\rceil! \leq n! \,.$$

(4.1) has to be maximized with respect to (4.7). Rewrite (4.7) a little bit:

$$(4.8) \qquad \alpha_1 \left\lfloor \frac{n}{2} \right\rfloor! \left\lceil \frac{n}{2} \right\rceil! + 2\alpha_2 \frac{n}{2} \left\lfloor \frac{n-1}{2} \right\rfloor! \left\lceil \frac{n-1}{2} \right\rceil! \leq n! \,.$$

Compare the coefficients of α_1 and $2\alpha_2$ in (4.8).

$$(4.9) \qquad \left\lfloor \frac{n}{2} \right\rfloor! \left\lceil \frac{n}{2} \right\rceil! \geq \frac{n}{2} \left\lfloor \frac{n-1}{2} \right\rfloor! \left\lceil \frac{n-1}{2} \right\rceil!$$

holds (with equality for even n). Replacing the coefficient of α_1 in (4.8) using (4.9), the inequality

$$(\alpha_1 + 2\alpha_2)\frac{n}{2} \left\lfloor \frac{n-1}{2} \right\rfloor! \left\lceil \frac{n-1}{2} \right\rceil! \leq n!$$

is obtained, what results in

$$|\mathcal{F}| = \alpha_1 + 2\alpha_2 \leq \frac{n!}{\frac{n}{2} \left\lfloor \frac{n-1}{2} \right\rfloor! \left\lceil \frac{n-1}{2} \right\rceil!} = 2\binom{n-1}{\left\lfloor \frac{n-1}{2} \right\rfloor}. \qquad \blacksquare$$

5. The Method, in General

Let $\mathcal{P}$ be the set of forbidden subposets. Let $\mathcal{F}$ be a family of subsets of $[n]$ such that the poset induced by $\mathcal{F}$ in B_n contains no member of $\mathcal{P}$ as a subposet. $\mathrm{La}(n, \mathcal{P})$ denotes the largest size of such a family. Consider the connected components of the poset induced by $\mathcal{F}$. The family of all possible components is denoted by $\mathcal{Q} = \mathcal{Q}(\mathcal{P})$.

In our Section 4 we had $\mathcal{P} = \{V_2, \Lambda_2\}$. Then $\mathcal{Q}(\{V_2, \Lambda_2\}) = \{P_1, P_2\}$.

If $Q \in \mathcal{Q}$ let Q_n^* be a realization of Q in the Boolean lattice B_n, that is, Q is embedded into B_n and a size (of a subsets) is associated with each element $q \in Q_n^*$. Here $Q \to Q_n^*$ denotes that Q_n^* is a realization of Q. In Section 4, for instance, P_2 is a path containing two elements, while P_2 is a labelled path, labelled with two integers a and b.

Furthermore $c(Q_n^*)$ denotes the number of chains going through Q_n^*. In our example these numbers are $a!(n-a)!$ and $a!(b-a)!(n-b)!$, respectively.

Let $\min_{Q \to Q_n^*} c(Q_n^*) = c_n^*(Q)$ be the smallest number of chains with respect to the realizations. In the example: $c_n^*(P_1) = \left\lfloor \frac{n}{2} \right\rfloor! \left\lceil \frac{n}{2} \right\rceil!$, $c_n^*(P_2) = n \left\lfloor \frac{n-1}{2} \right\rfloor! \left\lceil \frac{n-1}{2} \right\rceil!$.

Theorem 5.1.

$$\mathrm{La}(n, \mathcal{P}) \leq \frac{n!}{\inf_{Q \in \mathcal{Q}(\mathcal{P})} \frac{c_n^*(Q)}{|Q|}}.$$

Proof. Let $\mathcal{F}$ be a family without a copy of any of the posets in $\mathcal{P}$. The connected components of the poset induced by $\mathcal{F}$ all belong to $\mathcal{Q}(\mathcal{P})$. Since no chain can go through two distinct components, the sum of the numbers of chains cannot exceed the total number of chains.

$$(5.1) \qquad \sum_{Q \in \mathcal{Q}(\mathcal{P})} \sum_{Q_n^* : Q \to Q_n^*} c(Q_n^*) \le n! .$$

Since

$$Q \to Q_n^* \quad \text{implies} \quad c_n^*(Q) \le c(Q_n^*)$$

(5.1) can be replaced by

$$(5.2) \qquad \sum_{Q \in \mathcal{Q}(\mathcal{P})} \left| \{ Q_n^* : Q \to Q_n^* \} \right| c_n^*(Q) \le n! .$$

Easy manipulations on the left hand side give

$$\sum_{Q \in \mathcal{Q}(\mathcal{P})} \left| \{ Q_n^* : Q \to Q_n^* \} \right| |Q| \frac{c_n^*(Q)}{|Q|}$$

$$\ge \sum_{Q \in \mathcal{Q}(\mathcal{P})} \left| \{ Q_n^* : Q \to Q_n^* \} \right| |Q| \inf_{Q \in \mathcal{Q}(\mathcal{P})} \frac{c_n^*(Q)}{|Q|}$$

and a new form of (5.2):

$$(5.3) \qquad \inf_{Q \in \mathcal{Q}(\mathcal{P})} \frac{c_n^*(Q)}{|Q|} \sum_{Q \in \mathcal{Q}(\mathcal{P})} \left| \{ Q_n^* : Q \to Q_n^* \} \right| |Q| \le n! .$$

Here

$$\sum_{Q \in \mathcal{Q}(\mathcal{P})} \left| \{ Q_n^* : Q \to Q_n^* \} \right| |Q| = |\mathcal{F}|,$$

and (5.3) gives

$$|\mathcal{F}| \inf_{Q \in \mathcal{Q}(\mathcal{P})} \frac{c_n^*(Q)}{|Q|} \le n!$$

what proves the theorem. $\blacksquare$

6. THE UPPER ESTIMATE IN THEOREM 1.3

This theorem already has two different proofs in [11] and [4], however each of these proofs needed an *ad hoc* idea, our new method also works here. It needs some tedious calculations, but the principal idea is as easy as in Section 4. Especially if the concise form, Theorem 5.1 is used.

Suppose that $\mathcal{F}$ contains no V_2 as a subposet. Then it cannot contain a P_3 either. It is easy to deduce that the components of the poset spanned by $\mathcal{F}$ are all of type Λ_r where $0 \leq r$. This is a new phenomenon! The sizes of the components are unbounded. Yet, the method works. This is why we had to write "inf" in Theorem 5.1.

In terms of Section 5: $\mathcal{P} = \{V_2\}$ and $\mathcal{Q}(\mathcal{P}) = \{\Lambda_0, \Lambda_1, \Lambda_2, \ldots, \Lambda_r \ldots\}$. The following lemma gives a good lower estimate on $c_n^*(\Lambda_r)$. For the sake of completeness the proof from [10] is repeated.

Lemma 6.1. *Suppose* $6 \leq n, 1 \leq r$. *Then*

$$u^*!(n - u^*)! + r u^*! u^*(n - u^* - 1)! \leq c_n^*(\Lambda_r)$$

holds where $u^* = u^*(n) = \frac{n}{2} - 1$ *if* n *is even,* $u^* = \frac{n-1}{2}$ *if* n *is odd and* $r - 1 \leq n$, *while* $u^* = \frac{n-3}{2}$ *if* n *is odd and* $n < r - 1$.

In the case $r = 0$ *the inequality* $\left\lfloor \frac{n}{2} \right\rfloor! \left\lceil \frac{n}{2} \right\rceil! \leq c_n^*(\Lambda_0)$ *holds.*

Proof. By symmetry we can consider V_r instead of Λ_r. Since it was done in this form in [10] it is more convenient to use this form for the proof. Let $V(r; u, u_1, \ldots, u_r)$ $(u < u_1, \ldots, u_r)$ be a realization of Λ_r (in notation $V_r \to V(r; u, u_1, \ldots, u_r))$ where the subset of u elements is included in all other ones of sizes $u_1, \ldots, u_r$, respectively.

1. One can easily show by using the sieve that

$$c\big(V(r; u, u_1, \ldots, u_r)\big)$$

$$= u!(n - u)! + \sum_{i=1}^{r} u_i!(n - u_i)! - \sum_{i=1}^{r} u!(u_i - u)!(n - u_i)!.$$

This will actually be used in the form

$$(6.1) \quad c\big(V(r; u, u_1, \ldots, u_r)\big)$$

$$= \sum_{i=1}^{r} \left(\frac{1}{r} u!(n - u)! + u_i!(n - u_i)! - u!(u_i - u)!(n - u_i)! \right).$$

Dividing one term by $n!$ two useful forms are obtained for the summand in (6.1):

$$(6.2) \qquad \frac{1}{r\binom{n}{u}} + \frac{1}{\binom{n}{u_i}} - \frac{1}{\binom{n}{u_i}\binom{u_i}{u}} = \frac{1}{r\binom{n}{u}} + \frac{1}{\binom{n}{u_i}}\left(1 - \frac{1}{\binom{u_i}{u}}\right)$$

and

$$(6.3) \qquad \frac{1}{r\binom{n}{u}} + \frac{1}{\binom{n}{u_i}} - \frac{1}{\binom{n}{u}\binom{n-u}{n-u_i}} = \frac{1}{\binom{n}{u_i}} + \frac{1}{\binom{n}{u}}\left(\frac{1}{r} - \frac{1}{\binom{n-u}{n-u_i}}\right).$$

2. First we will show that (6.2)–(6.3) attains its minimum under the condition $u < u_i$ for some pair $u, u_i = u + 1$.

If $\frac{n}{2} - 1 \leq u$, fix u and consider changing u_i in (6.2). Here, $\binom{n}{u_i}$ is a decreasing fuction of u_i in the interval $\left[\lfloor \frac{n}{2} \rfloor, n\right]$, while $\binom{u_i}{u}$ is increasing. Therefore, one can suppose that $u_i = u + 1$, and we are done.

Else, $\frac{n}{2} - 1 > u$ and the method above in (6.2) leads to $u_i = \lfloor \frac{n}{2} \rfloor$. Fix this value and increase u using (6.3). It will not increase by moving u to $u = \lfloor \frac{n}{2} \rfloor - 1$.

Hence, we obtained the lower estimate

$$\min_u \left(\frac{1}{r}u!(n-u)! + (u+1)!(n-u-1)! - u!1!(n-u-1)!\right)$$

$$= \min_u \left(\frac{1}{r}u!(n-u)! + u!u(n-u-1)!\right)$$

for (6.2)–(6.3) and therefore we have

$$(6.4) \qquad \min_u \left(u!(n-u)! + ru!u(n-u-1)!\right) \leq c\big(V(r; u, u_1, \ldots, u_r)\big)$$

This minimum will be determined in the rest of the proof.

3. Suppose now that $2 \leq r$. Take the "derivative" of $f_r(u) = u!(n-u)! + ru!u(n-u-1)!$, that is, compare two consecutive places of $f_r(u)$. When does the inequality

$$(6.5) \qquad f_r(u-1) = (u-1)!(n-u+1)! + r(u-1)!(u-1)(n-u)!$$

$$< f_r(u) = u!(n-u)! + ru!u(n-u-1)!.$$

hold? It is equivalent to

$$0 < 2(r-1)u^2 - \big(n(r-3)+r-1\big)u - n^2 + (r-1)n.$$

The discriminant of the corresponding quadratic equation in u is

$$\big(n(r-3)+r-1\big)^2 + 8(r-1)\big(n^2-(r-1)n\big)$$

$$= (r+1)^2 n^2 - 2(r-1)(3r-1)n + (r-1)^2.$$

The latter expression can be strictly upper estimated by

$$\big((r+1)n-(r-1)\big)^2,$$

if $r+1 < 3r-1$ holds, that is, if $r > 1$. Hence, the larger root α_2 of the quadratic equation is less than

$$\frac{n(r-3)+r-1+(r+1)n-(r-1)}{4(r-1)} = \frac{n}{2}.$$

On the other hand, as it is easy to see, $\big(n(r+1)-3(r-1)\big)^2$ is a lower bound for the discriminant if $r-1 \le n$ holds. Using this estimate we obtain that $\frac{n-1}{2} \le \alpha_2$ in this case. Substituting this lower estimate into the formula for the smaller root α_1 we obtain $\alpha_1 \le 0$ when $n \ge r-1$. Since (6.5) holds exactly below α_1 and above α_2, we can state that $f_r(u)$ attains its minimum at $u = \lfloor \alpha_2 \rfloor$. By the inequalities above we can conclude that this is at $\frac{n}{2}-1$ if n is even and $\frac{n-1}{2}$ if n is odd. The statement of the lemma is proved in the case of $n \ge r-1$.

Else suppose $n < r-1$. The inequality $\alpha_2 < \frac{n-1}{2}$ can be proved in the same way as in the previous case. On the other hand, $6 \le n$ implies that $\big(n(r+1)-5(r-1)\big)^2$ is a lower estimate on the discriminant, hence we have $\frac{n}{2}-1 < \alpha_2$. This gives that $\alpha_1 < \frac{3}{2}$. The if n is even, $\lfloor \alpha_2 \rfloor$ is again $\frac{n}{2}-1$, while $\lfloor \alpha_2 \rfloor = \frac{n-3}{2}$ when n is odd. Although $f_r(0) < f_r(1)$ is allowed by this estimate, it is easy to check that $f_r(0) > f_r(1)$ holds in reality. By (6.4) the proof is finished for $r \ge 2$.

The case $r = 1$ is much easier. The comparison (6.5) leads to a linear inequality which is an equality for $u = \frac{n}{2}$. The formula $f_1(u)$ also has its minimum at $\lfloor \frac{n-1}{2} \rfloor$. (But it has the same value at $\frac{n}{2}-1$ and $\frac{n}{2}$.) $\blacksquare_L$

The inequality

$$(6.6) \quad u^*! u^*(n-u^*-1)! \le \frac{u^*!(n-u^*)! + r u^*! u^*(n-u^*-1)!}{r+1} \quad (0 \le r)$$

is a consequence of the lemma and the remark on the case $r = 0$.

Lemma 6.1 and (6.6) gives

$$(6.7) \qquad u^*!u^*(n - u^* - 1)! \leq \frac{c_n^*(\Lambda_r)}{|\Lambda_r|}.$$

Substituting this into Theorem 5.1 we obtain

$$|\mathcal{F}| \leq \frac{n!}{u^*!u^*(n - u^* - 1)!}.$$

This right hand side is equal to

$$\binom{n}{\lfloor \frac{n}{2} \rfloor} \frac{\frac{n}{2}}{\frac{n}{2} - 1}, \qquad \binom{n}{\lfloor \frac{n}{2} \rfloor} \frac{\frac{n+1}{2}}{\frac{n-1}{2}}, \qquad \binom{n}{\lfloor \frac{n}{2} \rfloor} \frac{\frac{n-1}{2}}{\frac{n-3}{2}}$$

in the cases $u^* = \frac{n}{2} - 1, \frac{n-1}{2}$ and $\frac{n-3}{2}$, respectively. These are all equal to

$$= \binom{n}{\lfloor \frac{n}{2} \rfloor} \left(1 + \frac{2}{n} + O\left(\frac{1}{n^2} \right) \right). \qquad \blacksquare$$

7. A CONSTRUCTION = A LOWER ESTIMATE

Although we concentrate in this paper on the upper estimates, it seems to be important to show the construction serving as a lower estimate in Theorem 1.3

The construction for a family avoiding a V_2 is the following. Take all the sets of size $\lfloor \frac{n}{2} \rfloor$ and a family $A_1, \ldots A_m$ of $\lfloor \frac{n}{2} \rfloor + 1$-element sets satisfying the condition $|A_i \cap A_j| < \lfloor \frac{n}{2} \rfloor$ for every pair $i < j$. It is easy to see that this family contains no V_2. We only have to maximize m. Since the $\lfloor \frac{n}{2} \rfloor$-element subsets of the A_is are all distinct, we have

$$m \left(\lfloor \frac{n}{2} \rfloor + 1 \right) \leq \binom{n}{\lfloor \frac{n}{2} \rfloor}.$$

This gives the upper estimate

$$(7.1) \qquad m \leq \binom{n}{\lfloor \frac{n}{2} \rfloor} \frac{2}{n}.$$

There is a very nice construction of such sets A_i with

$$(7.2) \qquad m = \binom{n}{\lfloor \frac{n}{2} \rfloor + 1}^{\frac{1}{n}} = \binom{n}{\lfloor \frac{n}{2} \rfloor}\left(\frac{1}{n} + \Omega\left(\frac{1}{n^2}\right)\right).$$

Let $\lfloor \frac{n}{2} \rfloor + 1$ be denoted by k in this proof for saving space. Consider the sets $\{x_1, x_2, \ldots, x_k\}$ where these xs are distinct integers in the interval $[n]$ satisfying

$$(7.3) \qquad x_1 + x_2 + \ldots + x_k \equiv a \pmod{n}$$

for some fixed $a \in [n]$. It is easy to see that the intersection of two such sets is $< k$, as we needed. The total number of k-element sets is $\binom{n}{k}$, therefore there is an a for which the number of sets satisfying (7.3) is at least $\frac{1}{k}\binom{n}{k}$. We found the necessary number of "good" sets.

This construction can be found in this form in [9] (the paper contains a much more general form), but the basic idea have appeared in [5] and [16], too.

It is a longstanding conjecture of coding theory what the right constant is here, 1 or 2. Or if the limit exists at all? This is why there is a disturbing factor 2 between the second terms of the lower and upper estimates in Theorem 1.3. This gap cannot be bridged without making progress in the problem in coding theory mentioned above.

8. Excluding the N

The poset N contains 4 distinct elements a, b, c, d satisfying $a < c$, $b < c$, $b < d$. In the Boolean lattice a subposet N consists of four distincts subsets satisfying $A \subset C$, $B \subset C$, $B \subset D$. It is somewhat surprising that excluding N the result is basically the same as in the case of V_2. The goal of the present section is to prove the following theorem.

Theorem 8.1 [10].

$$\binom{n}{\lfloor \frac{n}{2} \rfloor}\left(1 + \frac{1}{n} + \Omega\left(\frac{1}{n^2}\right)\right) \leq \mathrm{La}(n, N) \leq \binom{n}{\lfloor \frac{n}{2} \rfloor}\left(1 + \frac{2}{n} + O\left(\frac{1}{n^2}\right)\right)$$

holds.

The lower estimate is obtained from Theorem 1.3, since $\mathrm{La}(n, V_2) \leq \mathrm{La}(n, N)$.

The upper estimate will be proved by Theorem 5.1, again. Here $\mathcal{P} = \{N\}$.

Let $\mathcal{F}$ be a family of subsets of $[n]$ containing no four distinct members forming an N. Consider the poset $P(\mathcal{F})$ spanned by $\mathcal{F}$ in B_n. What can its components be? A component might be a P_3, but no component can contain a P_3 as a proper subposet, since adding one more element to P_3 an N is created no matter which element of P_3 is in relation with the new element. Let $a < b$ be two elements of a component. We claim that a and b cannot be both comparable within the component with some other distinct elements c, d (say, in this order), unless they are a part of a P_3. Indeed, the choices $c < a$ and $b < d$ lead to a P_3, therefore the only possibility is $a < c, d < b$. This is an N, contradicting the assumption. But one of them can be comparable with many others in the same direction. Therefore the following ones are the only possible components:

$$\mathcal{Q}(\mathcal{P}) = \{P_3, \Lambda_0, \Lambda_1, \Lambda_2, \ldots, \Lambda_r, \ldots, V_1, V_2, \ldots, V_r, \ldots\}.$$

In order to use Theorem 5.1 we have to give a good lower bound on the ratios

$$(8.1) \qquad \frac{c_n^*(P_3)}{3}, \qquad \frac{c_n^*(\Lambda_r)}{r+1}, \qquad \frac{c_n^*(V_r)}{r+1}.$$

(6.7) is a good lower estimate on the middle one. By symmetry, the same applies for the last one. The only unknown one is the first ratio. Its exact value is determined in [10] (Lemma 3.1). We do not repeat the proof, since both the statement and the proof are obvious.

Let $P(3; u, v, w)$ $(u < v < w)$ be a realization of P_3 by sets of sizes $u < v < w$.

Lemma 8.2. $c\big(P(3; u, v, w)\big)$ $(u < v < w)$ *takes its minimum for the values* $u = \lfloor \frac{n}{2} \rfloor - 1$, $v = \lfloor \frac{n}{2} \rfloor$, $w = \lfloor \frac{n}{2} \rfloor + 1$, *that is,*

$$\left(\left\lfloor \frac{n}{2} \right\rfloor - 1 \right)! \left(\left\lceil \frac{n}{2} \right\rceil - 1 \right)! \left(\left\lfloor \frac{n}{2} \right\rfloor^2 - n \left\lfloor \frac{n}{2} \right\rfloor + n^2 - 1 \right) \leq c\big(P(3; u, v, w)\big).$$

Hence we have

$$(8.2) \qquad \frac{1}{3} \left(\left\lfloor \frac{n}{2} \right\rfloor - 1 \right)! \left(\left\lceil \frac{n}{2} \right\rceil - 1 \right)! \left(\left\lfloor \frac{n}{2} \right\rfloor^2 - n \left\lfloor \frac{n}{2} \right\rfloor + n^2 - 1 \right) \leq \frac{c_n^*(P_3)}{3}$$

If we are lucky, the left hand side of (8.2) is not smaller than the left hand side of (6.7). Indeed, the inequality

$$u^*!u^*(n-u^*-1)! \le \frac{1}{3}\left(\left\lfloor\frac{n}{2}\right\rfloor - 1\right)!\left(\left\lceil\frac{n}{2}\right\rceil - 1\right)!\left(\left\lfloor\frac{n}{2}\right\rfloor^2 - n\left\lfloor\frac{n}{2}\right\rfloor + n^2 - 1\right)$$

can be easily checked (for $2 \le n$) by distinguishing the three cases of u^*.

Since $u^*!u^*(n-u^*-1)!$ is a lower estimate for all ratios in (8.1) the proof of Theorem 7.1 can be finished as in the case of Theorem 1.3. $\blacksquare$

Remarks. Knowing the estimate of Lemma 6.1 we obtained the proof of Theorem 7.1 almost free, we only had to show that P_3 does not decrease the infimum in the denominator of Theorem 5.1. Of course we cannot state that $\mathrm{La}(n, V_2) = \mathrm{La}(n, N)$, they are equal only asymptotically.

It is interesting to mention that the "La" function will jump if the excluded poset contains one more relation. The *butterfly* $\bowtie$ contains 4 elements: a, b, c, d with $a < c$, $a < d$, $b < c$, $b < d$.

Theorem 8.3. [4] *Let* $n \ge 3$. *Then* $\mathrm{La}(n, \bowtie) = \binom{n}{\lfloor n/2 \rfloor} + \binom{n}{\lfloor n/2 \rfloor + 1}$.

9. A FURTHER GENERALIZATION

Observe that the main part of a large family is near the middle, the total number of sets far from the middle is small. More precisely, let $0 < \alpha < \frac{1}{2}$ be a fixed real number. The total number of sets F (for a given n) of size satisfying

$$(9.1) \qquad |F| \notin \left[n\left(\frac{1}{2}-\alpha\right), n\left(\frac{1}{2}+\alpha\right)\right]$$

is very small. It is well-known (see e.g. [1], page 214) that for a fixed constant $0 < \beta < \frac{1}{2}$

$$\sum_{i=0}^{\lfloor \beta n \rfloor} \binom{n}{i} = 2^{n(h(\beta)+o(1))}$$

holds where $h(x) = -x\log_2 x - (1-x)\log_2(1-x)$. Therefore the total number of sets satisfying (9.1) is at most

$$(9.2) \qquad 2\sum_{i=0}^{\lfloor n(\frac{1}{2}-\alpha)\rfloor}\binom{n}{i} = 2^{n(h(\frac{1}{2}-\alpha)+o(1))} = \binom{n}{\lfloor\frac{n}{2}\rfloor}o\left(\frac{1}{n^2}\right)$$

where $0 < h\left(\frac{1}{2} - \alpha\right) < 1$ is a constant.

In view of this observation we can improve our main tool, Theorem 5.1. First we have to generalize $c_n^*(Q)$. Let $c_n^{*\alpha}(Q)$ denote $\min_{Q \to Q_n^*} c(Q_n^*)$ where only those realizations Q_n^* are considered whose member subsets are of size in the interval (9.1). It is obvious that $c_n^*(Q) \leq c_n^{*\alpha}(Q)$. We actually believe that they are equal for large n, but we cannot prove this statement.

Theorem 9.1. *Let* $0 < \alpha < \frac{1}{2}$ *be a real number. Then*

$$\mathrm{La}(n, \mathcal{P}) \leq \frac{n!}{\inf_{Q \in \mathcal{Q}(\mathcal{P})} \frac{c_n^{*\alpha}(Q)}{|Q|}} + \binom{n}{\lfloor \frac{n}{2} \rfloor} O\left(\frac{1}{n^2}\right).$$

Proof. Let $\mathcal{F}$ be a family containing no subposet belonging to $\mathcal{P}$. Define $\mathcal{F}_\alpha \subset \mathcal{F}$ consisting of the sets F of sizes in the interval (9.1). The rest, $\mathcal{F} - \mathcal{F}_\alpha$ is denoted by $\mathcal{F}_{\overline{\alpha}}$.

$$(9.3) \qquad |\mathcal{F}_{\overline{\alpha}}| \leq \binom{n}{\lfloor \frac{n}{2} \rfloor} O\left(\frac{1}{n^2}\right)$$

is a consequence of (9.2). We have to prove only that

$$(9.4) \qquad |\mathcal{F}_\alpha| \leq \frac{n!}{\inf_{Q \in \mathcal{Q}(\mathcal{P})} \frac{c_n^{*\alpha}(Q)}{|Q|}}.$$

The proof of Theorem 5.1 can be repeated. The only difference is that since all the sets in $\mathcal{F}_\alpha$ are of size in the interval (9.1), the components have the same property therefore c_n^* can be really replaced by $c_n^{*\alpha}$. The sum of (9.3) and (9.4) gives the statement of the theorem. ∎

Theorem 9.2. *Let* $1 \leq r$ *be a fixed integer, independent on* n. *Suppose that every element* $Q \in \mathcal{Q}(\mathcal{P})$ *has the following property: if* $a \in Q$ *then* a *covers at most* r *elements of* Q. *Then*

$$\mathrm{La}(n, \mathcal{P}) \leq \binom{n}{\lfloor \frac{n}{2} \rfloor} \left(1 + 2\frac{r}{n} + O\left(\frac{1}{n^2}\right)\right).$$

Proof. Theorem 9.1 will be used with $0 < \alpha < \frac{1}{8r+12}$. Suppose that $Q \in \mathcal{Q}(\mathcal{P})$, $Q \to Q_n^*$, that is, Q_n^* is a realization of Q and its member sets are of size in the interval (9.1). Using the first two terms of the sieve

(9.5)
$$c(Q_n^*) \geq \sum_{F \in Q_n^*} |F|!\,(n - |F|)! - \sum_{F,G \in Q_n^*, G \subset F} |G|!\,(|F| - |G|)!\,(n - |F|)!$$

$$= \sum_{F \in Q_n^*} \left(|F|!\,(n - |F|)! - \sum_{G \in Q_n^*:\, G \subset F} |G|!\,(|F| - |G|)!\,(n - |F|)! \right).$$

Consider one term of (9.5) and divide it by $n!$:

(9.6)
$$\frac{1}{\binom{n}{|F|}} \left(1 - \sum_{G \in Q_n^*:\, G \subset F} \frac{1}{\binom{|F|}{|G|}} \right).$$

$n\left(\frac{1}{2} - \alpha\right) \leq |G| < |F| \leq n\left(\frac{1}{2} + \alpha\right)$ implies $|F| - |G| \leq 2\alpha n$, therefore $|G|$ can be $|F|-1, |F|-2, \ldots, |F| - \lfloor 2\alpha n \rfloor$. The number of sets G with $|G| = |F|-1$ is at most r by the assumption of the theorem. It is easy to see that the number of sets G with $|G| = |F| - i$ is at most r^i. Hence the following lower estimate on (9.6) is obtained:

(9.7)
$$\frac{1}{\binom{n}{|F|}} \left(1 - \sum_{i=1}^{\lfloor 2\alpha n \rfloor} \frac{r^i}{\binom{|F|}{i}} \right).$$

We will show that these negative terms are increasing in absolute value, when $n \geq 4r$. Compare two neighboring terms.

$$\frac{r^i}{\binom{|F|}{i}} \geq \frac{r^{i+1}}{\binom{|F|}{i+1}}$$

holds iff

(9.8)
$$|F| \geq (r + 1)i + r.$$

Since $|F| \geq n\left(\frac{1}{2} - \alpha\right)$ and $i \leq 2\alpha n$, it is sufficient to show

$$n\left(\frac{1}{2} - \alpha\right) \geq (r + 1)2\alpha n + r$$

rather than (9.8). However this is an easy consequence of $0 < \alpha < \frac{1}{8r+12}$ and $n \geq 4r$. By the monotonity a new lower estimate is obtained for (9.7):

$$(9.9) \qquad \frac{1}{\binom{n}{|F|}} \left(1 - \frac{r}{|F|} - \frac{r^2}{\binom{|F|}{2}} - 2\alpha n \frac{r^3}{\binom{|F|}{3}}\right).$$

A further decrease can be obtained by choosing the appropriate, but different values of $|F|$ in the two factors of (9.9). Choose $|F| = \lfloor \frac{n}{2} \rfloor$ in the first factor and $|F| = \lceil n(\frac{1}{2} - \alpha) \rceil$ in the second one. The lower estimate

$$(9.10) \qquad \frac{1}{\binom{n}{\lfloor \frac{n}{2} \rfloor}} \left(1 - 2\frac{r}{n(1 - 2\alpha)} + O\left(\frac{1}{n^2}\right)\right)$$

is obtained.

(9.6)–(9.10) are lower estimates on one term of (9.5). Since (9.10) does not depend on F, we have the lower estimate

$$n!|Q_n^*|\frac{1}{\binom{n}{\lfloor \frac{n}{2} \rfloor}} \left(1 - 2\frac{r}{n(1 - 2\alpha)} + O\left(\frac{1}{n^2}\right)\right)$$

on (9.5). Using the trivial $|Q_n^*| = |Q|$

$$\frac{c_n^{*\alpha}(Q)}{|Q|} \geq |Q|\frac{n!}{\binom{n}{\lfloor \frac{n}{2} \rfloor}} \left(1 - 2\frac{r}{n(1 - 2\alpha)} + O\left(\frac{1}{n^2}\right)\right)$$

can be obtained what is independent on Q. Substitute this lower estimate into the first term of the statement of Theorem 9.1 to obtain an upper estimate:

$$\binom{n}{\lfloor \frac{n}{2} \rfloor}\frac{1}{1 - \frac{2r}{n(1-2\alpha)} - O\left(\frac{1}{n^2}\right)} = \binom{n}{\lfloor \frac{n}{2} \rfloor} \left(1 + \frac{2r}{n(1 - 2\alpha)} + O\left(\frac{1}{n^2}\right)\right).$$

Theorem 9.1 gives

$$\mathrm{La}(n, \mathcal{P}) \leq \binom{n}{\lfloor \frac{n}{2} \rfloor} \left(1 + 2\frac{r}{n(1 - 2\alpha)} + O\left(\frac{1}{n^2}\right)\right).$$

Since this holds for arbitrary small α, the proof is complete. $\blacksquare$

If the family $\mathcal{F}$ contains no V_{r+1} then the components cannot contain a set which is contained in $r + 1$ other sets. Therefore the conditions of Theorem 9.2 are satisfied (in the dual form). This is why Theorem 9.2 implies Theorem 1.5. However, as we will see in the next section, a stronger statement follows.

10. Excluding Induced Posets, Only

One can ask what happens if we exclude the posets R belonging to $\mathcal{P}$ only in a strict form, that is, there is no induced copy in the poset induced in B_n by the family. Given a "small" poset R, $\mathrm{La}^\sharp(n, R)$ denotes the maximum number of elements of $Y \subset 2^{[n]}$ (that is, the maximum number of subsets of $[n]$) such that R is not an induced subposet of the poset spanned by Y in B_n. This obviously generalizes for $\mathrm{La}^\sharp(n, \mathcal{P})$ where $\mathcal{P}$ is a set of posets.

For instance, calculating $\mathrm{La}(n, V_2)$ the path of length 3, P_3 is also excluded, while in the case of $\mathrm{La}^\sharp(n, V_2)$ this is allowed, three sets A, B, C are excluded from the family only when $A \subset B$, $A \subset C$ but B and C are incomparable. As we saw in the proof of Theorem 1.3, $\mathcal{Q}(V_2)$ consists of Λ_rs $(0 \le r)$. The set $\mathcal{Q}^\sharp(V_2)$ of possible components when only the induced V_2s are excluded is much richer. $\mathcal{Q}^\sharp(V_2)$ contains all posets whose graph is a "descending" tree with one maximal vertex. That is, not only the sizes of these posets are unbounded, but their depths, as well. Yet, this case can be also be treated, on the basis of Theorem 9.2. To be precise we have to modify the formulations of our previous theorems. These modifications need no proofs, since the original proofs did not really depended on $\mathcal{P}$, only on $\mathcal{Q}(\mathcal{P})$ and this is simply replaced by $\mathcal{Q}^\sharp(\mathcal{P})$.

Theorem 10.1.
$$\mathrm{La}^\sharp(n, \mathcal{P}) \le \frac{n!}{\inf_{Q \in \mathcal{Q}^\sharp(\mathcal{P})} \frac{c_n^*(Q)}{|Q|}}.$$

Theorem 10.2. *Let $0 < \alpha < \frac{1}{2}$ be a real number. Then*

$$\mathrm{La}^\sharp(n, \mathcal{P}) \le \frac{n!}{\inf_{Q \in \mathcal{Q}^\sharp(\mathcal{P})} \frac{c_n^{*\alpha}(Q)}{|Q|}} + \binom{n}{\lfloor \frac{n}{2} \rfloor} O\left(\frac{1}{n^2}\right).$$

Theorem 10.3. *Let $1 \le r$ be a fixed integer, independent on n. Suppose that every element $Q \in \mathcal{Q}^\sharp(\mathcal{P})$ has the following property: if $a \in Q$ then a covers at most r elements of Q. Then*

$$\mathrm{La}^\sharp(n, \mathcal{P}) \le \binom{n}{\lfloor \frac{n}{2} \rfloor} \left(1 + 2\frac{r}{n} + O\left(\frac{1}{n^2}\right)\right).$$

It is quite obvious that if $Q \in \mathcal{Q}^\sharp(V_{r+1})$ then no element of Q is covered by more than r other elements. Theorem 10.3 can be applied in a dual (upside-down) form.

Theorem 10.4.

$$\mathrm{La}^{\sharp}(n, V_{r+1}) \leq \binom{n}{\lfloor \frac{n}{2} \rfloor}\left(1 + 2\frac{r}{n} + O\left(\frac{1}{n^2}\right)\right).$$

This is a stronger form of Theorem 1.5. The special case $r = 1$ was solved in [6].

11. Concluding Remarks

1. We obtained Theorem 8.1 almost free after having the proof of Theorem 1.3 with our method. This probably will often happen. The solution for a given excluded configuration can be obtained by putting together estimates for "allowed" posets, which have been already solved for other excluded patterns.

2. In most of our results the extremal configuration consist of some full levels plus a thinned out level. Theorem 4.1 shows that this is not necessary true in all cases.

3. We do not see the limits of our method. We hope we will be able to generalize for more levels, like Theorems 8.3 and 1.2. However we are quite sure that it will not give the solution for every family $\mathcal{P}$ of posets.

4. The problem $\mathcal{P} = \{V_2, \Lambda_3\}$ is very instructive. It is easy to see that $\mathcal{Q}(\mathcal{P}) = \{P_1, P_2, \Lambda_2\}$. The cases of even and odd n are somewhat different. Consider first the case when n is even. Then $\frac{1}{3}c_n^*(\Lambda_2) = \frac{n}{2}!\frac{n}{2}!\left(1 - \frac{2}{3n}\right) < \frac{1}{2}c_n^*(P_2) = c_n^*(P_1)$. Therefore Theorem 5.1 implies

Theorem 11.1. *If n is even then*

$$\mathrm{La}(n, V_2, \Lambda_3) \leq \frac{n!}{\frac{n}{2}!\frac{n}{2}!\left(1 - \frac{2}{3n}\right)} = \binom{n}{\frac{n}{2}}\left(1 + \frac{2}{3n} + O\left(\frac{1}{n^2}\right)\right).$$

Observe that Theorem 9.2 would only give $\frac{2}{n}$ in the second term. The conclusion is that, although Theorem 9.2 is rather strong in general, it could be week in special cases. On the other hand, it seems that the obvious construction, taking all $\frac{n}{2}$-element subsets is not optimal. We found a better construction for $n = 4$: $\big\{\{1, 2\}, \{1, 3\}, \{1, 4\}, \{2, 4\}, \{3, 4\}, \{1, 2, 3\}, \{2, 3, 4\}\big\}$. But we do not know such a construction for infinitely many n.

Suppose now that n is odd. Then $\frac{1}{3}c_n^*(\Lambda_2) = \frac{n-1}{2}!\frac{n-1}{2}!\left(\frac{n}{2} - \frac{1}{6}\right) < \frac{1}{2}c_n^*(P_2) < c_n^*(P_1)$. Theorem 5.1 gives the following upper estimate. The lower estimate comes from Theorem 4.1.

Theorem 11.2. *If n is odd then*

$$\binom{n}{\frac{n-1}{2}}\left(1 + \frac{1}{n}\right) \leq \mathrm{La}(n, V_2, \Lambda_3) \leq \binom{n}{\frac{n-1}{2}}\left(1 + \frac{4}{3n - 1}\right).$$

Since the right hand side is an integer (42) for $n = 7$, it might cherish the hope that there are some nice constructions (similar to the construction of Theorem 4.1) giving equality in the upper bound, at least for some n. This equality can be achieved only by 14 Λ_2s: consisting of 14 four-element subsets $F_1, \ldots, F_{14}$ and their 3-element subset pairs: T_i^1, T_i^2 where these are 28 distinct 3-element subsets, $T_i^1, T_i^2 \subset F_i$ but $T_i^j \not\subset F_\ell$ holds whenever $i \neq \ell$. Let $R_1, \ldots, R_7$ denote the remaining seven 3-element sets. Since every F contains exactly four 3-element subsets, it must contain two of the Rs. Consequently there are 28 pairs $R \subset F$. Since each R has four 4-element supersets, they must be all in the family of Fs. However the minimum number of 4-element supersets of seven 3-element sets is 15. (This can be shown by taking the complements. The minimum size of the shadow of a family of seven 4-element sets can be obtained from $7 = \binom{5}{4} + \binom{3}{3} + \binom{2}{2}$. It is $\binom{5}{3} + \binom{3}{2} + \binom{2}{1} = 15$.) Our hopes failed.

But there might be other (perhaps infinitely many) n giving equality in the upper bound of Theorem 11.2. The next candidate is $n = 15$.

5. One feels that the optimal construction uses sets in the middle, only.

Conjecture 1. *For every poset $\mathcal{P}$ there is a constant $c(\mathcal{P})$ (independent on n) such that $\mathrm{La}(n, \mathcal{P})$ can be achieved by a family containing sets of size in the interval*

$$\left[\frac{n}{2} - c(\mathcal{P}), \frac{n}{2} + c(\mathcal{P})\right].$$

Let us remark that we have proved a much weaker version in Section 9: $\mathrm{La}(n, \mathcal{P})$ can be asymptotically achieved by sets from the interval

$$\left[n\left(\frac{1}{2} - \alpha\right), n\left(\frac{1}{2} + \alpha\right)\right]$$

where $0 < \alpha$ is arbitrarily small.

The conjecture is not true for the induced case. For instance if the induced Λ_2 is excluded (P_3 is allowed, determination of $\mathrm{La}^\sharp(n, \Lambda_2)$) then the optimal construction always contains the $\emptyset$.

References

[1] Noga Alon and Joel H. Spencer, *The Probabilistic Method,* John Wiley & Sons (1992).

[2] B. Bollobás, On generalized graphs, *Acta. Math. Acad. Sci. Hungar.,* **16** (1965), 447–452.

[3] Annalisa de Bonis, Gyula O.H. Katona, Forbidden r-forks, to appear in *Order.*

[4] Annalisa De Bonis, Gyula O.H. Katona, Konrad J. Swanepoel, Largest family without $A \cup B \subseteq C \cup D$, *J. Combin. Theory Ser. A,* **111**, (2005), 331–336.

[5] R. C. Bose and T. R. N. Rao, *On the theory of unidirectional error correcting codes,* Technical report, SC-7817, Sept. 1978, Dept. Comp. Sci., Southern Methodist University, Dallas, TX.

[6] Teena Carroll, Gyula O.H. Katona, *Largest family without an induced $A \subset B$, $A \subset C$,* submitted.

[7] Konrad Engel, *Sperner Theory,* Encyclopedia of Mathematics and its Applications **65** (1997), Cambridge University Press.

[8] P. Erdős, On a lemma of Littlewood and Offord, *Bull. Amer. Math. Soc.,* **51** (1945) 898–902.

[9] R. L. Graham and H. J. A. Sloane, Lower bounds for constant weight codes, *IEEE IT,* **26**, 37–43.

[10] Jerrold R. Griggs and Gyula O.H. Katona, No four sets forming an N, to appear in *J. Combin Theory.*

[11] G.O.H. Katona and T. G. Tarján, Extremal problems with excluded subgraphs in the n-cube, *Lecture Notes in Math.,* **1018**, 84–93.

[12] D. Lubell, A short proof of Sperner's lemma, *J. Combin. Theory,* **1** (1966), 299.

[13] L.D. Meshalkin, A generalization of Sperner's theorem on the number of subsets of a finite set, *Teor. Verojatnost. i Primen.,* **8** (1963), 219-220 (in Russian with German summary).

[14] E. Sperner, Ein Satz über Untermegen einer endlichen Menge, *Math. Z.,* **27** (1928), 544–548.

[15] Hai Tran Thanh, An extremal problem with excluded subposets in the Boolean lattice, *Order,* **15** (1998), 51–57.

[16] R. R. Varshamov and G. M. Tenengolts, A code which corrects a single asymetric error (in Russian), *Avtom. Telemech.,* **26**(2) (1965), 288–292.

[17] K. Yamamoto, Logarithmic order of free distributive lattices, *J. Math. Soc. Japan,* **6** (1954), 347–357.

Gyula O.H. Katona

Rényi Institute
Budapest, Hungary

e-mail: `ohkatona@renyi.hu`

BOLYAI SOCIETY
MATHEMATICAL STUDIES, 17

Horizons of Combinatorics
Balatonalmádi
pp. 141–161.

Subsums of a Finite Sum and Extremal Sets of Vertices of the Hypercube[*]

DEZSŐ MIKLÓS

We will investigate the maximum size of the subsets of the vertices of a hypercube satisfying the property that the subspace (or cone) spanned by them will not intersect (contain) a given – other – subset of the vertices of the cube. It will turn out that the two cases when, on one side, we consider the spanned subspaces over $GF(2)$, i.e. work only inside the hypercube, and, on the other side, consider subspaces over $\mathbb{R}$, will yield different results. In the first case the maximal subsets with the required property will naturally form a subspace (unless we assume some further properties about the chosen vertices) by themselves – and we may not speak about a cone over $GF(2)$) –, while in the second case they will not necessarily do. For the more detailed, real case an interesting connection to the maximal size of subsums of a (positive) sum which are equal to 0 (in which case the numbers are supposed to be non-zero) or which are also positive, will be pointed out as well as a conjecture given related to the Littlewood–Offord and Erdős–Moser problems.

1. Introduction

Consider the different special cases of the following general question:

Question 1.1. How many vertices (maybe of a certain further property, e.g. of fixed weight) of the n-dimensional hypercube can be picked such that their span, that is the subspace spanned by them – either over $GF(2)$ or over $\mathbb{R}$ – (or, in case of $\mathbb{R}$ the convex or positive span, that is, the

*The work of the author was supported by the Hungarian National Foundation for Scientific Research grant numbers NK062321 and NK048826 and by the following projects of the European Community: INTAS 04-77-7171, COMBSTRU-HPRN-CT-2002-000278, FIST-MTKD-CT-003006.

cone spanned by them), does not contain or does not intersect certain configurations of the hypercube (vertieces, vertices of certain weight(s), subspaces or hyperplanes)?

Throughout the paper, C_n will denote the set of the vertices of the n-dimensional hypercube, $M \subset C_n$ will be a subset of it, $\mathrm{span}_2(M)$ will denote the subspace (of C_n) spanned by M over $GF(2)$, $\mathrm{span}_{\mathbb{R}}(M)$ will denote the subspace of $\mathbb{R}^n$ spanned by M (which is naturally not a subset of the hypercube, but contains some vertices, at least the vertices belonging to M) and $\mathrm{cone}\,(M) = \mathrm{cone}_{\mathbb{R}}(M)$ will denote the cone spanned by M (over $\mathbb{R}$), that is the collection of points of $\mathbb{R}^n$ obtained as a linear combination of the points from M with all positive coefficients. The elements (vertices) of the hypercube C_n will be considered as vectors (many times identified with the subset of $[n]$ having them as their characteristic vectors) and will be denoted by $\underline{x}_i$ with the vectors of exactly one, the ith coordinate equal to 1 denoted by $\underline{e}_i = \{0, \ldots, 0, 1, 0, \ldots, 0\}$, and the whole 1 vector (vertex) denoted by $\underline{1} = \{1, 1, \ldots, 1\}$.

The most obvious forms of the above general question are these: How big M can be such that $\underline{1} \notin \mathrm{span}\,(M)$ or $\underline{e}_1 = (1, 0, \ldots, 0) \notin \mathrm{span}\,(M)$ or none of the vectors (vertices of the hypercube) $\underline{e}_i$ are in $\mathrm{span}\,(M)$, where span can mean either span_2 or $\mathrm{span}_{\mathbb{R}}$, or $\underline{1} \notin \mathrm{cone}\,(M)$? Moreover, naturally, the k-weighted versions of these questions can be asked as well, that is when we restrict the choices of the vertices of C_n only to those having weight k, i.e. exactly k coordinates equal to 1, all others being equal to 0.

It will turn out that the two most interesting questions of the above ones are equivalent to the following ones, respectively

Question 1.2. How should we give a set of n real numbers $x_1, x_2, \ldots, x_n$ none of them being equal to 0, such that they maximize the number of sums $\sum_{i \in B} x_i$ equal to 0, where the B's are subsets of $\{1, 2, \ldots, n\}$ (or B's are subsets of $\{1, 2, \ldots, n\}$ of cardinality k or at most k)?

Question 1.3. Let $x_1, x_2, \ldots, x_n$ be given numbers such that $\sum_{i=1}^{n} x_i > 0$. What is maximum number of negative subsums (or the minimum number of positive subsums) of exactly k of these numbers?

In section 2 of the paper we will explore the non-restricted cases (when the chosen vertices are not restricted to weight k vertices only) of the above questions and point out the connection of it to the Littlewood–Offord problem (basically to Question 1.2 above). It will also be shown that Question 1.2 is equivalent to the following question:

Question 1.4. Let $x_1, x_2, \ldots, x_n$ be given numbers. What is the maximum number of the subsums $\sum x_{j_i}$ of them which can be disclosed without the disclosure of the values x_i, that is such a way that knowing those subsums nobody could calculate the values of the x_i's?

We will also discuss in this chapter the relation of $\mathrm{span}_2(M)$ and $\mathrm{span}_{\mathbb{R}}(M)$ for a given subset M of vertices of the hypercube.

In section 3 the k-weight cases will be inspected, in particular the relation of them to Question 1.3 and to the following question as well:

Question 1.5. Let $x_1, x_2, \ldots, x_n$ be given numbers. What is the maximum number of the subsums $\sum_{i=1}^{k} x_{j_i}$ (of exactly k of or at most k of these numbers) which can be given without the disclosure of the values of x_i's?

It will be pointed out that the answers to the questions have many times kind of "phase transition" phenomena, that is in lower dimension they will be different from the ones in higher dimension.

Finally, in section 4 we will return to the non-restricted case and investigate, or at least pose some further questions, where the subsets of the vertices of the hypercube to be avoided will be of other nature, like all vertices of weight 2, or similar. An interesting conjecture will be formulated for this case, related to the Erdős–Moser problem.

It was brought to the attention of the author during the preparation of this survey paper that during recent years R. Ahlswede, H. Aydinian, and L. H. Khachatrian wrote a series of papers [1, 2, 3, 4, 5] dealing with similar questions and especially [2, 3] contain results stated in this paper as well, though mostly from a different approach and most of the time with different proofs.

2. THE GENERAL QUESTIONS

Proposition 2.1. *If for an $M \subset C_n$ the size of $M > 2^{n-1}$ then $(1, 0, \ldots, 0)$ $\in \mathrm{span}\,(M)$, and therefore $(1, 1, \ldots, 1) \in \mathrm{span}\,(M)$.*

Proof. It is easy to see that $|M| > 2^{n-1}$ implies that two "complementary" pair of vertices, that is, two vertices whose sum as vectors equals $(1, 1, \ldots, 1)$, should be in M, and therefore $(1, 1, \ldots, 1) \in \mathrm{span}\,(M)$. The fact that

$|M| > 2^{n-1}$ implies $(1, 0, \ldots, 0) \in \text{span}\,(M)$ will be shown by induction on n. The base cases ($n = 2$ or 3) are easy to check.

Divide M into two disjoint subsets, M_1 being the set of vertices with last coordinates 0, and M_2 the set of vertices with last coordinates 1. Since $|M| > 2^{n-1}$, either $|M_1| > 2^{n-2}$ or $|M_2| > 2^{n-2}$. In the first case, by the induction hypothesis, the vector $(1, 0, \ldots, 0)$ of length $n - 1$ is in the span of the vectors obtained by truncating the last 0 coordinate from the vertices of M_1, and therefore the vector $\underline{e}_1 = (1, 0, \ldots, 0)$ of length n is in the span of M_1. In the second case, knowing already that $\underline{1} = (1, 1, \ldots, 1) \in \text{span}\,(M)$, subtract the vectors of M_2 from this vector, obtaining more than 2^{n-2} vectors in span M with last coordinate equal to 0, thus reducing this case to the first one.

Note that the above arguments work for both of span_2 and $\text{span}_{\mathbb{R}}$, therefore this is proposition is valid, independently whether we work over $GF(2)$ or $\mathbb{R}$.

Also, the same arguments show the validity of the following two propositions as well (in case of Proposition 2.2, again, independently whether we take the span over $GF(2)$ or $\mathbb{R}$).

Proposition 2.2. *If for an* $M \subset C_n$ *the size of* $|M| > 2^{n-1}$ *then* $\text{span}\,(M) \supset C_n$.

Proposition 2.3. *If for an* $M \subset C_n$ *the size of* $|M| > 2^{n-1}$ *then* $(1, 1, \ldots, 1) \in \text{cone}\,(M)$.

Remark 2.4. All the above bounds are sharp, since for any $x \in C_n$, $x \neq \underline{0}$ one can take a non-zero coordinate of x, and M as the set of all (2^{n-1}) vertices of C_n which have 0 at this coordinate. The span of this M will obviously not contain x (neither in $GF(2)$ nor in $\mathbb{R}$). This construction is valid for the vertex $(1, 1, \ldots, 1)$ and for positive span (cone) as well.

Remark 2.5. The question about the maximum size of M with span M not containing completely any given subset B of C_n is handled by the above propositions: Any vertex of C_n, in particular any vertex of B can be "avoided" by an M of size 2^{n-1} (that is, span M will not contain that vertex, therefore will not contain B completely). On the other hand, any M of size greater than 2^{n-1} will have span M containing the whole C_n, in particular, span M will contain B completely. Again, in the argument above span may be meant both as span_2 or $\text{span}_{\mathbb{R}}$.

The situation becomes more diverse in case we want to avoid with the span of M *all* vertices $\underline{e}_i$.

Theorem 2.6. *The maximum size of a subset M of the vertices of C_n such that none of the vertices $\underline{e}_i = (0,0,\ldots,0,1,0,\ldots,0)$ are in $\mathrm{span}_{\mathbb{R}}(M)$ is*
$$\binom{n}{\lfloor n/2 \rfloor} = \sum_{i=0}^{\lfloor n/2 \rfloor} \binom{\lceil n/2 \rceil}{i}\binom{\lfloor n/2 \rfloor}{i} = \sum_{i=0}^{\lfloor n/2 \rfloor} \binom{\lceil n/2 \rceil}{i}\binom{\lfloor n/2 \rfloor}{\lfloor n/2 \rfloor - i}$$

(as shown by taking all vertices having the same number of 1 coordinates among the first $\lfloor n/2 \rfloor$ and last $\lceil n/2 \rceil$ coordinates).

In case of span_2 the above bound may not be valid (since in that case we have M a subspace of C^n, definitely having size of a power of 2), and so the situation is quite different, as the following remark shows.

Remark 2.7. The maximum size of a subset M of the vertices of C_n such that none of the vertices $(0,0,\ldots,0,1,0,\ldots,0)$ are in $\mathrm{span}_2(M)$ is still 2^{n-1}, as shown by the following example: divide the set of coordinates into two arbitrary subsets A and B (in extreme case $|B| = 1$) and take all of those vertices of C_n which have the same parity of 1 coordinates in A and B (in case of $|A| = 2^{n-1}$ and $|B| = 1$ it simply means taking all the vertices with arbitrary coordinates in A and the only coordinate belonging to B being the parity check bit of them). All linear combinations over $GF(2)$ will preserve the parity of the 1 coordinates both in A and B, therefore no vertices with exactly one coordinate equal to 1 will be among them. The previous propositions show that this is the best possible construction.

Remark 2.8. Though it is clear that for any subset M of the vertices of C_n $\dim \mathrm{span}_2 M \leq \dim \mathrm{span}_{\mathbb{R}} M$, we can not state any containment relation between $\mathrm{span}_2 M$ and $\mathrm{span}_{\mathbb{R}} M \cap C_n$. If one chooses M_1 to be the set of vertices (of dimension 4) $\{(0000), (1111), (1010), (1001), (0110), (0101), (1100), (0011)\}$, then – as it can be easily seen – $\mathrm{span}_2 M_1 = M_1$ over $GF(2)$, while $\mathrm{span}_{\mathbb{R}} M_1 = \mathbb{R}^4$, showing therefore that $\mathrm{span}_2 M \not\supseteq \mathrm{span}_{\mathbb{R}} M \cap C_n$ (in general). On the other hand, choosing $M_2 = \{(0000), (1111), (1010), (1001), (0110), (0101)\}$, we get that $\mathrm{span}_2 M_2 = M_1$, while $\mathrm{span}_{\mathbb{R}} M_2$ will not contain any further vertices of C_n, therefore showing that $\mathrm{span}_2 M \not\subseteq \mathrm{span}_{\mathbb{R}} M \cap C_n$ (in general).

Clearly, Theorem 2.6 is equivalent to the following theorem:

Theorem 2.9 (Miller et al. 1991 [14, 15], Griggs, 1997 [13]).
Let $x_1, x_2, \ldots, x_n$ be given numbers. The maximum number of the subsums

$\sum_{i=1}^{k} x_{j_i}$ which can be given without the disclosure of the values x_i is
$\sum_{i=0}^{\lfloor n/2 \rfloor} \binom{\lceil n/2 \rceil}{i}\binom{\lfloor n/2 \rfloor}{i} = \sum_{i=0}^{\lfloor n/2 \rfloor} \binom{\lceil n/2 \rceil}{i}\binom{\lfloor n/2 \rfloor}{\lfloor n/2 \rfloor - i} = \binom{n}{\lfloor n/2 \rfloor}$

(as shown by taking all subsums having the same number of elements among the first $\lfloor n/2 \rfloor$ and last $\lceil n/2 \rceil$ elements).

Indeed, if we take the characteristic vectors of the subsums in Theorem 2.9 (and view them as vertices of C_n), their span$_{\mathbb{R}}$ must avoid all vertices of C_n with exactly one 1 coordinate (otherwise the corresponding $\underline{e}_i$ would be a linear combination of the disclosed sum, therefore easily computable). On the other hand, if we have a subset M of the vertices of C_n such that span$_{\mathbb{R}} M$ does not contain any $\underline{e}_i \in C_n$, consider a solution to the equation $\underline{x}\mathbf{M} = \underline{b}$ where $\mathbf{M}$ is the matrix consisting of columns formed by the element of M as vectors, $\underline{b}$ is the vector consisting of the supposedly disclosed values of the subsums corresponding to the vectors in M and therefore $\underline{x}$ consisting of coordinates which might be the possible values of the x_i's, giving exactly these values of the subsums. Now, $\underline{e}_i = \{0, 0, \ldots, 1, \ldots, 0\}$, having exactly and only the ith coordinate equal to 1, is not in span$_{\mathbb{R}} M$, therefore $\underline{e}_i'$, the component of it perpendicular to the subspace span$_{\mathbb{R}} M$ has non-zero value in the ith coordinate. That is, $(\underline{x} + \underline{e}_i')$ is another solution of the equation $(\underline{x} + \underline{e}_i')\mathbf{M} = \underline{b}$ and has a different value in the ith coordinate, showing that it is impossible to calculate the value of x_i from the disclosed values of the subsums.

In [7] or [8, 9] it was proven that the above two theorems are equivalent to the following one as well.

Theorem 2.10. *Given a set of n real numbers $x_1, x_2, \ldots, x_n$ none of them being equal to 0, the maximum number of sums $\sum_{i \in B} x_i$ equal to 0, where the B's are subsets of $\{1, 2, \ldots, n\}$ is*

$$\sum_{i=0}^{\lfloor n/2 \rfloor} \binom{\lceil n/2 \rceil}{i}\binom{\lfloor n/2 \rfloor}{i} = \sum_{i=0}^{\lfloor n/2 \rfloor} \binom{\lceil n/2 \rceil}{i}\binom{\lfloor n/2 \rfloor}{\lfloor n/2 \rfloor - i} = \binom{n}{\lfloor n/2 \rfloor}$$

(as shown by taking $\lfloor n/2 \rfloor$ 1's and $\lceil n/2 \rceil$ -1's).

This theorem has a similarity with the famous Littlewood–Offord problem:

Problem 2.11 (Littlewood–Offord). How do we select – not necessarily distinct – complex numbers $x_1, x_2, \ldots x_n$, with $|x_i| \geq 1$ and an open unit diameter ball B, such that they maximize the number out of the 2^n sums $\sum_{i \in I} a_i$, $I \subseteq [n]$, lying inside B?

The following simpler version of it, considering only reals, instead of complex numbers, i.e., vectors of dimension two, was first solved by Erdős [10]. His argument can be used to prove Theorem 2.10, more precisely, the following, a bit more general statement:

Theorem 2.12. *Given a set of n real numbers $x_1, x_2, \ldots, x_n$ none of them being equal to 0, the maximum number of sums $\sum_{i \in B} x_i$ equal to a fixed t (where the B's are subsets of $\{1, 2, \ldots, n\}$) is $\binom{n}{\lfloor n/2 \rfloor}$.*

Proof (though it can be found at several papers, we include here the following, simplest argument, based on Griggs, copying proof of Erdős' for the Littlewood–Offord problem and even further simplified by Cameron [6]). Divide the set of indices $\{1, 2, \ldots, n\}$ into two parts, the indices of the negative and the positive x_i's: $M = N \cup P$. Assign to any subset of the indices B a new subset $B' = (N \cap B) \cup (\overline{N} \cap P)$. Clearly, $B_1' \subset B_2'$ yields that $\sum_{i \in B_1} x_i < \sum_{i \in B_2} x_i$, therefore the subset of the indices yielding the same subsum value form a Sperner system, their number may not exceed $\binom{n}{\lfloor n/2 \rfloor}$.

Remark 2.13. The questions about the maximum size of M with cone M not containing any or all of the vertices of the form $\underline{e}_i = (0, \ldots, 0, 1, 0, \ldots, 0)$ are trivial, therefore not interesting. (A cone spanned by the vertices of the cube will contain a vertex of type $\underline{e}_i = (0, \ldots, 0, 1, 0, \ldots, 0)$ iff the vertex $\underline{e}_i$ is among the spanning vertices.)

3. The weight k restricted case

Again, as throughout in the paper, let C_n denote the vertices of the n dimensional hypercube and let $M_k \subset C_n$ be a subset of it consisting of vertices of weight k only (vertices with exactly k coordinates equal to 1). In this chapter we will discuss the following general question:

Question 3.1. How big M_k can be such that $(1, 0, 0, \ldots, 0) \notin \operatorname{span}(M_k)$ or $(1, 1, \ldots, 1) \notin \operatorname{span}(M_k)$ or $(1, 1, \ldots, 1) \notin \operatorname{cone}(M_k)$ (again, span denoting the general question or answer for both of the cases over $GF(2)$ or $\mathbb{R}$, while span_2 or $\operatorname{span}_{\mathbb{R}}$ will stand for the specific cases, but the cone case is considered only over $\mathbb{R}$).

To obtain lower bounds, consider the following sets M_k of vertices of C_n of weight k

1. vertices with first coordinate $= 0$ (the number of them is $\binom{n-1}{k}$ and their span (cone, in case of $\mathbb{R}$) will definitely contain neither $(1, 0, 0, \ldots, 0)$ nor $(1, 1, \ldots, 1)$).

2. vertices with last coordinate $= 1$ (the number of them is $\binom{n-1}{k-1}$ and their span over $\mathbb{R}$ or even over $GF(2)$ will not contain $(1, 0, 0, \ldots, 0)$, while their span over $\mathbb{R}$ will not contain $(1, 1, \ldots, 1)$ either).

3. vertices with exactly one of the last two coordinates $= 1$, and having the remaining $k - 1$ 1 coordinates chosen from the remaining $n - 2$ positions (the number of them is $2\binom{n-2}{k-1}$ and their span over $\mathbb{R}$ or $GF(2)$ will not contain $(1, 0, 0, \ldots, 0)$, and their span over $\mathbb{R}$ will not contain $(1, 1, \ldots, 1)$ either, provided $n \neq 2k$).

In case 2, assume that we are given values $x_1, \ldots x_n$ and some subsums of k of them – having characteristic vectors equal to the given vertices in M_k – are disclosed, all containing x_n. Increase the value of x_n by $k - 1$ and decrease all others by 1, therefore the given subsums will remain, though each of the values and the sum of them are changed, that is, neither x_1 nor $\sum_{i=0}^{n} x_i$ can be calculated from the given subsums, and therefore none of $(1, 0, 0, \ldots, 0)$ and $(1, 1, \ldots, 1)$ can be in $\mathrm{span}_{\mathbb{R}} M_k$. A similar argument works for case 3, increasing the values of the variables corresponding to the first two coordinates each by $k - 1$ and decreasing all other values by 1. However, only in case of $n \neq 2k$ will this last argument work for the vector $(1, 1, \ldots, 1)$ since decreasing the first $2k - 2$ elements by 1 and increasing the last two by $k - 1$ will not change the total sum of the numbers in case of $n = 2k$.

In case of $GF(2)$ and case 2, assuming that $(1, 0, 0, \ldots, 0) = \underline{e}_1 \in \mathrm{span}_2 M_k$ the number of vectors in the linear combination giving $\underline{e}_1$ must be even (since the last coordinate must be 0 in the linear combination), resulting an even number of 1's in total, not allowing $\underline{e}_1$ in $\mathrm{span}_2 M_k$. A similar argument works for case 3, only now the number of vectors in the linear combination yielding $\underline{e}_1$ having last coordinate equal to 1 must be even, as well as the number of vectors having the last but one coordinate equal to 1. Since we only consider vectors with exactly one of the last two coordinates equal to 1, these two collections of vectors are disjoint, giving in total an even number vectors, and therefore an even number of "1"'s in the linear combination, which, therefore, may not give $\underline{e}_1$. Therefore, for every n and k one can choose a set of at least $\max\left\{\binom{n-1}{k}, \binom{n-1}{k-1}, 2\binom{n-2}{k-1}\right\}$ vectors of weight k (to form our M_k) such that neither $\operatorname{span}_2 M_k$ nor $\operatorname{span}_{\mathbb{R}} M_k$ contain $\underline{e}_1$. Also, we can choose a set of at least $\binom{n-1}{k}$ vectors of weight k (to form our M_k) such that $\operatorname{span}_2 M_k$ does not contain $\underline{1} = \{1, 1, \ldots, 1\}$ by construction number 1 above. However, construction number 2, giving a lower bound of $\binom{n-1}{k-1}$ as well, may or may not work, depending on the actual values of n and k. For example, if the parity of n and k are different, a parity argument will show that the span_2 of the set M_k obtained by construction 2 will still not contain $\underline{1}$, while if both $\binom{n-1}{k-1}$ and $\binom{n-2}{k-2}$ are odd, simply the sum of all vectors with last coordinate equal to 1 will give $\underline{1}$.

Remark 3.2.

$$
\max\left\{\binom{n-1}{k}, \binom{n-1}{k-1}, 2\binom{n-2}{k-1}\right\} =
\begin{cases}
\binom{n-1}{k-1} & \text{for } n \leq 2k-2 \\[2mm]
\binom{n-1}{k-1} = 2\binom{n-2}{k-1} & \text{for } n = 2k-1 \\[2mm]
2\binom{n-2}{k-1} & \text{for } n = 2k \\[2mm]
\binom{n-1}{k} = 2\binom{n-2}{k-1} & \text{for } n = 2k+1 \\[2mm]
\binom{n-1}{k} & \text{for } n \geq 2k+2
\end{cases}
$$

and

$$
\max\left\{\binom{n-1}{k}, \binom{n-1}{k-1}\right\} =
\begin{cases}
\binom{n-1}{k-1} & \text{for } n \leq 2k-1 \\[2mm]
\binom{n-1}{k-1} = \binom{n-1}{k} & \text{for } n = 2k \\[2mm]
\binom{n-1}{k} & \text{for } n \geq 2k+1
\end{cases}
$$

Let $m_1 = m_1^{(n,k)} = \max\left\{\binom{n-1}{k}, \binom{n-1}{k-1}, 2\binom{n-2}{k-1}\right\}$ and $m_2 = m_2^{(n,k)} = \max\left\{\binom{n-1}{k}, \binom{n-1}{k-1}\right\}$. These two numbers will be the exact bounds for the $\mathbb{R}$ case. Unfortunately the case of $GF(2)$ is much more complicated, where further parity constraints must be considered, since, e.g., in case of k being even there is no way to get a linear combination of vectors of weight k giving $\underline{e}_1$, a vector of weight 1, that is odd weight.

Theorem 3.3. *If for an $n \geq k$ and $M_k \subset C_n$ the size of $M_k > m_1$ then $(1, 0, \ldots, 0) \in \operatorname{span}_{\mathbb{R}}(M_k)$, and therefore $\operatorname{span}_{\mathbb{R}}(M_k) = C_n$.*

Proof goes by induction on n, and then for a fixed n, by induction on k, that is, to prove the validity of the statement for (n, k), we will assume it is true for every (n', k') with either $n' < n$ or in case of $n' = n$ with $k' < k$. The base cases are $k = 1$ (trivial) and then for every bigger k we will also need the $n = k+1$ case, when trivially $\binom{n-1}{k-1} = \binom{k}{k-1} = k$ is the right bound (having more than that many vertices would include *all* vertices of weight k and therefore any vertex of weight 1 – and, as a consequence, all other vertices – would be in $\text{span}_{\mathbb{R}}$).

For the inductional step (in general), assume that $|M_k| > m_1 = m_1^{(n,k)}$ and will prove that $\underline{e}_1 \in \text{span}_{\mathbb{R}} M_k$. Let M_k be partitioned into two subsets, M_k^i, the sets of vertices from M_k having their last coordinates equal to i, $i = 0, 1$. In case $|M_k^0| > m_1^{(n-1,k)}$ we will trivially have $(1, 0, \ldots, 0) \in \text{span}(M_k)$. Similarly, in case of $|M_k^1| > m_1^{(n-1,k-1)}$ a linear combination of the vertices obtained from the vertices of M_k^1 considering only the first $n - 1$ coordinates (and having $k - 1$ 1 coordinates there) will be of the form $(1, 0, \ldots, 0)$. The same linear combination on the last, n^{th} coordinates (all of them being 1) will result a_1 and therefore we will have a vertex of the form $\underline{v}_1 = (1, 0, \ldots, 0, a_1) \in \text{span}(M_k)$. Similarly, for every $1 \leq i \leq n - 1$ we will have a vertex of form $\underline{v}_i = (0, \ldots, 0, 1, 0, \ldots, 0, a_i) \in \text{span}(M_k)$, where the only 1 coordinate is at position i. An easy counting of the total weights of the linear combination on the first $n - 1$ and, independently, on the last, n^{th} coordinate will give that $a_i = \frac{1}{k-1}$ for every i. Summing up the appropriate choice of $k - 1$ of these vertices will give us all of those vertices of weight k which have the last coordinate equal to 1, in $\text{span}_{\mathbb{R}} M_k$. Assume there is another vertex of M_k, that is, another one from M_k^0, with last coordinate equal to 0, say $\underline{v}$. Change the first 1 coordinate – assume it's position is at i – of $\underline{v}$ to 0, and the very last coordinate (which was supposed to be 0) to 1, resulting a vertex already in $\text{span}_{\mathbb{R}} M_k$. Take the difference of them, $(0, \ldots, 0, 1, 0 \ldots, 0, -1)$ and consider the earlier $v_i = (0, \ldots, 0, 1, 0, \ldots, 0, a_i)$, an appropriate linear combination of which giving $(0, \ldots, 0, 1)$, or, equivalently, $\left(0, \ldots, 0, \frac{1}{k-1}\right)$. Subtracting this from $\underline{v}_1$ will result $\underline{e}_1 = (1, 0, \ldots, 0)$, the sought vertex in $\text{span}_{\mathbb{R}} M_k$.

Therefore, we only need to check that for every $n \geq k + 2$ and set of vertices M_k of dimension n and weight k with $|M_k| > m_1 = m_1^{(n,k)}$ either *(a)* $|M_k^0| > m_1^{(n-1,k)}$ or *(b)* $|M_k^1| > m_1^{(n-1,k-1)}$ *and we have a vertex in M_k^0 as well.* For that, it will be enough to prove that $m_1^{(n,k)} \geq m_1^{(n-1,k)} + m_1^{(n-1,k-1)}$, since the second condition is automatically ensured by the fact that $|M_k^1| \leq$

$\binom{n-1}{k-1} \leq m_1^{(n,k)} < |M_k|$, and therefore a vertex from M_k must be outside of M_k^1, that is, in M_k^0.

The inequality

$$(1) \qquad m_1^{(n,k)} \geq m_1^{(n-1,k)} + m_1^{(n-1,k-1)}$$

is almost always true and should be checked for all possible values of (n,k), together with the missing inductional steps when (1) does not hold. Again, assume that $|M_k| > m_1 = m_1^{(n,k)}$

(i) $n \leq 2k-2$ in which case $m_1^{(n,k)} = \binom{n-1}{k-1}$, $m_1^{(n-1,k)} = \binom{n-2}{k-1}$ and $m_1^{(n-1,k-1)} = \binom{n-2}{k-2}$ and therefore (1) is true, $\underline{e}_1 \in \mathrm{span}_{\mathbb{R}} M_k$.

(ii) $n = 2k-1$ in which case consider $\overline{M_k} = \{\underline{1} - a \colon a \in M_k\}$, a set of vertices of the hypercube C_{2k-1} of weight $k-1$. By the induction hypothesis we know that if $\big(|M_k| = \big)|\overline{M_k}| > m_1^{(2k-1,k-1)} = \binom{2k-2}{k-1} = m_1^{(2k-1,k)}$ then both $\underline{e}_1$ and $\underline{1}$ are in $\mathrm{span}_{\mathbb{R}} \overline{M_k}$, that is, there are linear combinations of the vectors from M_k such that $\sum d_i(\underline{1} - \underline{a}_i) = \underline{1}$ and $\sum c_i(\underline{1} - \underline{a}_i) = \underline{e}_1$. From the first equation we have $\sum d_i \underline{a}_i = \big(\sum d_i - 1\big)\underline{1}$, where again an easy calculation of the weights shows that $\sum d_i = \frac{n}{n-k}$ and therefore $\underline{1} \in \mathrm{span}_{\mathbb{R}} M_k$. From the second equation we get that $\sum c_i \underline{1} - \sum c_i \underline{a}_i = \underline{e}_1$, that is $\underline{e}_1 = \sum -c_i \underline{a}_i + \big(\sum c_i\big)\underline{1}$, yielding $\underline{e}_1 \in \mathrm{span}_{\mathbb{R}} M_k$.

(iii) $n = 2k$ in which case $m_1^{(n,k)} = 2\binom{n-2}{k-1}$, $m_1^{(n-1,k)} = \binom{n-2}{k-1}$ and $m_1^{(n-1,k-1)} = \binom{n-2}{k-1}$ and therefore (1) is true, $\underline{e}_1 \in \mathrm{span}_{\mathbb{R}} M_k$.

(iv) $n = 2k+1$, in which case we assume that $|M_k| > m_1 = m_1^{(2k+1,k)} = \binom{2k}{k}$. First assume that $|M_k^1| > m_1(2k, k-1) = \binom{2k-1}{k-1}$, in which case the general induction step described at the beginning of the prof works. Otherwise we may assume that $|M_k^0| \geq \binom{2k-1}{k} = \frac{1}{2}\binom{2k}{k}$, which is unfortunately not enough to prove that $\underline{e}_1 \in \mathrm{span}_{\mathbb{R}} M_k$, but gives us two "complementary" vectors in the first $2k$ coordinates, giving $\underline{b} = (1,1,1,\ldots,1,1,0) \in \mathrm{span}_{\mathbb{R}} M_k$. Consider $M'^1_k = \{\underline{b} - \underline{a} \colon \underline{a} \in M_k^1\} \subset \mathrm{span}_{\mathbb{R}} M_k$, a set of vectors of dimension $2k+1$ with exactly $k+1$ "1" coordinates among the first $2k$ coordinates, -1 as the last coordinate and all other coordinates equal to 0. By induction we know that in case the number of these vectors (equal to $|M'^1_k| = |M_k^1|$)

is more than $m_1(2k, k+1) = \binom{2k-1}{k+1}$, an argument similar to the general inductional step will show that for every $1 \le i \le n-1$ there is a vertex of form $\underline{v}_i = (0, \ldots, 0, 1, 0, \ldots, 0, a_i) \in \text{span}_{\mathbb{R}}(M_k)$, where the only 1 coordinate is at position i, and a similar counting of the total weights of the linear combination on the first $n-1$ and, independently, on the last, n^{th} coordinate will give that $a_i = \frac{-1}{k+1}$ for every i. By an earlier comment still there must be a vector, say $\underline{b} \in M_k^0$. Take all of those $\underline{v}_i$'s which have "1" coordinates at the positions of the "1" coordinates of $\underline{b}$, sum them up, resulting a vector only different from $\underline{b}$ in the last coordinate (where $\underline{b}$ has 0, while the sum obviously $-\frac{k}{k+1}$). It gives that $\left\{0, 0, \ldots, -\frac{k}{k+1}\right\} \in \text{span}_{\mathbb{R}} M_k$, which together with $\underline{v}_i = (0, \ldots, 0, 1, 0, \ldots, 0, a_i) \in \text{span}_{\mathbb{R}}(M_k)$ gives that $\underline{e}_i = (0, \ldots, 0, 1, 0, \ldots, 0, 0) \in \text{span}_{\mathbb{R}}(M_k)$ for every i.

(v) $n \ge 2k+2$ in which case $m_1^{(n,k)} = \binom{n-1}{k}$, $m_1^{(n-1,k)} = \binom{n-2}{k}$ and $m_1^{(n-1,k-1)} = \binom{n-2}{k-1}$ and therefore (1) is true, $\underline{e}_1 \in \text{span}_{\mathbb{R}} M_k$. $\blacksquare$

Theorem 3.4. *If for an $M_k \subset C_n$ the size of $M_k > m_2$ then $(1, 1, \ldots, 1) \in \text{span}(M_k)$.*

Proof. Note that $m_1(n, k) = m_2(n, k)$ almost always, with the only exception of $n = 2k$. However, in this last case $m_2(2k, k) = \binom{2k-1}{k} = \binom{2k-1}{k-1} = \frac{1}{2}\binom{2k}{k}$ and assuming $|M_k| > m_2(n, k)$ in this case gives us two "complementary" vectors in M_k, sum of which is exactly equal to $(1, 1, \ldots, 1)$. This, together with the constructions at the beginning of this chapter plus Theorem 3.2 completes the proof. $\blacksquare$

It is natural to ask whether the result of Theorem 3.3 remains valid if we restrict ourselves to convex combinations of the vectors in M_k, that is, which value of $|M_k|$ will surely result that $(1, 1, \ldots, 1, 1) \in \text{cone} \, M_k$. Note here that this question is only meaningful over $\mathbb{R}$ and even in that case meaningless for $(1, 0, \ldots, 0, 0)$, since a cone contains a vertex of type $(1, 0, \ldots, 0, 0)$ iff the vertex is among the vertices spanning the cone.

The situation becomes more complicated, as the following example shows that for certain $n > 2k$ we have a set of k-uniform subsets (set of vertices of weight k) $M_k \subset C_n$ of size bigger than $m_2(n, k) = \binom{n-1}{k}$ with $(1, 1, \ldots, 1, 1) \notin \text{cone} \, M_k$.

Let $n = 3k + 1$ and consider M_k, the set all of those vertices of C_n of weight k which have at least one of the first three coordinates equal to 1. In this case we have $\binom{3k-2}{k}$ vertices of weight k having the first three coordinates 0, which is less than $\binom{3k}{k-1}$, and therefore the number of vertices having at least one of the first three entries equal to 1, $\binom{3k+1}{k} - \binom{3k-2}{k}$, is more than $\binom{3k+1}{k} - \binom{3k}{k-1} = \binom{3k}{k} = \binom{n-1}{k}$.

We claim that the cone spanned by M_k will not contain $(1, 1, \ldots, 1)$. The simplest way to see it is to give an alternative form of the question above and consider the example in that framework:

Question 3.5. Let $x_1, x_2, \ldots, x_n$ be given numbers such that $\sum_{i=1}^{n} x_i > 0$. What is maximum number of negative subsums (or the minimum number of positive subsums) of exactly k of these numbers?

It is obvious that for certain choices of $x_1, x_2, \ldots, x_n$ one will have at least $\binom{n-1}{k}$ negative k-subsums, shown by the example of many small (absolute value) negative numbers and one big (absolute value) positive number, e.g. $\{-1, \ldots, -1, n\}$. Therefore, the minimum number of positive subsums is at most $\binom{n-1}{k-1}$ and in Question 3.5 the bound is at least as big as $\binom{n-1}{k}$, but maybe bigger sometimes, shown by the following example, equivalent to the above one about the cone not containing $(1, 1, \ldots, 1, 1)$:

Consider $3k + 1$ numbers: $\{2 - 3k, 2 - 3k, 2 - 3k, 3, 3, \ldots, 3\}$ whose sum is 1. In this case there are $\binom{3k-2}{k}$ positive subsums, which is less than $\binom{3k}{k-1}$, and therefore $\binom{3k+1}{k} - \binom{3k-2}{k}$ negative subsums, which is more than $\binom{3k+1}{k} - \binom{3k}{k-1} = \binom{3k}{k} = \binom{n-1}{k}$.

To see the analogy between the two questions one may use a linear algebraic argument similar to the earlier cases: Assume the cone spanned by the vertices of C_n of weight k which have at least one of the first three coordinates equal to 1. To each of these vertices assign the k-subsum of the set $\{2 - 3k, 2 - 3k, 2 - 3k, 3, 3, \ldots, 3\}$ with having the vertex as it's characteristic vector. Since all of these sumbsums are negative, it may not happen that a linear combination of them will give the vector $(1, 1, \ldots, 1, 1)$ corresponding to the total, therefore positive sum.

Although the previous example shows that the bound might be bigger than $\binom{n-1}{k}$ in some cases, the following theorem still shows that the general bound will be this number.

Theorem 3.6 (Manickam, Miklós, 1989). *Let $x_1, x_2, \ldots, x_n$ be given numbers such that $\sum_{i=1}^{n} x_i > 0$. The minimum number of positive subsums of exactly k of these numbers is $\binom{n-1}{k-1}$ if $n > n_1(k)$ or k divides n.*

Corollary 3.7. *If for an $M_k \subset C_n$ the size of $M_k > \binom{n-1}{k}$ and either $n > n_1(k)$ or k divides n, then $(1, 1, \ldots, 1) \in \mathrm{cone}\,(M_k)$.*

Next, we may ask the question analogous to the one asked for the non-restricted case (and we will restrain only to the real case here as well), that is

Question 3.8. What is the maximum size of a subset M_k of the vertices of the n-dimensional hypercube (all of weight k or weight at most k) such that none of the vertices of the form $(0, 0, \ldots, 0, 1, 0, \ldots, 0)$ are in $\mathrm{span}_{\mathbb{R}}(M_k)$?

Similar to the non-restricted case, this question is proven to be equivalent in [8, 9] to the following two questions:

Question 3.9. Let $x_1, x_2, \ldots, x_n$ be given numbers. What is the maximum number of the subsums $\sum_{i=1}^{k} x_{j_i}$ (of exactly k of or at most k of these numbers) which can be given without the disclosure of the values x_i?

and

Question 3.10. Let $x_1, x_2, \ldots, x_n$ be given numbers. What is the maximum number of the subsums $\sum_{i=1}^{k} x_{j_i}$ (of exactly k of or at most k of these numbers) being equal to 0?

Let the answer to these 3 questions be defined by $M(n, k)$ (in case of M_k consisting of vectors all of weight k), and $N(n, k)$ (in case of vectors of weight at most k), resp.

One may see that any number of the form $m_1(n_1, n_2, k_1, k_2) = \binom{n_1}{k_1}\binom{n_2}{k_2}$ with $n_1 + n_2 = n$ and $k_1 + k_2 = k$, and, independently,

$$m_2(n, k) = \sum_{i=1}^{\frac{k}{2}} \binom{\left\lceil \frac{n}{2} \right\rceil}{i}\binom{\left\lfloor \frac{n}{2} \right\rfloor}{i}$$

are lower bounds for $N(n, k)$, and $m_1(n_1, n_2, k_1, k_2)$ is a lower bound for $M(n, k)$ as well. For the bound $m_1(n_1, n_2, k_1, k_2)$ and Question 3.8 take M_k as the set of vertices of the hypercube containing exactly k_1 "1" coordinates among the first n_1 coordinates (and therefore exactly k_2 "1" coordinates among the last n_2 coordinates, for Question 3.9 subsums containing exactly k_1 x_i's among the first n_1 given numbers (and therefore exactly k_2 x_i's among the last n_2 given numbers), and for Question 3.10 consider n_1 copies of k_2 and n_2 copies of $-k_1$ (the checking of the validity of the fact these samples will have the required properties is based on some previous argument in this paper and left to the reader). For the bound $m_2(n,k)$ and Question 3.8 take M_k as the set of vertices of the hypercube containing exactly i "1" coordinates both among the first $\lceil \frac{n}{2} \rceil$ and last $\lfloor \frac{n}{2} \rfloor$ coordinates for $1 \le i \le \lfloor \frac{k}{2} \rfloor$, for Question 3.9 subsums containing exactly i x_i's both among the first $\lceil \frac{n}{2} \rceil$ and last $\lfloor \frac{n}{2} \rfloor$ given numbers (again, for $1 \le i \le \lfloor \frac{k}{2} \rfloor$), and for Question 3.10 consider $\lceil \frac{n}{2} \rceil$ copies of 1 and $\lfloor \frac{n}{2} \rfloor$ copies of -1. The checking is again easy and left to the reader.

In case of $m_1(n_1, n_2, k_1, k_2)$ these numbers give many different lower bounds. In order to get the best one, we need to maximize $\binom{n_1}{k_1}\binom{n_2}{k_2}$ for $n_1 + n_2 = n$ and $k_1 + k_2 = k$. A somewhat surprising result – since it shows that the maximum is reached at a rather marginal point – is in [8, 9]:

Theorem 3.11 (Demetrovics, Katona, Miklós, 2004). *Suppose* $4 \le k$, $n_1(k) \le n$. *The maximum size of a subset* M_k *of the vertices of the n-dimensional hypercube (all of weight at most k) such that none of the vertices of the form* $(0, 0, \ldots, 0, 1, 0, \ldots, 0)$ *are in* $\mathrm{span}_{\mathbb{R}}(M_k)$ *is*

$$N(n,k) = \left(\left\lfloor \frac{\left\lfloor \frac{(n+1)(k-1)}{k} \right\rfloor}{k-1} \right\rfloor \right) \left(n - \left\lfloor \frac{(n+1)(k-1)}{k} \right\rfloor \right)$$

which answer is of the form $\binom{n_1}{k_1}\binom{n_2}{k_2}$ *with* $n_1 = \left\lfloor \frac{(n+1)(k-1)}{k} \right\rfloor$, $n_2 = n - \left\lfloor \frac{(n+1)(k-1)}{k} \right\rfloor$, $k_1 = k - 1$, $k_2 = 1$, *therefore also is the answer for* $M(n,k)$, *the case with vectors all of weight exactly equal to k.* This later result, even without the assumption that $n_1(k) \le n$ was also obtained by Ahlswede, Aydinian, and Khachatrian in [3].

On the other hand, the assumption of $n_1(k) \le n$ is necessary for the case of $N(n,k)$, when the sizes of the chosen subsets (or, equivalently, weight of the chosen vertices) are only bounded above by k, not necessarily equal to k. For example, as Theorems 2.9 shows,

$$N(n,n) = \binom{n}{\lfloor n/2 \rfloor} = \sum_{i=0}^{\lfloor n/2 \rfloor} \binom{\lceil n/2 \rceil}{i}\binom{\lfloor n/2 \rfloor}{i} = m_2(n,n)$$

and not $m_1(n_1, n_2, k_1, k_2)$ for ceratin values of the parameters.

It is expected that the same construction remains the best if n is not much larger than k, that is (assuming for convenience that k is even)

$$N(n,k) = \sum_{i=1}^{\frac{k}{2}} \binom{\lceil \frac{n}{2} \rceil}{i}\binom{\lfloor \frac{n}{2} \rfloor}{i}.$$

For example, it is known that $N(12,6) = \binom{6}{3}\binom{6}{3} + \binom{6}{2}\binom{6}{2} + \binom{6}{1}\binom{6}{1} = m_2(12,6)$, however, $N(20,6) = M(20,6) = \binom{17}{5} \cdot 3 = m_1(17,3,5,1) = \max m_1(n_1,n_2,k_1,k_2)$ with $n_1 + n_2 = 20$ and $k_1 + k_2 = 6$, like in Theorem 3.11.

4. FURTHER QUESTIONS – BACK TO THE UNRESTRICTED CASE

Returning to the unrestricted case, we may ask, for example, for the maximum size of M such that $\operatorname{span}_{\mathbb{R}} M$ avoids all vertices of weight $(n-1)$, or, using the same translation we had earlier, given n real numbers $x_1, x_2, \ldots, x_n$, find the maximum number of subset sums $\sum_{i \in B} x_i$ equal to 0, where the B's are subsets of $\{1, 2, \ldots, n\}$, with the condition that none of the sums $\left(\sum_{i=1}^{n} x_i\right) - x_j$ are equal to 0, that is no sum of $n-1$ of the given numbers is equal to zero. Or, in general, find the maximum number of subset sums $\sum_{i \in B} x_i$ equal to 0, for any set of n real numbers $x_1, x_2, \ldots, x_n$, with the condition that none of the sums $\sum_{j=1}^{r} x_{i_j}$ are equal to 0 (for a given r). (For $r = n$ the answer is 2^{n-1} by Proposition 2.1 and for $r = 1$ it is $\binom{n}{\lfloor n/2 \rfloor}$ by Theorem 2.6). This later will correspond to finding the maximum size of M such that $\operatorname{span}_{\mathbb{R}} M$ avoids all vertices of weight r.

We prove the following theorem which can also be found in [3] with a different proof.

Theorem 4.1. *For $n \geq 8$ the maximum size of M such that $\operatorname{span}_{\mathbb{R}} M$ avoids all vertices of weight $(n-1)$, (and, equivalently, for given n real numbers $x_1, x_2, \ldots, x_n$, the maximum number of subset sums $\sum_{i \in B} x_i$ equal to 0, where no sum of $n-1$ of the given numbers is equal to zero) is 2^{n-2}.*

We will prove the following, somewhat stronger and more precise theorem:

Theorem 4.2. *The maximum number of subset sums $\sum_{i \in B} x_i$ equal to 0, where no sum of $n-1$ of the given numbers is equal to zero is $\max\left(\binom{n}{\lfloor n/2 \rfloor}, 2^{n-2}\right)$. More precisely, if we further assume that none of the numbers is equal to zero we have at most $\binom{n}{\lfloor n/2 \rfloor}$ 0 subset sums, while if we further assume that the total sum of the numbers is not zero we have at most 2^{n-2} 0 subset sums.*

Equivalently, the maximum size of M such that $\mathrm{span}_\mathbb{R}\, M$ avoids all vertices of weight $(n-1)$ is $\max\left(\binom{n}{\lfloor n/2 \rfloor}, 2^{n-2}\right)$, with being at most $\binom{n}{\lfloor n/2 \rfloor}$ if all vertices of weight 1 are avoided further, and at most 2^{n-2} if the vertex $\underline{1} = \{1, 1, \ldots, 1\}$ is avoided as well.

Proof. Note that this later theorem states only – not necessarily sharp – upper bounds. The bounds here can be reached. In case when none of the numbers are 0 – or no vertex of weight 1 is contained in $\mathrm{span}_\mathbb{R}\, M$ – by the original construction of Theorem 2.6 (but this construction will fulfill our main assumption only when n is even, and therefore the bound here might not be sharp) and for the other case by the numbers $1, 1, 0, 0, 0, \ldots, 0$, or, equivalently, choosing all vertices to be in M whose first two coordinates are equal to 0. This later construction always works, so the maximum number we are looking for is always at least 2^{n-2}.

The case when no chosen number can be zero or no vertex of weight 1 can be in $\mathrm{span}_\mathbb{R}\, M$ is the immediate consequence of Theorem 2.6.

The other upper bound (2^{n-2}) will be proven by induction on n, with $n = 3$ being trivial. We will distinguish two cases, if there is a 0 among the chosen numbers (a vertex of weight 1 in $\mathrm{span}_\mathbb{R}\, M$) or if not. In the first case leave that 0 out, resulting $n - 1$ numbers, such that the sum of them is not 0 (since it is an $n - 1$ subsum of the original numbers) and no subsum of $n - 2$ of them is 0 either (since that, together with the left element, would also result an $n - 1$ subsum of the original numbers equal to 0). Therefore these $n - 1$ numbers will satisfy the assumption and therefore there are at most 2^{n-3} 0 subsums of them. If we add back the left element, being it 0, it can be added to the already 0 subsums, thus doubling the total number of 0-subsums from the original n numbers. The remaining case, that is when there is no 0 among the chosen numbers, will be handled by the following lemma.

Lemma 4.3. *The maximum number of subset sums $\sum_{i \in B} x_i$ equal to 0, where none of the given n numbers is equal to 0, none of the sum of any $n - 1$ of them is equal to zero and the total sum of them is neither 0 is 2^{n-2}.*

Note that the lemma itself is a trivial consequence of Theorem 2.6 for most of the values of n, since the condition that none of these numbers are equal to zero already implies the upper limit of $\binom{n}{\lfloor n/2 \rfloor}$. However, we need it for small values of n as well (when $\binom{n}{\lfloor n/2 \rfloor} > 2^{n-2}$ to have our induction step worked completely).

Proof of the Lemma will be by induction on n again, being trivial for $n = 2, 3$. If chosen numbers have all the same value, that yields no 0 subsum at all, so we may assume there are two different ones of them, x_1 and x_2. Take all the 4 subsets B_1, B_2, B_3 and B_4 of $\{x_1, x_2\}$ and consider the 2^{n-3} pairs of complementary subsets of the remaining $n - 2$ numbers. We claim that for any A_1, A_2 of these pairs at most two of the 8 subsets $B_i \cup A_k$ can give 0 sum, therefore the total number of 0 subsums is at most $2 \times 2^{n-3} = 2^{n-2}$.

If $x_1 + x_2 = 0$ then at most one of the A_k's, say A_1 gives a 0 sum (both may not, since than, together with $x_1 + x_2 = 0$ the total sum would be 0), and then A_1 and $A_1 \cup \{x_1, x_2\}$ would give 0 sum. Since the $x_i \neq 0$, the sets $A_1 \cup \{x_i\}$ will not give 0 sum, since A_2 does not give a zero sum, neither does $A_2 \cup \{x_1, x_2\}$ and $A_2 \cup \{x_i\}$ may not give 0 either, since in that case $A_1 \cup A_2 \cup \{x_i\}$ would give a subsum of $n - 1$ numbers equal to 0.

If $x_1 + x_2 = 0$ and none of the A_k's give a 0 sum, we can add at most one of x_1 and x_2 to the sum of A_1 to obtain 0, and similarly for A_2, giving again at most 2 0 sums of the form $B_i \cup A_k$.

If $x_1 + x_2 \neq 0$ then the 4 numbers 0, x_1, x_2 and $x_1 + x_2$ are all different, and so at most one of them can be added to A_1 or A_2 to yield a sum equal to 0. ∎

Now, Theorem 4.1 is an immediate consequence of Theorem 4.2 observing that for $n \geq 8$ we have $\binom{n}{\lfloor n/2 \rfloor} \leq 2^{n-2}$.

The next natural question is to find the maximum size of M such that $\mathrm{span}_{\mathbb{R}} M$ avoids all vertices of weight 2, or, given n real numbers $x_1, x_2, \ldots, x_n$ ($\neq 0$), to find the maximum number of subsums $\sum_{i \in B} x_i = 0$, where the B's are subsets of $\{1, 2, \ldots, n\}$, with none of the sums $x_i + x_j = 0$. This one has a close resemblance to the following well known problem of Erdős and Moser asked after finding the solution of the Littlewood–Offord problem (see Problem 2.11) and it's high symmetry:

Problem 4.4 (Erdős–Moser, [11]). How do we select *distinct* nonzero real numbers $x_1, x_2, \ldots x_n$ and a target sum t to maximize the number of subset sums $= t$?

which is equivalent to

Problem 4.5. How do we select nonzero real numbers $x_1, x_2, \ldots x_n$ to maximize the number of subset sums $= 0$, with all $x_i - x_j \neq 0$?

A possible candidate for the largest such set of numbers is family of the n distinct integers closest to 0 and the target $t = 0$. This was proven to be the best construction by Stanley in [17].

State formally our similar problem in the same language:

Problem 4.6. How do we select (nonzero) real numbers $x_1, x_2, \ldots x_n$ to maximize the number of subset sums $= 0$, with all $x_i + x_j \neq 0$, and what is this maximum?

Note that here the nonzero assumption does not really change the problem, since the condition $x_i + x_j \neq 0$ implies that at most one of the chosen numbers can be equal to zero. If so, remove it and then the remaining $n - 1$ numbers will satisfy the original condition together with the further assumption that none of them are equal to zero. Taking here the best construction, one can always add the last, 0 element to every 0-sum, doubling the number of zero sums. That is, if $m_1(n)$ denotes the maximum number of 0-sums above without assuming that the given numbers are nonzeros and $m_2(n)$ with the additional assumption that they may not be equal to zero, we have that $m_1(n) = 2 \times m_2(n - 1)$.

Both the author, and, independently, Ahlswede, Aydinian, and Khachatrian in [3] conjecture that

Conjecture 4.7. Given n real numbers $x_1, x_2, \ldots x_n$, with all $x_i + x_j \neq 0$ and further, no $x_i = 0$, the maximum number of subset sums $= 0$ ($=m_2(n)$ with the above notation), or, equivalently, the maximum size of M such that $\operatorname{span}_\mathbb{R} M$ avoids all vertices of weight 2 and 1 of the hypercube is $\binom{\lceil 2n/3 \rceil}{2} \lfloor n/3 \rfloor$. This bound can be reached by the choice of

$$\{-1, -1, \ldots, -1, 2, \ldots 2\}$$

with $\lceil 2n/3 \rceil$ copies of -1 and $\lfloor n/3 \rfloor$ copies of 2 in the subsum "language" or by choosing M as the set of all vertices of the hypercube which have exactly two 1's among the first $\lceil 2n/3 \rceil$ coordinates and one 1's among the last $\lfloor n/3 \rfloor$ coordinates.

Note that the linear algebra argument already used several times shows here as well that the later choice of M will have the required property: Assume that we are given n numbers $x_1, x_2, \ldots x_n$ and disclose the value of the sum of any three of them, such that two are chosen from the first $\lceil 2n/3 \rceil$ ones and the third from the last $\lfloor n/3 \rfloor$ ones. Increasing the value of the first $\lceil 2n/3 \rceil$ by 1 and decreasing the value of the last $\lfloor n/3 \rfloor$ by 2 will leave the value of all the disclosed subsums unchanged, while the value of any of these numbers and any sum of 2 of these numbers will be changed. Therefore, no linear combination of these triple sums can be equal to any sums of 2 of the numbers.

In the general framework stated in the first paragraph of this chapter, further questions might be asked about the size of M if $\mathrm{span}_\mathbb{R} M$ does not contain any vertices of a given constant weight r, or, similarly, about the maximum number of subset sums $\sum_{i \in B} x_i$ equal to 0, for any set of n real numbers $x_1, x_2, \ldots, x_n$, with the condition that none of the sums $\sum_{j=1}^{r} x_{i_j}$ are equal to 0 for a given r. Assuming the validity of Conjecture 4.5 it would not be difficult to prove that for $r = 3$ the best bound is the same as for $r = 1$, that is $\binom{n}{\lfloor n/2 \rfloor}$ with the same construction. Similarly, the $r = 4$ and 5 cases (provided n is big enough compared to k) would be easy to handle with the conjecture, for the even case similar to the $r = 2$ case and the odd to the $r = 1$ case. However, in case of $k = 6$ the construction in the above conjecture does not work, and we have not even a conjecture for the best structure.

Further, asymptotic results might be found in [2] and [3].

A further possible direction of this research to investigate the case when the span of the subset of the vertices is taken over $GF(2)$, for both the unrestricted and restricted cases. Then the subset sum equivalency does not work, we are completely left on linear algebra and other direct tools and many times the answer may depend on the parity of the parameters involved. We have considered only a few of these cases in this paper.

REFERENCES

[1] R. Ahlswede, H. Aydinian and L. H. Khachatrian, Intersection theorems under dimension constraints, *J. Combin. Theory Ser. A*, **113** (2006), 483–519.

[2] R. Ahlswede, H. Aydinian and L. H. Khachatrian, Forbidden $(0, 1)$-vectors in hyperplanes of $\mathbb{R}^n$: the restricted case, *Designs, Codes and Cryptography*, **29** (2003), 17–28.

[3] R. Ahlswede, H. Aydinian and L. H. Khachatrian, Forbidden $(0, 1)$-vectors in hyperplanes of $\mathbb{R}^n$: the unrestricted case, *Designs, Codes and Cryptography*, **37** (2005), 151–167.

[4] R. Ahlswede, H. Aydinian and L. H. Khachatrian, Extremal problems under dimension constraints, *Discrete Math.*, **273** (2003), 9–21.

[5] R. Ahlswede, H. Aydinian and L. H. Khachatrian, Maximum numbers of constant weight vertices of the unit n-cube contained in a k-dimensional subspace, *Combinatorica*, **23** (2003), 5–22.

[6] P. Cameron, personal communication.

[7] F. Y. Chin and G. Ozsoyoglu, Auditing and inference control in statistical databases, *IEEE Transactions on Software Engineering*, **SE-8** (1982), 574–582.

[8] J. Demetrovics, G. O. H. Katona and D. Miklós, On the security of individual data, *Ann. Math. Artif. Intell.*, **46** (2006), 98–113.

[9] J. Demetrovics, G. O. H. Katona and D. Miklós, On the Security of Individual Data, in: *Foundations of Information and Knowledge Systems*, Lecture Notes in Computer Science 2942 Springer (2004), pp. 49–58.

[10] P. Erdős, On a lemma of Littlewood and Offord, *Bull. Amer. Math. Soc.*, **51** (1945), 898–902.

[11] P. Erdős, Extremal Problems in number theory, in: *Theory of Munbers*, Amer. Math. Soc., Providence, (1965), pp. 181–189.

[12] J. R. Griggs, Concentrating subset sums at k points, *Bull. Inst. Comb. Applns.*, **20** (1997) 65–74.

[13] J. R. Griggs, Database security and the distribution of subset sums in $\mathbb{R}^m$, in: *Graph Theory and Combinatorial Biology* (Balatonlelle, 1996), Bolyai Soc. Math. Stud., **7**, Budapest, (1999), pp. 223–252.

[14] M. Miller, I. Roberts and I. Simpson, Application of Symmetric Chains to an Optimization Problem in the Security of Statistical Databases, *Bull. ICA*, **2** (1991), 47–58.

[15] M. Miller, I. Roberts and I. Simpson, Preventation of Relative Compromise in Statistical Databases Using Audit Expert, *Bull. ICA*, **10** (1994), 51–62.

[16] N. Manickam and D. Miklós, On the number of nonnegative partial sums of a nonnegative sum, in: *Combinatorics* (Eger, 1987), Colloq. Math. Soc. Jnos Bolyai, **52**, North-Holland, Amsterdam (1988), pp. 385–392.

[17] P. M. Stanley, Weyl groups, the hard Lefschetz theorem, and the Sperner property, *SIAM J. Alg. Discr. Math.*, **1** (1980), 168–184.

Dezső Miklós

Alfréd Rényi Institute of Mathematics
Hungarian Academy of Sciences
Budapest P.O.B. 127
H-1364 Hungary

e-mail: dezso@renyi.hu

BOLYAI SOCIETY
MATHEMATICAL STUDIES, 17

Horizons of Combinatorics
Balatonalmádi
pp. 163–177.

Combinatorial Conditions for the Rigidity of Tensegrity Frameworks

ANDRÁS RECSKI[*]

1. Introduction

Rigidity of *bar-and-joint frameworks* has been studied for centuries. If the exact positions of the joints of such a framework are known, the rank of the so called rigidity matrix determines whether the framework is rigid[1]. If the underlying graph is given only, the rigidity of the framework cannot always be determined: if certain conditions (depending on the dimension of the space) are not satisfied then the framework cannot be rigid, no matter what the actual positions of the joints are, otherwise rigidity can be realized by some (in fact, almost all) positions of the joints. These graph theoretic conditions can be checked in polynomial time for the 1- and 2-dimensional frameworks, while the complexity questions are mainly open for higher dimensions. For surveys of such results the reader is referred to [6, 13, 16].

Unlike combinatorialists, civil engineers are interested in the actual stresses of the bars as well; in particular, the question whether a given bar is under compression or under tension is of great importance. A bar which is always under tension (or compression) can theoretically be replaced by a cable (by a strut, respectively), leading to the concept of *tensegrity*

---

[*]Supported by Grant No. OTKA T67651 of the Hungarian National Science Fund and the National Office for Research and Technology. The useful remarks of the referee are gratefully acknowledged.

[1]For brevity, we use the word "rigid" instead of the more precise "infinitesimally rigid", see the definitions in the next section.

frameworks where some of the joints are connected by bars (with constant length), some others by cables (where the length is only bounded from above) or by struts (with bounds from below only).

In what follows, we survey some – partly new – combinatorial results concerning tensegrity frameworks, emphasizing issues of computational complexity as well. The reader is also referred to previous surveys like [4, 15].

2. Bar-and-joint Frameworks

Let $G = (V, E)$ be a simple graph with vertex set V (or $V(G)$) and edge set E (or $E(G)$). A *bar-and-joint framework* or shortly *framework* in the d-dimensional space is a pair $F = (G, p)$ consisting of a graph G and a map $p\colon V \to \mathbb{R}^d$. In this context, vertices and edges are also called joints and bars (or rods), respectively. The *rigidity matrix* $R(G, p)$ or shortly $R(F)$ of the framework has $|E|$ rows and $d|V|$ columns: if an edge $\{v_i, v_j\} \in E$ corresponds to a row then the entries in the d columns corresponding to vertices v_i and v_j contain the coordinates of $p(v_i) - p(v_j)$ and $p(v_j) - p(v_i)$, respectively, while the remaining entries are zero.

Example 1. Consider the complete graph K_4 as the graph of a 2-dimensional framework. Its rigidity matrix in general looks like

$$
\begin{bmatrix}
x_1 - x_2 & x_2 - x_1 & 0 & 0 & y_1 - y_2 & y_2 - y_1 & 0 & 0 \\
x_1 - x_3 & 0 & x_3 - x_1 & 0 & y_1 - y_3 & 0 & y_3 - y_1 & 0 \\
x_1 - x_4 & 0 & 0 & x_4 - x_1 & y_1 - y_4 & 0 & 0 & y_4 - y_1 \\
0 & x_2 - x_3 & x_3 - x_2 & 0 & 0 & y_2 - y_3 & y_3 - y_2 & 0 \\
0 & x_2 - x_4 & 0 & x_4 - x_2 & 0 & y_2 - y_4 & 0 & y_4 - y_2 \\
0 & 0 & x_3 - x_4 & x_4 - x_3 & 0 & 0 & y_3 - y_4 & y_4 - y_3
\end{bmatrix}
$$

The rigidity matrix of the framework $F_1 = (K_4, p_1)$ with the mapping $p_1\colon 1 \to (0,0);\ 2 \to (1,0);\ 3 \to (0,1);\ 4 \to (1,1)$ and that of $F_2 = (K_4, p_2)$ with the mapping $p_2\colon 1 \to (0,0);\ 2 \to (1,0);\ 3 \to \left(\frac{1}{2}, \frac{1}{2}\right);\ 4 \to (1,1)$ are as follows:

$$
\begin{bmatrix}
-1 & 1 & 0 & 0 & 0 & 0 & 0 & 0 \\
0 & 0 & 0 & 0 & -1 & 0 & 1 & 0 \\
-1 & 0 & 0 & 1 & -1 & 0 & 0 & 1 \\
0 & 1 & -1 & 0 & 0 & -1 & 1 & 0 \\
0 & 0 & 0 & 0 & 0 & -1 & 0 & 1 \\
0 & 0 & -1 & 1 & 0 & 0 & 0 & 0
\end{bmatrix}
;\quad
\begin{bmatrix}
-1 & 1 & 0 & 0 & 0 & 0 & 0 & 0 \\
-\frac{1}{2} & 0 & \frac{1}{2} & 0 & -\frac{1}{2} & 0 & \frac{1}{2} & 0 \\
-1 & 0 & 0 & 1 & -1 & 0 & 0 & 1 \\
0 & \frac{1}{2} & -\frac{1}{2} & 0 & 0 & -\frac{1}{2} & \frac{1}{2} & 0 \\
0 & 0 & 0 & 0 & 0 & -1 & 0 & 1 \\
0 & 0 & -\frac{1}{2} & \frac{1}{2} & 0 & 0 & -\frac{1}{2} & \frac{1}{2}
\end{bmatrix}
$$

The null space of $R(F)$ can be interpreted as the space of the velocities of the joints during a motion. A map $p'\colon V \to \mathbb{R}^d$ is called an *infinitesimal motion* if $(p_i - p_j)(p_i' - p_j') = 0$ holds for every edge $\{v_i, v_j\} \in E$.

Any congruent motion of $\mathbb{R}^d$ can be considered as an infinitesimal motion $p' = Sp + t$ (where S is a fixed $d \times d$ skew-symmetric matrix and t is a fixed d-dimensional vector), hence the rank of the rigidity matrix of an arbitrary d-dimensional framework (G, p) with $n = |V(G)| \geq d+1$ cannot be greater than $nd - \binom{d+1}{2}$.

A d-dimensional framework (G, p) with $n = |V| \geq d+1$ is *infinitesimally rigid* if the rank of its rigidity matrix is $nd - \binom{d+1}{2}$. Throughout, we shall use the shorter expression *rigid*; however, note that there are several other rigidity concepts as well, see [6, 16], for example. This implies, in particular, that a rigid framework in the 1-, 2- and 3-dimensional space must contain at least $n - 1, 2n - 3$ and $3n - 6$ edges, respectively.

For example, both frameworks in Example 1 are rigid (the rank of both matrices is 5). F_1 remains rigid if we delete any one of the edges (any row of the left hand side matrix is a linear combination of the other five). In case of F_2 if we delete the first, the fourth or the fifth row of the matrix, its rank decreases to 4, if we delete any other row then the rank remains 5.

A continuous function $\pi(q, t)\colon V \times [0, 1) \to \mathbb{R}^d$ is a *deformation* of the framework $F = (G, p)$ if $\pi(q, 0) = p$ and $\|\pi(v_i, t) - \pi(v_j, t)\| = \|p(v_i) - p(v_j)\|$ holds for every edge $\{v_i, v_j\} \in E(G)$ and for every value of t. A congruent motion of the framework is a trivial deformation (the above equality holds for every pair of vertices, whether they are adjacent or not); some frameworks have nontrivial deformations as well, like the one in Example 1 if one deletes the diagonals $\{1, 4\}$ and $\{2, 3\}$.

Rigid frameworks cannot have nontrivial deformations, see [16], Theorem 49.1.4, but this is only necessary for rigidity (delete, for example, edge $\{2, 3\}$ from F_2 of Example 1).

Clearly, if a rigid d-dimensional framework F with n joints has exactly $nd - \binom{d+1}{2}$ bars, it is *minimally rigid,* that is, it does not remain rigid if we remove any of its bars. In this case the rows of its rigidity matrix are linearly independent. More generally, the row space matroid $\mathcal{M}(F)$ of the rigidity matrix of F contains the information about the redundancy of the bars: the rigidity of F is preserved after the removal of a bar if and only if the corresponding row was contained in a circuit of $\mathcal{M}(F)$. For example, $\mathcal{M}(F_1)$ for our first example is a circuit of size 6 while $\mathcal{M}(F_2)$ is the direct sum of three bridges (corresponding to the bars adjacent to joint 2) and a circuit of size 3.

3. Tensegrity Frameworks

Civil engineers are interested in the actual stresses of the bars of a rigid framework under a given load; in particular, the question whether a given bar is under compression or under tension is of great importance. A bar which is always under tension (or compression) can theoretically be replaced by a cable (by a strut, respectively), leading to the concept of tensegrity frameworks where some of the joints are connected by bars (with constant length), some others by cables (where the length is only bounded from above) or by struts (with bounds from below only). Formally, a *tensegrity framework* is a pair $T = (G, p)$ consisting of a graph $G = (V, E)$ with a tripartition $E = B \cup C \cup S$ and a map $p \colon V \to \mathbb{R}^d$. The edges in the three subsets are called *bars, cables* and *struts*, respectively, together we shall refer to them as *tensegrity elements.*

A map $p' \colon V \to \mathbb{R}^d$ is called an *infinitesimal motion* of the tensegrity framework $T = (G, p)$ if

$$(p_i - p_j)(p_i' - p_j') \leq 0 \quad \text{holds for every cable} \quad \{v_i, v_j\} \in C,$$
$$(p_i - p_j)(p_i' - p_j') = 0 \quad \text{holds for every bar} \quad \{v_i, v_j\} \in B, \quad \text{and}$$
$$(p_i - p_j)(p_i' - p_j') \geq 0 \quad \text{holds for every strut} \quad \{v_i, v_j\} \in S.$$

A d-dimensional tensegrity framework $T = (G, p)$ is *infinitesimally rigid* (or, in this paper, shortly *rigid*) if every infinitesimal motion of it is a congruent motion of $\mathbb{R}^d$.

A continuous function $\pi(q, t) \colon V \times [0, 1) \to \mathbb{R}^d$ is a *deformation* of the tensegrity framework $T = (G, p)$ if $\pi(q, 0) = p$ and, for every value of t,

$$\|\pi(v_i, t) - \pi(v_j, t)\| \leq \|p(v_i) - p(v_j)\| \quad \text{holds for every cable} \ \{v_i, v_j\} \in C,$$
$$\|\pi(v_i, t) - \pi(v_j, t)\| = \|p(v_i) - p(v_j)\| \quad \text{holds for every bar} \ \{v_i, v_j\} \in B \text{ and}$$
$$\|\pi(v_i, t) - \pi(v_j, t)\| \geq \|p(v_i) - p(v_j)\| \quad \text{holds for every strut} \ \{v_i, v_j\} \in S.$$

If each cable and strut of a tensegrity framework T is replaced by a bar then the resulting bar-and-joint framework $F(T)$ is called the *underlying bar-and-joint framework* of T. If, for some tensegrity framework T with $C \cup S \neq \emptyset$, this $F(T)$ is minimally rigid then T could not be rigid; a necessary condition for the rigidity of a d-dimensional tensegrity framework (containing at least one cable or strut) with n joints is that it must have at least $nd - \binom{d+1}{2} + 1$ tensegrity elements. For example, F_1 of Example 1 remains rigid if we replace bars $\{1,4\}$ and $\{2,3\}$ by cables and the other four bars by struts or *vice versa*. Similarly, F_2 of the same example remains rigid if we replace bars $\{1,3\}$ and $\{3,4\}$ by cables and $\{1,4\}$ by strut or *vice versa* (while the bars incident to joint 2 remain bars).

In general, the following theorem holds:

Theorem 1 ([15], see also [13], Proposition 18.3.1). *Suppose that a rigid 2-dimensional framework F has n joints and $2n - 2$ bars, hence there exists a nonzero vector w satisfying $wR(F) = 0$, where $R(F)$ denotes the rigidity matrix of F. If we replace the bars, corresponding to the positive (negative) entries of w by cables (struts, respectively) and keep the bars if the corresponding entry is zero, then the resulting tensegrity framework will be rigid.*

Observe that the rank of the rigidity matrix in the above theorem is $2n - 3$, hence this vector w is unique up to a nonzero constant multiplier. Since this constant can also be negative, we obtain a special case of the following more general theorem:

Theorem 2 ([16], Theorem 49.1.31). *Reversal theorem: A tensegrity framework is rigid if and only if its reversal (interchanging cables and struts) is rigid.*

For example, in case of F_1 in Example 1, a possible choice of the vector w may be $(1, 1, -1, -1, 1, 1)$. A slight perturbation of the coordinates of the joints would change the entries but would keep the sign pattern of this vector.

Example 2. As a less trivial example, consider Figure 1A. This drawing[2] can be considered as a 2-dimensional rigid framework. A routine but fairly long calculation (or some physical intuition) gives that (apart from its reversal) there is a unique way to replace every bar with cable or strut

---

[2]A coloured version of this drawing was the symbol of the Sixth Czech-Slovak International Symposium on Combinatorics, Graph Theory, Algorithms and Applications, Prague, 2006. Just like that conference was honouring the 60th birthday of Jarik Nešetřil, let us dedicate this example to him.

if we wish to preserve rigidity, see Figure 1B. Throughout this paper we follow the usual convention that bars, cables and struts are indicated by continuous, by dotted and by double lines, respectively.

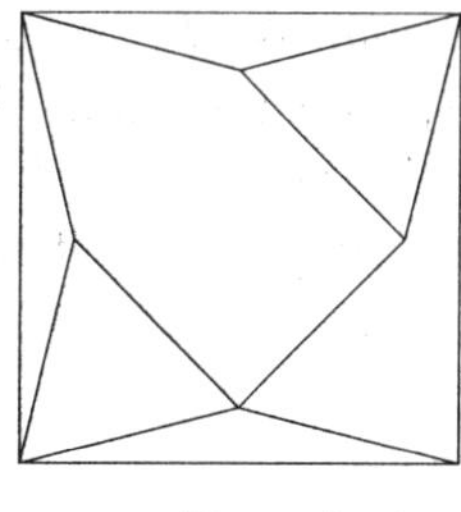

Fig. 1A

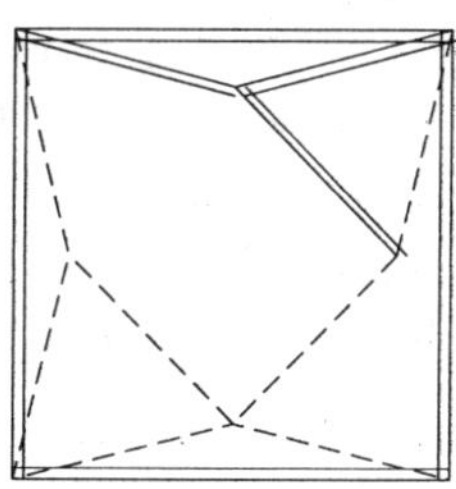

Fig. 1B

As one can expect, if a rigid 2-dimensional bar-and-joint framework with n joints has more than $2n-2$ bars then there is a larger degree of freedom for replacing the bars with cables or struts. Such an example will be presented below (Figure 3B for Example 3).

We mention two further useful relations between a tensegrity framework and its underlying bar-and-joint framework:

Theorem 3 ([15], Corollary 5.3). *If we remove one cable or strut from a rigid tensegrity framework and then replace all the remaining tensegrity elements by bars then the resulting bar-and-joint framework will be rigid.*

Suppose that (G, p) is rigid in $\mathbb{R}^d$ and $(G-e, p)$ is not, for some $e \in E(G)$. Then e is called a *critical* edge of G (in dimension d). Clearly, the rod corresponding to e in the framework (G, p) cannot be replaced by a cable or a strut if we wish to preserve the rigidity in $\mathbb{R}^d$. The converse is also true:

Theorem 4 [9]. *A graph can be realized in $\mathbb{R}^d$ as a rigid tensegrity framework without bars (i.e. with cables and struts only) if and only if it has no critical edges in dimension d.*

4. Genericity, the Theorems of Laman and Lovász–Yemini

The graph G of a framework $F = (G, p)$ alone cannot determine whether F is rigid. For example, if we delete an edge from K_4, the resulting graph can belong to a rigid 2-dimensional framework (delete any edge from F_1 of Example 1) or to a non-rigid one (delete $\{2,3\}$ from F_2 of the same example). The reason for this difference is that joints $1, 3, 4$ happen to be collinear in the latter case, leading to an infinitesimal motion. Another, less trivial example is the Kuratowski-graph $K_{3,3}$ which leads to a rigid 2-dimensional framework if and only if the six joints do not lie on a conic section, see [3].

In general, if (G, p_1) is rigid and (G, p_2) is not then the coordinates of the joints of the latter framework satisfy some additional algebraic relations, leading to some cancellations while the rank of the rigidity matrix is calculated. One can prove that if a realization of a graph leads to a rigid framework then "almost all" realizations have this property. Formally a graph G will be called *generic rigid in the d-dimensional space (as a bar-and-joint framework)* if there is a map $p \colon V \to \mathbb{R}^d$ so that the framework (G, p) is rigid. The set of these "rigid points" is open and its complement is of zero measure.

In the one-dimensional space a bar-and-joint framework is rigid if and only if its graph is connected, see [11]. Hence, in particular, a 1-dimensional minimally rigid framework with n joints must have $n - 1$ bars.

We have seen that a 2-dimensional minimally rigid bar-and-joint framework with n joints must have $2n - 3$ bars. However, a graph with n vertices and $2n - 3$ edges is not necessarily generic rigid in the plane: if we add a path of 3 edges between two vertices of K_4, the resulting graph has 6 vertices and 9 edges, yet, any realization of it leads to a non-rigid framework, since one part of it is "overbraced". The classical result of Laman states that essentially this is the only possible reason:

Theorem 5 [10]. *A simple graph G with n vertices and $2n - 3$ edges is generic rigid in the plane as a bar-and-joint framework if and only if $|E'| \leq 2|V'| - 3$ holds for every subgraph $G' = (V', E')$ of G with at least two vertices.*

Checking the condition of Laman's theorem for all subgraphs would require exponential time; however, the following equivalent condition by Lovász and Yemini can be checked in polynomial time, by using the matroid partition algorithm [5]:

Theorem 6 [11]. *A simple graph G with n vertices and $2n - 3$ edges is generic rigid in the plane as a bar-and-joint framework if and only if, for any*

$e \in E(G)$, the edge set of $G + e$ is the union of two edge-disjoint spanning trees, where $G + e$ means that edge e is replaced by a pair of parallel edges.

It is a very easy exercise in graph theory to show that this condition is equivalent to the following, seemingly stronger one:

Theorem 7 ([12], Proposition 1). *A simple graph G with n vertices and $2n - 3$ edges is generic rigid in the plane as a bar-and-joint framework if and only if, for any $x, y \in V(G)$, the edge set of $G + \{x, y\}$ is the union of two edge-disjoint spanning trees, where $G + \{x, y\}$ means that G is extended by a new edge between the (not necessarily adjacent) vertices x and y.*

For $d > 2$ the straightforward modification of Laman's condition is necessary for the generic minimal rigidity of the graph but it is not sufficient [1]. No polynomial time algorithm is known to decide whether a graph is generic rigid in $\mathbb{R}^d$ for $d > 2$ as a bar-and-joint framework.

5. What are the Analogous Results for Tensegrity Frameworks?

The planar tensegrity frameworks of Figures 2A-B (and their reversals) are rigid. A small perturbation of joint 4 does not destroy rigidity, provided that this joint remains inside the shaded areas. The set of the "rigid points" p of a rigid tensegrity framework (G, p), if the graph G and the tripartition of $E(G)$ fixed, is open (just like in case of bar-and-joint frameworks) but its complement is of positive measure.

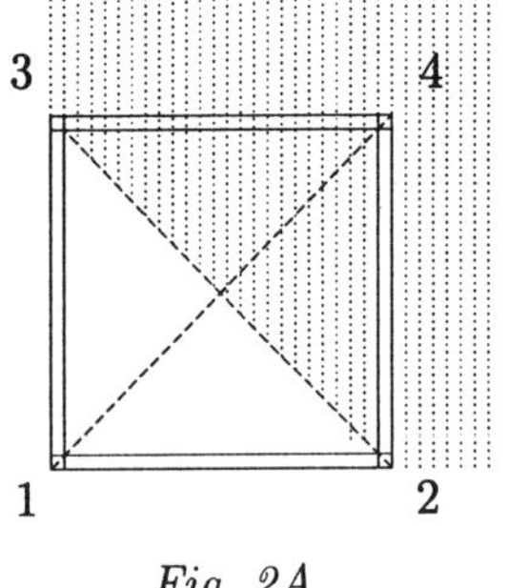

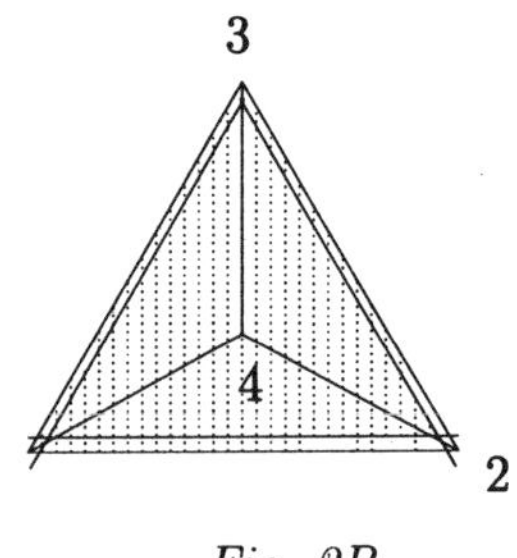

Fig. 2A   Fig. 2B

Hence we keep the formal definition that a graph G and a tripartition $E = B \cup C \cup S$ of its edge set will be called *generic rigid in the d-dimensional space (as a tensegrity framework)* if there is a map $p\colon V \to \mathbb{R}^d$ so that the resulting tensegrity framework (G, p) is rigid; however, it will not be true any more that if such a p exists then "almost all" choices are good for the given tripartition.

Before formulating the analogues of the theorems of the previous section, we introduce the more restrictive concept of *r-tensegrity frameworks* where all the bars must be replaced by cables or struts. A graph G will be called *generic rigid in the d-dimensional space (as an r-tensegrity framework)* if there is a map $p\colon V \to \mathbb{R}^d$ and a bipartition $E = C \cup S$ of its edge set of G so that the resulting r-tensegrity framework (G, p) is rigid.

Theorem 8. *A simple graph G with n vertices and $2n - 2$ edges is generic rigid in the plane as an r-tensegrity framework if and only if $|E'| \leq 2|V'| - 3$ holds for every proper subgraph $G' = (V', E')$ of G with at least two vertices.*

Proof. *Necessity.* If G is generic rigid in the plane as an r-tensegrity framework then, by Theorem 3, $G - e$ is generic rigid in the plane as a bar-and-joint framework for every $e \in E(G)$. Hence, for every proper subgraph $G' = (V', E')$ of G consider an edge $e \in E - E'$ and recall (Theorem 5) that $|E'| \leq 2|V'| - 3$ holds.

Sufficiency. If G has the required property then $G - e$ is generic rigid in the plane as a bar-and-joint framework for every $e \in E(G)$. Hence G satisfies the condition of Theorem 1. All we have to prove that if we apply Theorem 1, the result will be an r-tensegrity framework, that is, the vector w will have nonzero entries only. But it is obvious since if entry w_i (corresponding to edge e_i) were zero then $G - e_i$ would not be rigid as a bar-and-joint framework. $\blacksquare$

Suppose that a simple graph G has n vertices, $2n - 2$ edges, and $E(G)$ is the union of two edge-disjoint spanning trees. An edge $e \in E(G)$ is *replaceable* if this latter property remains true for $(G - e) + \{x, y\}$ for any two vertices $x, y \in V(G)$.

Theorem 9. *A simple graph G with n vertices and $2n - 2$ edges is generic rigid in the plane as an r-tensegrity framework if and only if $E(G)$ is the union of two edge-disjoint spanning trees and all the edges of G are replaceable.*

Theorem 10. *Given a graph G with n vertices and $2n - 2$ edges. If we wish to realize it as a rigid tensegrity framework in the plane so that a particular* *edge e_0 must not be a bar then it is possible if and only if $E(G)$ is the union of two edge-disjoint spanning trees and e_0 is replaceable.*

The *proof* of these theorems are straightforward if one uses the equivalence of the conditions of Theorems 6, 7 with those of Theorem 5.

However, observe that using the matroid partition algorithm, Theorem 9 only leads to a yes or no answer, in polynomial time, to the question whether G is generic rigid in the plane as an r-tensegrity framework but it does not construct an appropriate bipartition $E = C \cup S$. This can also be obtained in polynomial time, see [9].

6. ANOTHER APPROACH

Using the terminology of complexity theory, one may formulate the following decision problems:

$\mathbf{P}_{\mathrm{bar}}(d)$
Input: A graph G with n vertices and $nd - \binom{d+1}{2}$ edges
Question: Is G generic rigid in the d-dimensional space as a bar-and-joint framework?

$\mathbf{P}_{r\text{-ten}}(d)$
Input: A graph G with n vertices and $nd - \binom{d+1}{2} + 1$ edges
Question: Is G generic rigid in the d-dimensional space as an r-tensegrity framework?

Both problems are in NP for any value of d (this is obvious for the first problem and follows from the theory of linear programming in case of the second one). Theorem 5 states that $\mathbf{P}_{bar}(2)$ is in co-NP, Theorem 6 states that $\mathbf{P}_{bar}(2)$ is in P and Theorem 9 states that $\mathbf{P}_{r\text{-ten}}(2)$ is in P. We have mentioned in Section 4 that $\mathbf{P}_{bar}(1)$ is also in P and so is $\mathbf{P}_{r\text{-ten}}(1)$ as well, see [14].

Recall that the input of $\mathbf{P}_{r\text{-ten}}(d)$ consists of a graph G only; asking the question whether it can be realized as a rigid "general" tensegrity framework (with all three types of tensegrity elements permitted) would be uninteresting since the "surest" thing is to use bars only. Rather, the following, more difficult decision problem should be introduced:

$\mathbf{P}^*_{\text{ten}}(d)$

Input: A graph G with n vertices and $nd - \binom{d+1}{2} + 1$ edges and a tripartition of its edge set $E(G)$

Question: Does there exists a rigid tensegrity framework in the d-dimensional space, with this underlying graph and with this tripartition?

For example, the answer to Problem $\mathbf{P}_{r\text{-ten}}(2)$ is "yes" for the input graph $G = K_4$, see Figures 2A–B. The answer to Problem $\mathbf{P}^*_{\text{ten}}(2)$ for the same input graph and a tripartition $E(G) = B \cup C \cup S$ may be "yes" (for example, if $|C| = 2$ and these two edges are independent, that is, they together cover all the vertices, see Figure 2A, or if $|C| = 3$ and these three edges form a star or a circuit, see Figure 2B) and may be "no" (for example, if $|C| = 2$, $|S| = 4$ but the two edges of C form a path, or if $|C| = |S| = 3$ but both subsets form a path).

As another illustration of the difference between the two problems, recall (Figures 2A–B) that if a graph with a given tripartition is generically rigid in $\mathbb{R}^d$ as a tensegrity framework then the set of "bad points" may form a set with positive measure. This is not the case if the tripartition is not *a priori* given:

Theorem 11. *Let a graph G and a tripartition $E = B \cup C \cup S$ of its edge set with $C \cup S \neq \emptyset$ be generic rigid in $\mathbb{R}^d$ as a tensegrity framework. Then the set of "rigid points" $\{p \colon p \in \mathbb{R}^d$ and there exists a (possibly different) tripartition $E = B' \cup C' \cup S'$ of $E(G)$ with $C' \cup S' \neq \emptyset$ so that (G, p) is rigid$\}$ is open and its complement is of zero measure.*

Proof. If $e \in C \cup S$ then, by Theorem 3, the graph $G - e$ is generic rigid in $\mathbb{R}^d$ as a bar-and-joint framework. Hence almost all of its realizations are rigid (the measure of the set of the "bad" points is zero) and then any of these rigid realizations can be extended to a rigid realization of G as a tensegrity framework with $C' \cup S' \neq \emptyset$, for example with the trivial choice $C' = \{e\}$, $S' = \emptyset$, $B' = E - \{e\}$. ∎

Theorem 12. $\mathbf{P}^*_{\text{ten}}(1)$ *can be answered in polynomial time.*

Proof. The following algorithm (see [14]) answers the question even in the more general case when the number of edges of the input graph is arbitrary:

Replace each $e \in B$ with a pair of parallel edges and put one of the new edges to C and the other to S. The resulting graph G' is not simple any more and its edge set has a bipartition $C' \cup S'$. The answer to the original question is "yes" if and only if G' is 2-edge-connected and each of its 2-vertex-connected components intersects both C' and S'. These latter conditions can clearly be checked in polynomial time. ∎

Conjecture 13. $\mathbf{P}^{*}_{\mathrm{ten}}(2)$ *can be answered in polynomial time.*

This conjecture seems to be open and the following example (an extension of [15], Example 6.5) shows that the situation is rather complex for the 2-dimensional case.

Example 3. Figure 3A shows various rigid planar realizations of a graph consisting of a circuit of size 6 and four diagonals. Imagine that joint A of the first framework "slowly moves to the left", that is, towards joint D, while the position of the other five joints is fixed. The first (i.e. leftmost) framework is in generic position. When the six joints lie on a conic section then the only edge which destroys bipartiteness must become a bar (see the second framework, it "separates" the first and the third ones which are both in generic position). In general, the odd numbered frameworks are in generic position and are separated by nongeneric ones; if a tensegrity element "jumps" from cable to strut or *vice versa* then it is temporarily a bar in the nongeneric position.

The way how the seventh and the ninth generic frameworks are separated is more complex: since joints A and D coincide, the rank of the rigidity matrix decreases by two rather than by one. Hence, even if we decide that, say, the edges adjacent to joint F are replaced by two cables and a strut, we have a further degree of freedom, see all the frameworks of Figure 3B. Here we use a similar convention (odd numbered r-tensegrity frameworks are separated by even numbered ones which contain bars as well). Observe that the tensegrity element CD is redundant (can be removed without destroying rigidity) in frameworks 3, 4 and 5, CE is redundant in frameworks 5, 6 and 7, and DE is redundant in the last three frameworks.

7. Rigidity and Connectivity

Recall that a graph is generic rigid in the 1-dimensional space as a bar-and-joint framework if and only if it is connected. Rigidity of graphs as r-

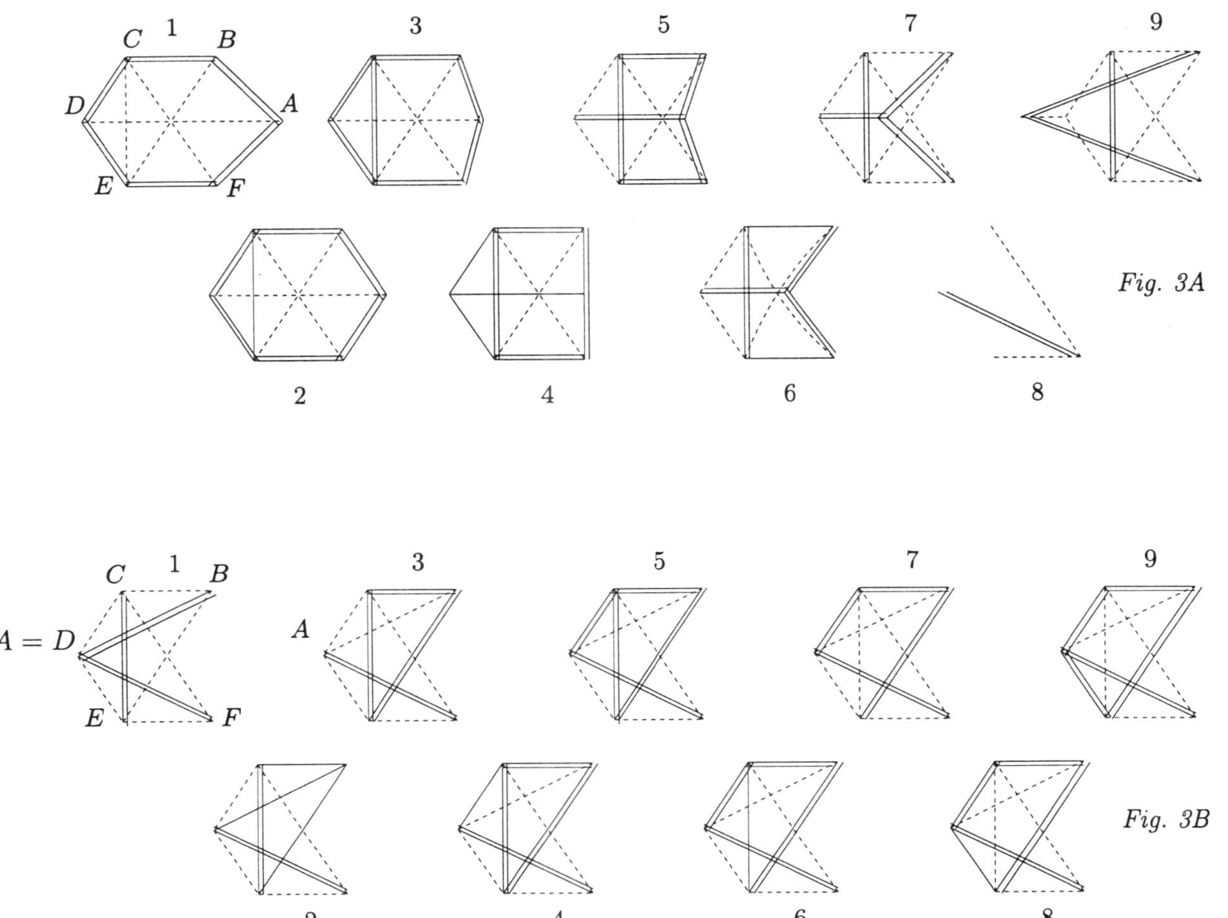

C 1 B
D A
E F
3 5 7 9
2 4 6 8
Fig. 3A
A = D
C 1 B
A
E F
3 5 7 9
2 4 6 8
Fig. 3B

tensegrity frameworks and/or in higher dimensional spaces are also related to the higher vertex- and edge-connectivity numbers of the graphs, see also Section 4.7 of [6].

One can easily prove (implicitly contained in [14], Theorem 1) that a graph is generic rigid in the 1-dimensional space as an r-tensegrity framework if and only if it is 2-edge-connected.

2-vertex-connectivity is a necessary condition for a graph to be generic rigid in the plane as a bar-and-joint framework, but it is not sufficient. 6-vertex-connectivity is already sufficient (and 5 is not enough), see [11]. Results for higher dimensions have recently been obtained by Bill Jackson and Tibor Jordán [8].

For the generic rigidity in the plane as an r-tensegrity framework a graph should be not only 2-vertex-connected but 3-edge-connected as well. However, neither 3-vertex-connectivity nor 4-edge-connectivity is required, see the rigid tensegrity framework of Figure 4.

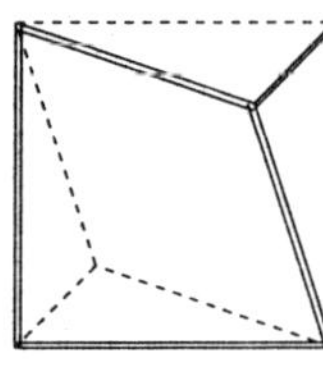

Fig. 4

References

[1] L. Asimow and B. Roth, The rigidity of graphs, *Trans. Amer. Math. Soc.*, **245** (1978), 279–289.

[2] L. Asimow and B. Roth, The rigidity of graphs II., *SIAM J. Appl. Math.*, **68** (1979), 171–190.

[3] E. D. Bolker and B. Roth, When is a bipartite graph a rigid framework? *Pacific J. Math.*, **90** (1980), 27–44.

[4] R. Connelly and W. Whiteley, Second-order rigidity and prestress stability for tensegrity frameworks, *SIAM J. Discrete Math.*, **9** (1996), 453–491.

[5] J. Edmonds, Minimum partition of a matroid into independent subsets, *J. Res. Natl. Bur. Stand.*, **69B** (1965), 67–72.

[6] J. Graver, B. Servatius and H. Servatius, *Combinatorial Rigidity*, American Mathematical Society (Providence, 1993).

[7] B, Jackson and T. Jordán, Connected rigidity matroids and unique realizations of graphs, *J. Combinatorial Theory, Ser. B.,* **94** (2005), 1–29.

[8] B. Jackson and T. Jordán, The d-dimensional rigidity matroid of sparse graphs, *J. Combinatorial Theory, Ser. B.,* **95** (2005), 118–133.

[9] T. Jordán, A. Recski and Z. Szabadka, *Rigid tensegrity labellings of graphs,* to appear.

[10] G. Laman, On graphs and rigidity of plane skeletal structures, *Eng. Math.,* **4** (1970), 331–340.

[11] L. Lovász and Y. Yemini, On generic rigidity in the plane, *SIAM J. Algebraic and Discrete Methods,* **3** (1982), 91–98.

[12] A. Recski, A network theory approach to the rigidity of skeletal structures, Part II. Laman's theorem and topological formulae, *Discrete Applied Math.,* **8** (1984), 63–68.

[13] A. Recski, *Matroid Theory and Its Applications in Electric Network Theory and in Statics,* Springer (Berlin, 1989).

[14] A. Recski and O. Shai, One-dimensional synthesis of graphs as tensegrity frameworks, in: *Proc. 4th Japanese-Hungarian Symposium on Discrete Math. and Its Appl.,* Budapest (2005), pp. 284–288.

[15] B. Roth and W. Whiteley, Tensegrity frameworks, *Trans. Amer. Math. Soc.,* **265** (1981), 419–446.

[16] W. Whiteley, Rigidity and Scene Analysis, in: *Handbook of Discrete and Computational Geometry* (J. E. Goodman and J. O'Rourke, eds.), CRC Press (Boca Raton, 1997), pp. 892–916.

András Recski

Department of Computer Science and Information Theory, Budapest University of Technology and Economics, H-1521 Budapest, Hungary

e-mail: recski@cs.bme.hu

BOLYAI SOCIETY
MATHEMATICAL STUDIES, 17

Horizons of Combinatorics
Balatonalmádi
pp. 179–188.

POLYGONAL GRAPHS

ÁKOS SERESS[*]

A near-polygonal graph is a graph Γ with a distinguished set $\mathcal{C}$ of cycles of common length m such that each path of length two lies in a unique element of $\mathcal{C}$. If m is the girth of Γ then the graph is called polygonal. We describe various constructions of polygonal and near-polygonal graphs, and some attempts toward their classification.

1. INTRODUCTION

Polygonal and near-polygonal graphs were defined in the abstract. We need one more notion: a polygonal graph Γ of girth m is called *strict polygonal* if the set $\mathcal{C}$ contains all cycles of Γ of length m. Some examples of strict polygonal graphs are K_n (of girth 3) and the 1-skeleton, i.e., the vertex- and edge-set, of a cube or a dodecahedron (of girth $m = 4$ and 5, respectively). In the Petersen graph P, every 2-path is contained in two 5-cycles so P is not strict polygonal. However, for a 5-cycle C in P and the subgroup $H \leq \mathrm{Aut}\,(P)$, $H \cong A_5$, the set of cycles $\mathcal{C} := C^H = \{C^h \mid h \in H\}$ covers each 2-path exactly once and so P is polygonal.

Polygonal and near-polygonal graphs are Manley Perkel's invention [13], [19] and for about a decade he was the only mathematician working in this subject. Recent activity is more widespread, partly due to the fact that polygonal graphs occur both in topological and algebraic graph theory and in geometries related to finite simple groups (see for example [15]). The purpose of this short survey is to popularize polygonal graphs; I find them fascinating as they introduce a design-theoretic flavor into graph theory.

[*]Supported in part by the NSF and the NSA.

The theory of polygonal graphs is in its infancy. We are still in the data-collecting phase, as most results are only constructions. Moreover, there is only a scarce supply of known polygonal graphs of girth at least six. There are two major construction ideas: one is topological (graph embeddings into surfaces and covers), while the other one is algebraic (coset graphs of groups). The two main sections of this paper are devoted to these two kinds of constructions.

The definition of polygonal graphs does not mention any requirements on the automorphism group of the graph, and it is not clear how much symmetry a polygonal graph must posess. It is easy to prove (see [14, Lemma 2.1]) that a connected near-polygonal graph is regular: indeed, let Γ be a near-polygonal graph with a set $\mathcal{C}$ of special cycles and let $\alpha, \beta \in V(\Gamma)$ be adjacent and of valency $d(\alpha)$ and $d(\beta)$, respectively. Then the number of 2-paths with middle point α and containing the edge $e = \{\alpha, \beta\}$ is $d(\alpha) - 1$, implying that exactly $d(\alpha) - 1$ cycles in $\mathcal{C}$ contain e. Similarly, the number of 2-paths with middle point β and containing the edge e is $d(\beta) - 1$, so there are exactly $d(\beta) - 1$ cycles in $\mathcal{C}$ containing e. Thus $d(\alpha) = d(\beta)$.

As a starting point toward the classification of polygonal graphs, it would be interesting to determine the set of pairs (r, m) for which r-regular polygonal graphs of girth m exist.

2. Topological Constructions

Trivalent graphs. Most topological constructions of polygonal graphs deal with the trivalent case. If a trivalent graph Γ of girth m can be embedded on a (not necessarily orientable) surface such that each face is an m-gon then the set of faces form the distinguished cycle set $\mathcal{C}$ of a polygonal graph. Conversely, if Γ is a trivalent polygonal graph with distinguished cycle set $\mathcal{C} = \{C_1, \ldots, C_k\}$ then attaching a 2-cell to each C_i we get a 2-dimensional topological space which can be embedded on some surface. Hence the construction of trivalent polygonal graphs is reduced to determining whether an appropriate embedding on some surface exists.

The observation in the previous paragraph readily implies that some Platonic solids are polygonal graphs: the tetrahedron, cube, and dodecahedron. More generally, Perkel [13] worked out which graph embeddings in [6, Table 8] yield trivalent polygonal graphs. These embeddings give graphs of girth up to 9, but only finitely many of them are of girth at least 7 (the examples with girth at least 7 are also listed in [17]). Hence the questions remained: are there infinitely many trivalent polygonal graphs of girth at least 7, and can the girth be arbitrarily large?

The first of these questions was answered by Perkel [18]. Given any triangle-free, trivalent graph Γ_0 of order n embedded on an orientable surface of genus $g > 0$, an idea of Brown and Connelly [3] can be applied to construct a trivalent polygonal graph of girth 7 and of order $14n$, embedded on a surface of genus $14(g - 1) + 1$. Iterating this construction, we obtain infinitely many trivalent polygonal graphs of girth 7.

Seven years later, Archdeacon proved the following remarkable theorem.

Theorem 2.1 [1]. *For all n divisible by 4, there exist trivalent polygonal graphs of girth n.*

The proof starts with a graph embedding interesting in its own right: Stahl and White [23] give an embedding of $\Gamma := K_{n/2,n/2}$ on an orientable surface S such that each face is a Hamiltonian cycle. Then [1] constructs a cover $\hat{\Gamma}$ of Γ, embedded on a surface $\hat{S}$, which keeps the property that every face is a cycle of length n and that the graph is $n/2$-regular, but in addition the girth of $\hat{\Gamma}$ is $n/2$. The next step is to add a new vertex ν_f on each face f of $\hat{\Gamma}$, and connect ν_f to the vertices on f. Since each face f was an n-gon and $n/2$ faces met at each vertex of $\hat{\Gamma}$, we obtain a regular graph Δ of degree n with all triangular faces. Finally, we take the dual Δ^* of Δ. The vertices of Δ^* are the triangles of Δ, with two triangles connected if and only if they share a common edge. Hence the graph Δ^* is trivalent, and the girth condition on $\hat{\Gamma}$ ensures that the girth of Δ^* is n. Also, Δ^* is polygonal, because the n-cycles defined by the triangles incident to a fixed vertex of Δ cover each 2-path in Δ^* exactly once.

We note that the construction of polygonal graphs of girth 7 mentioned earlier [18] also proceeds by first building a triangulation of a surface and then taking the dual graph. Both in [1] and in [18] the valency of the triangular graphs is relatively small, compared to the number of vertices. This corresponds to our intuition based on the fact that a triangulation is locally planar (the subgraph spanned by the neighborhood of any vertex is a cycle, which is a planar graph with a few edges). However, there are triangulations with many edges: in [22], for all $\varepsilon > 0$ a triangular graph on n vertices and with more than $n^{2-\varepsilon}$ edges is constructed, for some sufficiently large n. In a sense, this result is best possible because a triangular graph of order n must have $o(n^2)$ edges [5]. Unfortunately, the graphs in [22] cannot be used for the construction of near-polygonal graphs because the valencies of vertices take three different values.

We finish this subsection by commenting on the attempts to classify trivalent polygonal graphs. It is easy to see that the only polygonal graphs of girth three and four are the tetrahedron and cube, respectively and with some work it can also be shown that the only trivalent strict polygonal graph of girth 5 is the dodecahedron. There are infinitely many strict polygonal graphs of girth 6 (just wrap large portions of a hexagonal plain tessellation on a torus). Negami [12] classified trivalent polygonal graphs of girth 6 in the sense that they all must be duals of triangulations of the torus or the Klein bottle. The case of girth at least 7 is wide open.

Doubling the girth. Much less is known about polygonal graphs of valency at least 4. In a topological construction, we have to consider surface embeddings in higher dimensions because on a two-dimensional surface not all pairs of edges incident to a fixed vertex lie on a common face. Examples are the k-dimensional hypercubes C_k (of valency k and girth 4) and the graphs D_k obtained by identifying antipodal vertices of C_k (also of valency k and girth 4 for $k \geq 4$). There are also two sporadic examples. One of them is the 1-skeleton of a 4-dimensional polytope called the 120-cell. The 120-cell has 120 three-dimensional faces isomorphic to dodecahedrons, and the 1-skeleton is a strict polygonal graph of girth 5 and valency 4, on 600 vertices. The other example is obtained by identifying antipodal vertices of the 120-cell. That graph is also strict polygonal of girth 5 and valency 4 [16].

The main result of this subsection is the following theorem of Archdeacon and Perkel [2].

Theorem 2.2 [2]. *If there exists a polygonal graph Γ of valency r and girth m then there exists a polygonal graph Δ of valency r and girth $2m$. Moreover, if Γ is strict polygonal then Δ is strict polygonal as well.*

Given a polygonal graph $\Gamma(V, E)$ of valency r and girth m, [2] constructs Δ as a cover of Γ using the voltage graph method [9]. We define an elementary abelian 2-group G of order $2^{|E|}$, with a generating set $\{g_e \mid e \in E\}$. For each $e \in E$, we assign the group element g_e to e, and call it the voltage assignment. The vertex set of Δ is $V(\Delta) := V \times G$ and $(\alpha_1, h_1), (\alpha_2, h_2) \in V(\Delta)$ are defined to be adjacent if and only if $\{\alpha_1, \alpha_2\} \in E$ and $h_2 = h_1 + g_{\{\alpha_1, \alpha_2\}}$. Note that since G is an elementary abelian 2-group, the definition of edges does not depend on the ordering of the pairs (α_i, h_i). For fixed $\alpha_1, \alpha_2 \in V$, if $\{\alpha_1, \alpha_2\} \in E$ then Δ contains a perfect matching between the fibers $\{(\alpha_1, h) \mid h \in G\}$ and $\{(\alpha_2, h) \mid h \in G\}$, while if $\{\alpha_1, \alpha_2\} \notin E$ then there are no edges between these two fibers. Hence Δ is r-regular.

We indicate why the girth of Δ is $2m$ and why Δ is polygonal. Let $\varphi : V(\Delta) \to V$, $(\alpha, h) \mapsto \alpha$ be the natural projection. For any cycle C of Γ, $\varphi^{-1}(C)$ is a 2-regular graph on a vertex set of size $2^{|E|}|C|$, so $\varphi^{-1}(C)$ is the union of disjoint cycles. We claim that each cycle has length $2|C|$. Indeed, let $(\alpha, k) \in \varphi^{-1}(C)$ and let $\hat{C}$ be the unique cycle in $\varphi^{-1}(C)$ containing (α, k). Walking along $\hat{C}$ starting at (α, k), we return to the fiber $F_\alpha := \{(\alpha, h) \mid h \in G\}$ after $|C|$ steps, at a point $(\alpha, k + \ell)$ for some $\ell \in G$. Then ℓ is the sum of the voltage assignments along the path we walked on, so $\ell \neq 0$. Continuing the walk along $\hat{C}$, we return to the fiber F_α after $|C|$ further steps, at the point $(\alpha, k + \ell + \ell) = (\alpha, k)$. Hence $|\hat{C}| = 2|C|$. Also, given any cycle D in Δ, $\varphi(D)$ is a closed walk $(\alpha_0, \alpha_1, \ldots, \alpha_n = \alpha_0)$ in Γ, with the property that $\alpha_i \neq \alpha_{i+2}$ for all $i \leq n - 2$ (using the terminology of Section 3, $\varphi(D)$ is a closed $|D|$-arc). We can choose the starting point α_0 such that the second occurrence of α_0 is at α_j for some j, with the additional property that $\alpha_1, \ldots, \alpha_{j-1}$ are pairwise different. Walking along D starting at (α_0, k), we return to the fiber F_{α_0} after j steps, at the vertex $(\alpha_j, k + \ell) = (\alpha_0, k + \ell)$. Since the girth of Γ is m, we have $j \geq m$. Moreover, $\ell \neq 0$ because it is a sum of voltage assignments along a cycle of Γ. The remaining part of D projects on a closed arc of Γ, so its length is at least m. Adding the two estimates, we obtain $|D| \geq 2m$. Finally, it is clear that if C is the distinguished set of m-cycles in the polygonal graph Γ then $\varphi^{-1}(C)$ covers each 2-path in Δ exactly once.

3. Algebraic Constructions

The basic idea is to define an appropriate cycle C in a graph Γ and a subgroup $G \leq \mathrm{Aut}\,(\Gamma)$ such that $\mathcal{C} := \{C^g \mid g \in G\}$ is the distinguished cycle set of a near-polygonal graph. The difficulty is that G must be large enough that the G-images of C cover each 2-path, but C must be chosen carefully so that the 2-paths are covered only once. There are two sufficient conditions known to ensure these properties; not surprisingly, both of them can be applied to 2-arc transitive graphs. First we give the necessary definitions.

An s-*arc* in a graph Γ is a sequence $(\alpha_0, \alpha_1, \ldots, \alpha_s)$ of vertices such that $\alpha_i \neq \alpha_{i+2}$ for all $i \leq s - 2$. In particular, a 2-arc is a 2-path with a distinguished starting point. For $G \leq \text{Aut}\,(\Gamma)$, we say that Γ is (G, s)-arc transitive if G acts transitively on the s-arcs of Γ: that is, for any two s-arcs $(\alpha_0, \alpha_1, \ldots, \alpha_s)$, $(\beta_0, \beta_1, \ldots, \beta_s)$ of Γ, there exists $g \in G$ such that $\alpha_i^g = \beta_i$ for $0 \leq i \leq s$. The theory of s-arc transitive graphs is a major area of algebraic graph theory, and is much more advanced than the theory of polygonal graphs (see [21] and its references).

Suppose that a group G has a core-free subgroup H (that means that H contains no nontrivial $N \lhd G$) and an element $g \in (G \setminus H)$ satisfying $g^2 \in H$. Then we define the *coset graph* $\text{Cos}\,(G, H, HgH)$ to have vertex set $[G : H]$, the right cosets of H in G. Two cosets Hx, Hy are adjacent if and only if $yx^{-1} \in HgH$. It is easy to prove (see e.g. [21, Lemma 1.2]) that coset graphs $\text{Cos}\,(G, H, HgH)$ are undirected and $(G, 1)$-arc transitive. Conversely, any $(G, 1)$-arc transitive graph Γ is isomorphic to a coset graph $\text{Cos}\,(G, H, HgH)$ for some $H \leq G$ and $g \in (G \setminus H)$ satisfying $g^2 \in H$. We also have the following characterization of 2-arc transitive graphs. A coset graph $\text{Cos}\,(G, H, HgH)$ is $(G, 2)$-arc transitive if and only if for all vertices α, the point stabilizer G_α acts 2-transitively on the neighborhood $N(\alpha)$ of α. For the proof, just consider transitivity on the 2-arcs with middle point α.

Now we are ready to describe the sufficient conditions that ensure the existence of near-polygonal graphs. The first one is due to Perkel [19].

Lemma 3.1. [19] *Suppose that $\Gamma = \text{Cos}\,(G, H, HgH)$ is $(G, 2)$-arc transitive and let α, β, γ be vertices with the properties*

(i) $\quad \beta, \gamma \in \Gamma(\alpha)$;

(ii) *The only fixed points of the action of $G_{\alpha\beta\gamma}$ on $N(\alpha)$ are β and γ; and*

(iii) *There exists $h \in G$ normalizing $G_{\alpha\beta\gamma}$ and satisfying $\alpha^h = \beta$.*

Moreover, let C be the connected component containing α of the restriction of Γ to the fixed point set of $G_{\alpha\beta\gamma}$. Then C is a cycle, and Γ is near-polygonal with distinguished cycle set $\mathcal{C} = \{C^x \mid x \in G\}$.

Perkel used Lemma 3.1 to construct two infinite families of near-polygonal graphs. Let $p \neq 5$ be an odd prime and let H be a subgroup of $G^* := \text{PGL}\,(2, p^2)$ isomorphic to A_5. Moreover, let $g_\ell \in G^*$, for $\ell = 5, 6$, be an involution such that $|H \cap H^{g_\ell}| = 60/\ell$. Then for the group $G := \langle H, g_\ell \rangle$, the coset graph $\text{Cos}\,(G, H, Hg_\ell H)$ is ℓ-regular and satisfies the hypotheses of Lemma 3.1. We note that G is isomorphic to $\text{PSL}\,(2, p)$, $\text{PGL}\,(2, p)$, $\text{PSL}\,(2, p^2)$, or $\text{PGL}\,(2, p^2)$, depending on the residue class $p \bmod 40$.

What is the length m of the special cycles in these near-polygonal graphs? The paper [19] defines sequences that relate p and m in some mysterious way. It is easier to state the result in the valency 6 case. Let $a_1 = 2$, $a_2 = -4$, and $a_i = -2a_{i-1} - 5a_{i-2}$ for $i \geq 3$. Then, if G is defined over a field of characteristic p and the special cycles in the coset graph $\mathrm{Cos}\,(G, A_5, A_5 g_6 A_5)$ have length m then $p \mid a_m$. For example, $a_9 = -2 \cdot 359$ and indeed the special cycles in $\mathrm{Cos}\,(\mathrm{PSL}\,(2, 359), A_5, A_5 g_6 A_5)$ have length 9.

It is difficult to determine which of these near-polygonal graphs are polygonal because there is no easy method to determine the girth of these graphs. With some heroic hand calculations, in the 1970's Perkel proved that the valency 5 coset graphs obtained from $\mathrm{PSL}\,(2, 31)$ and $\mathrm{PSL}\,(2, 41)$ are strict polygonal of girth 5 and 7, respectively. Later, using Cayley, Perkel found four more polygonal examples from groups over the fields with 19, 49, 139, and 169 elements. Recently, using *GAP* [7], we found many more examples. All of these results are summarized in the following table.

G	valency	girth	strict polygonal?
$\mathrm{PSL}\,(2, 3^2)$	5	3	strict
$\mathrm{PSL}\,(2, 31)$	5	5	strict
$\mathrm{PSL}\,(2, 41)$	5	7	strict
$\mathrm{PSL}\,(2, 13^2)$	5	7	strict
$\mathrm{PSL}\,(2, 431)$	5	9	strict
$\mathrm{PSL}\,(2, 199)$	5	11	strict
$\mathrm{PSL}\,(2, 43^2)$	5	11	strict
$\mathrm{PSL}\,(2, 1951)$	5	13	strict
$\mathrm{PSL}\,(2, 36209)$	5	17	strict
$\mathrm{PSL}\,(2, 522919)$	5	19	strict
$\mathrm{PSL}\,(2, 459649)$	5	21	strict
$\mathrm{PSL}\,(2, 1747^2)$	5	23	strict
$\mathrm{PSL}\,(2, 5336599)$	5	23	strict
$\mathrm{PGL}\,(2, 11)$	5	4	strict
$\mathrm{PGL}\,(2, 19)$	5	6	non-strict
$\mathrm{PGL}\,(2, 379)$	5	14	non-strict
$\mathrm{PSL}\,(2, 19)$	6	5	non-strict
$\mathrm{PSL}\,(2, 139)$	6	7	strict
$\mathrm{PSL}\,(2, 359)$	6	9	strict
$\mathrm{PSL}\,(2, 1321)$	6	11	strict
$\mathrm{PSL}\,(2, 16901)$	6	13	strict
$\mathrm{PSL}\,(2, 12239)$	6	17	strict
$\mathrm{PSL}\,(2, 74761)$	6	21	non-strict

G	valency	girth	strict polygonal?
PGL $(2, 3^2)$	6	4	non-strict
PGL $(2, 7^2)$	6	8	non-strict
PGL $(2, 13^2)$	6	12	non-strict

The graphs with girth 23 in this table are the current champions among known 2-arc transitive polygonal graphs. It would be interesting to know how many polygonal graphs are in these two near-polygonal families. Another problem is whether the girth of these graphs tends to infinity as the defining characteristic of G increases. This question arose in the study of limits of vertex-primitive graphs, see [8].

Another sufficient condition for the existence of near-polygonal graphs was given by Li and Seress [11].

Lemma 3.2 [11]. *Let Γ be a connected sharply $(G, 2)$-arc transitive graph, i.e., for any two 2-arcs there is a unique element of G mapping the first 2-arc to the other one. Assume that for an arc (α, β) of Γ there exists an involution $g \in G$ such that $(\alpha, \beta)^g = (\beta, \alpha)$. Then Γ is near-polygonal.*

Note that the hypothesis of Lemma 3.2 is in a sense the opposite of the hypothesis of Lemma 3.1. For $\beta, \gamma \in \Gamma(\alpha)$, in Lemma 3.1 the 3-point stabilizer $G_{\alpha\beta\gamma}$ fixes only the vertices β and γ in $\Gamma(\alpha)$, while in Lemma 3.2 $G_{\alpha\beta\gamma}$ fixes pointwise the entire neighborhood $\Gamma(\alpha)$.

In graphs Γ satisfying the hypothesis of Lemma 3.2, we can construct the special cycle set the following way. Let $(\alpha_0, \alpha_1, \alpha_2, \alpha_3)$ be a 3-path in Γ, and let $h \in G$ be the unique group element such that $\alpha_i^h = \alpha_{i+1}$ for $0 \le i \le 2$. Then $C := \alpha_0^{\langle h \rangle}$ is a cycle in Γ, and the set of cycles $\mathcal{C} := \{C^\ell \mid \ell \in G\}$ covers each 2-path of Γ exactly once. We say that C is obtained by *spinning the 2-path* $(\alpha_0, \alpha_1, \alpha_2)$.

In certain cases we can control the length of the spun cycle. In [11], the following is proven for arbitrary $m \ge 5$. If $q = p^e$ is a prime power with $q \equiv \pm 1 \pmod{m}$ and $G = \mathrm{PGL}\,(2, q) \times \mathrm{PGL}\,(2, q)$, then there exists a near-polygonal coset graph Γ of G of valency q with $q(q-1)(q+1)^2$ vertices such that the length of special cycles is m. Moreover, in the cases $m = 5, 6, 7$, if q satisfies some additional mild arithmetic conditions then Γ can be chosen to be polygonal. These examples include the first infinite family of polygonal graphs with a 2-arc transitive group of automorphisms and of girth 7.

We finish this survey with some remarks about attempts toward the algebraic classification of polygonal graphs. As noted in the introduction, it is not clear how much symmetry a polygonal graph must posess. The trivalent examples of Section 2 obtained from the surface embeddings of [6] are 2-arc transitive, but it is not known which symmetry properties of the input graphs are preserved in the constructions in [18] and in Theorems 2.1, 2.2. There is no classification result assuming only the 2-arc transitivity, equivalently the 2-transitivity of Aut $(\Gamma)_\alpha$ on $\Gamma(\alpha)$, of a polygonal graph Γ. Classifications were obtained assuming that Aut $(\Gamma)_\alpha$ acts as the full symmetric or alternating group on $\Gamma(\alpha)$ and the girth is 4 or 5 [4], [10]. In [20], polygonal graphs of girth 6 are classified under the assumption that Aut $(\Gamma)_\alpha$ acts 3-homogeneously (i.e., transitively on the 3-element subsets) on $\Gamma(\alpha)$ and in addition the action of Aut $(\Gamma)_\alpha$ on the distance-3 neighborhood of α satisfies a mild arithmetic condition.

REFERENCES

[1]   D. Archdeacon, Densely embedded graphs, *J. Comb. Theory B*, **54** (1992), 13–36.

[2]   D. Archdeacon and M. Perkel, Constructing polygonal graphs of large girth and degree, *Congr. Num.*, **70** (1990), 81–85.

[3]   M. Brown and R. Connelly, On graphs with a constant link, in: *New Directions in the Theory of Graphs* (F. Harary, ed.), Academic Press (New York, 1973), pp. 19–51.

[4]   P. J. Cameron, Suborbits in transitive permutation groups, in: *Combinatorial Group Theory*, Math. Centre Tracts, **57** (1974), pp. 98–129.

[5]   L. H. Clark, R. C. Entringer, J. E. McCanna and L. A. Székely, Extremal problems for local properties of graphs, *Australasian J. Combin.*, **4** (1991), 25–31.

[6]   H. S. M. Coxeter and W. O. J. Moser, *Generators and Relations for Discrete Groups*, volume 14 of *Ergeb. Math. Grenzgeb.*, Springer-Verlag (Berlin, Göttingen, Heidelberg, 4th edition, 1957).

[7]   The GAP Group, *GAP – Groups, Algorithms, and Programming, Version* 4.4 (Aachen–St. Andrews, 2004).

[8]   M. Giudici, C. H. Li, C. E. Praeger, Á. Seress and V. I. Trofimov, On limit graphs of finite vertex-primitive graphs, *J. Comb. Theory A*, **114** (2007), 110–134.

[9]   J. L. Gross and T. W. Tucker, *Topological Graph Theory*, John Wiley & Sons (New York, 1987).

[10]   A. A. Ivanov, On 2-transitive graphs of girth 5, *European J. Comb.*, **8** (1987), 393–420.

[11]   C. H. Li and Á. Seress, Symmetrical path-cycle covers of a graph and polygonal graphs, *J. Comb. Theory A*, **114** (2007), 35–51.

[12] S. Negami, Uniqueniss and faithfulness of embeddings of graphs into surfaces, Ph. Thesis, Tokyo Inst. of Technology (1985).

[13] M. Perkel, Ph. D. Thesis, Univ. Michigan (1977).

[14] M. Perkel, Bounding the valency of polygonal graphs with odd girth, *Canad. J. Math.*, **31** (1979), 1307–1321.

[15] M. Perkel, A characterization of J_1 in terms of its geometry, *Geom. Dedicata*, **9** (1980), 291–298.

[16] M. Perkel, Polygonal graphs of valency four, *Congr. Num.*, **35** (1982), 387–400.

[17] M. Perkel, Trivalent polygonal graphs, *Congr. Num.*, **45** (1984), 45–70.

[18] M. Perkel, Trivalent polygonal graphs of girth 6 and 7, *Congr. Num.*, **49** (1985), 129–138.

[19] M. Perkel, Near-polygonal graphs, *Ars Comb.*, **26A** (1988), 149–170.

[20] M. Perkel and C. E. Praeger. On narrow hexagonal graphs with a 3-homogeneous suborbit, *J. Algebraic Comb.*, **13** (2001), 257–273.

[21] Á. Seress, Toward the classification of s-arc transitive graphs, in: *Groups St. Andrews* 2005, (C. Campbell, E. Robertson, ed.), volume 340 of London Math. Soc. Lecture Note Series, Cambridge Univ. Press (2007), pp. 401–414.

[22] Á. Seress and T. Szabó, Dense graphs with cycle neighborhoods, *J. Comb. Theory B*, **63** (1995), 281–293.

[23] S. Stahl and A. T. White, Genus embeddings for some complete tripartite graphs, *Discrete Math.*, **14** (1976), 279–296.

Ákos Seress

The Ohio State University
Dept. of Mathematics
Columbus, OH 43210
U.S.A.

BOLYAI SOCIETY
MATHEMATICAL STUDIES, 17

Horizons of Combinatorics
Balatonalmádi
pp. 189–213.

INFINITE COMBINATORICS: FROM FINITE TO INFINITE

LAJOS SOUKUP*

We investigate the relationship between some theorems in finite combinatorics and their infinite counterparts: given a "finite" result how one can get an "infinite" version of it? We will also analyze the relationship between the proofs of a "finite" theorem and the proof of its "infinite" version.

Besides these comparisons, the paper gives a proof of a theorem of Erdős, Grünwald and Vázsonyi giving the full descriptions of graphs having one/two-way infinite Euler lines. The last section contains some new results: an infinite version of a multiway-cut theorem is included.

1. INTRODUCTION

The introduction should be started with a negative statement: this paper is not a survey of the most important results of infinite combinatorics. Some surveys can be found in [15], [7] or in [8].

In this paper we intend to investigate the relationship between some theorems in finite combinatorics and their infinite counterparts: given a "finite" theorem how one can get a "infinite" version of it? So we study the methods of generalizations. We will survey some problems from finite combinatorics and we will analyze the relationship between their proofs and the proofs of their "infinite" versions.

---

*The preparation of this paper was supported by the Hungarian National Foundation for Scientific Research grant no. 61600 and 68262

Although this paper is not a guide how to get new "infinite" results we will give examples of applications of some basic proof methods from infinite combinatorics.

Beside the investigation of these connections, in section 3.1 we will recall some "forgotten" results of Erdős, Grünwald and Vázsonyi, (see [9] and [10]) with full proof because these theorems are not easily accessible in the literature (originally they were published in Hungarian, [9], then in German, [10]). Moreover, in section 4, we give an infinite version of a theorem from [11] concerning the minimal size of multi-way cuts.

The results from section 1–3 are folklore if no references are given. Only section 4 contains a new result of the author.

Our notation is standard. See e.g. [4].

The set of neighbouring vertices of a vertex v in a graph G is denoted by $\Gamma_G(v)$. If A is a set of vertices then $\Gamma_G(A) = \cup\{\Gamma_G(v)\colon v \in A\}$. The degree $d_G(v)$ of a vertex v is $\left|\Gamma_G(v)\right|$.

A *trail* T in a graph G is a sequence $T = \langle x_0, x_1 \ldots, x_n \rangle$ of vertices such that $E(T) = \langle x_i x_{i+1} \colon i < n \rangle$ is a family of pairwise different edges of G. The vertices x_0 and x_n are the *end-vertices* of the trail. A *circuit* is a trail whose end-vertices coincide. A *path* is a trail with distinct vertices.

A graph is *connected* iff there is a path between any two of its vertices. The maximal connected subgraphs of a graph are the *components* of the graph.

Directed trails and *directed paths* are defined similarly in directed graphs (digraphs, in short).

If G is a directed graph and $p = \langle x_0, x_1, \ldots, x_n \rangle$ is a directed path in G then we write $\mathrm{first}(p) = x_0$, $\mathrm{last}(p) = x_n$ and $\mathrm{E}(p) = \{x_0 x_1, \ldots, x_{n-1} x_n\}$.

If $G = (V, E)$ is a directed graph and $A \subset V$ then

$$\mathrm{In}(A) = \{v\colon \exists a \in A \; va \in E\}$$

and $\mathrm{in}(A) = \left|\mathrm{In}(A)\right|$; similarly,

$$\mathrm{Out}(A) = \{v\colon \exists a \in A \; av \in E\}$$

and $\mathrm{out}(A) = \left|\mathrm{Out}(A)\right|$.

Since we will discuss theorems in finite combinatorics and their infinite counterparts side by side we introduce the following terminology: theorems in finite combinatorics will be enumerated as *Finite Theorem 1*, *Finite Theorem 2*, etc, and the corresponding results from infinite combinatorics will be enumerated as *Infinite Theorem 1*, *Infinite Theorem 2*, etc.

2. METHOD OF PROOFS, TRANSFER PRINCIPLES

The first example illustrates the simplest case: there is no difference between the finite and infinite theorems, moreover the same proof works in both cases, all we should do is to remove the word "finite" from both the theorem and from its proof.

2.1. Connectedness

Finite Theorem 1. *A finite graph $G = (V, E)$ is connected iff given any partition (V_0, V_1) of the vertices into two non-empty sets there is an edge between V_0 and V_1.*

Proof. A connected graph clearly has this property.

To see the other direction let $x \in V$ and put

$$A = \{z \in V : \text{there is an } x\text{-}z\text{-path in } G\}.$$

Since there is no edge between A and $V \setminus A$ and $a \in A$, we have $A = V$. Thus from x there is a path to each vertex of G. ∎

Infinite Theorem 1. *A graph $G = (V, E)$ is connected iff given any partition (V_0, V_1) of the vertices into two non-empty sets there is an edge between V_0 and V_1.*

The same proof works.

The next example is – at least for the first sight – very similar.

2.2. Spanning trees

Finite Theorem 2. *Every finite connected graph $G = (V, E)$ has a spanning tree.*

We "know" that the same statement holds for arbitrary graphs:

Infinite Theorem 2. *Every connected graph $G = (V, E)$ has a spanning tree.*

But, as we will see soon, the relationship between theirs proofs is more delicate. The "finite theorem" has (at least) two different proofs:

First Proof. Let $T = (V, F)$ be a minimal connected subgraph of G. Then T can not contain a circle, so it is a spanning tree. ∎

The method of this proof can not be applied to get the "infinite" version because it is not easy to guarantee that there is a minimal connected subgraph of an infinite graph: an infinite graph G may contain a decreasing chain $G_0, G_1, \ldots$ of connected subgraphs of G such that $V(G_i) = V(G)$ but $\cap_{i \in \mathbb{N}} E(G_i) = \emptyset$.

Now consider the second proof of the finite theorem.

Second Proof. Let $T = (V', E')$ be a maximal subtree of G. Since there is no edge between V' and $V \setminus V'$ we have $V' = V$. Hence T is a spanning tree. ∎

This proof can be modified to get the infinite theorem:

Proof. Let $\mathcal{T}$ be the family of subtrees of G. For $T, T' \in \mathcal{T}$ write $T \prec T'$ iff T is a subtree of T'.

Since $\mathcal{T}$ is closed under increasing union, $\langle \mathcal{T}, \prec \rangle$ has a maximal element $T = (V', E')$ by Zorn's Lemma. Since there is no edge between V' and $V \setminus V'$ we have $V' = V$. Hence T is a spanning tree. ∎

So almost the same proof works, but we used Zorn's Lemma (i.e. Axiom of Choice) It is a natural question whether we really need the Axiom of Choice? The next theorem gives the answer:

Theorem 2.1 (ZF). *If every connected graph has a spanning tree then Axiom of Choice holds.*

Proof. Let $\mathcal{A} = \{A_i : i \in I\}$ be a family of non-empty sets. We want to find a choice function.

First we can assume that the elements of $\mathcal{A}$ are pairwise disjoint.

Construct a graph $G = (V, E)$ as follows. Let

$$V = \{x\} \cup \{y_i, z_i : i \in I\} \cup \cup\{A_i : i \in I\},$$

where $\{x\} \cup \{y_i, z_i : i \in I\}$ are new, pairwise different vertices, and put

$$E = \{xy_i : i \in I\} \cup \cup_{i \in I}\{z_i a, a y_i : a \in A_i\}.$$

Then G is connected, so, by the assumption, it has a spanning tree $T = (V, F)$. Then

(i) $\{xy_i : i \in I\} \subset F$,

(ii) for each $i \in I$ there is exactly one $a_i \in A_i$ such that $z_i a_i, a_i y_i \in F$,

(iii) for each $a \in A_i \setminus \{a_i\}$ we have $z_i a \in F$ iff $ay_i \notin F$.

Thus $f(i) = a_i$ is a choice function for $\mathcal{A}$ and f is definable using T. ∎

So it was a case when we have the same theorem for finite and infinite. Even the proofs are almost the same, but in the infinite case we should use Axiom of Choice to get some maximal structure.

Next we will see an example when the finite case has a straightforward generalization for the countable case, but there is no way to get some similar result for uncountable graphs.

2.3. Normal spanning tree

A *normal spanning tree* (or *depth-first search tree*) of a connected graph $G = (V, E)$ is a rooted subtree T of G such that for each edge $xy \in E$ the endpoints x and y are comparable in the rooted tree order.

Finite Theorem 3. *Every finite connected graph has a normal spanning tree.*

Proof. Apply the depth-first algorithm to construct a normal spanning tree. ∎

What about the infinite graphs? Well, the complete graph on $\aleph_1$, $K_{\aleph_1}$, does not have a normal spanning tree. On the other hand, we have

Infinite Theorem 3. *Every countable connected graph has a normal spanning tree.*

The simple greedy depth-first algorithm may not work even in $K_{\aleph_0}$ because it may find an infinite path which does not contain all the vertices. However an inductive algorithm may works: using a carefully chosen ordering we can guarantee that all the vertices is included in some finite step into the spanning tree.

So far we have seen examples when either we have the same statement for finite and infinite graph, or it was clear that certain statements simply fail for uncountable graphs.

2.4. Pseudo-winners in tournaments

Given a directed graph $G = (V, E)$ for $A \subset V$ let $\mathrm{Out}_1(A) = A \cup \mathrm{Out}(A)$ and $\mathrm{Out}_n(A) = \mathrm{Out}_1\big(\mathrm{Out}_{n-1}(A)\big)$ for $n > 1$, i.e. $v \in \mathrm{Out}_n(A)$ iff there is a path of length at most n which leads from some elements of A to v. Similarly, let $\mathrm{In}_1(A) = A \cup \mathrm{In}(A)$ and $\mathrm{In}_n(A) = \mathrm{In}_1\big(\mathrm{In}_{n-1}(A)\big)$ for $n > 1$. If $A = \{v\}$ write $\mathrm{Out}_n(v)$ for $\mathrm{Out}_n\big(\{v\}\big)$, and $\mathrm{In}_n(v)$ for $\mathrm{In}_n\big(\{v\}\big)$.

Let $T = (V, E)$ be a tournament and let $t \in V$. We say that t is a *pseudo-winner* iff $\mathrm{Out}_2(t) = V$.

Finite Theorem 4. *Every finite tournament has a pseudo-winner.*

Proof. If t has maximal out-degree then t is a pseudo-winner.

Indeed, let $v \in V$. If $tv \in E$ then $v \in \mathrm{Out}(t) \subset \mathrm{Out}_2(t)$.

If $vt \in E$ then $t \in \mathrm{Out}(v) \setminus \mathrm{Out}(t)$, so there is $s \in \mathrm{Out}(t) \setminus \mathrm{Out}(v)$ because $\big|\mathrm{Out}(t)\big|$ was maximal. But then tsv is a directed path of length 2, and so $v \in \mathrm{Out}_2(t)$. ∎

Now consider the infinite case. The simplest generalization fails because

Observation 2.2. *There is no pseudo-winner in the tournament $(\mathbb{Z}, <)$.*

However, in $(\mathbb{Z}, <)$ we have $\mathbb{Z} = \mathrm{In}_1(0) \cup \mathrm{Out}_1(1)$. As it turns out, this behavior of $\mathbb{Z}$ is not exceptional.

Infinite Theorem 4. *A tournament $T = (V, E)$ contains a pseudo-winner or there are $x \neq y \in V$ such that $V = \mathrm{Out}_1(x) \cup \mathrm{In}_1(y)$.*

Proof. Indeed, if y is not a pseudo-winner witnessed by x, i.e. $x \notin \mathrm{Out}_2(y)$, then $V = \mathrm{Out}_1(x) \cup \mathrm{In}_1(y)$. ∎

One can find other interesting results concerning the structure of infinite tournaments, see e.g. [13]:

Theorem 2.3. (1) *Let $T = (V, E)$ be an infinite tournament. If $V = \mathrm{Out}_n(v)$ for some $n \geq 3$ and $v \in V$ then $V = \mathrm{Out}_3(w)$ for some $w \in V$. (2) There is an infinite tournament $T = (V, E)$ such that $V = \mathrm{Out}_3(v)$ for some $v \in V$ but $V \neq \mathrm{Out}_2(w)$ for any $w \in V$.*

This was an example when the finite and the infinite theorems are quite different. But the infinite case is also easy provided you know what you have to prove.

The next subsection contains an example when the infinite theorem is still open.

2.5. Quasi-kernels in digraphs

Let $G = (V, E)$ be a digraph, $A, B \subset V$. An independent set $A \subset V$ is a *quasi-kernel (quasi-sink)* iff $V = \mathrm{Out}_2(A)$ $(V = \mathrm{In}_2(A))$.

Finite Theorem 5 (Chvátal, Lovász, [5]). *Every finite digraph has a quasi-kernel.*

The simple generalization fails even for infinite tournaments: the tournament $(\mathbb{Z}, <)$ is a counterexample.

For infinite tournament it was easy to find an infinite version of this theorem but what is the right generalization for infinite digraphs?

Theorem 4 implies that *if $G = (V, E)$ is an infinite tournament then there are point $x \neq y \in V$ s.t. $V = \mathrm{Out}_2(\{x\}) \cup \mathrm{In}_2(\{y\})$.* One can guess that this formulation gives us the right infinite version of the theorem of Chvátal and Lovász, namely we conjectured that *every digraph contains two disjoint independent sets, A and B such that $V = \mathrm{Out}_2(A) \cup \mathrm{In}_2(B)$.*

For a while we tried to find a counterexample, but at some point we've found a quite easy way to show that:

Infinite Theorem 5 (P. L. Erdős, A. Hajnal, –, [13]). *Every digraph contains two disjoint independent sets, A and B such that $V = \mathrm{Out}_2(A) \cup \mathrm{In}_2(B)$.*

However, during hunting counterexamples we realized that all the digraphs we could construct have a much stronger property which led us to the formulation of the following conjecture.

Kernel–Sink Conjecture. *Given a directed graph $G = (V, E)$ there is a partition (W_0, W_1) of V such that $G[W_0]$ has a quasi-kernel and $G[W_1]$ has a quasi-sink.*

Let us remark that theorem 4 implies this statement for infinite tournaments. In [13] we prove that this conjecture holds for different classes of infinite graphs, but the conjecture is still open.

In the next subsection we will see a problem when the finite case is trivial, the general infinite case is hard but solved, however the countable case is completely open.

2.6. Unfriendly partitions

Let $G = (V, E)$ be a graph. A partition (A, B) of V is called *unfriendly* iff every vertex has at least as many neighbor in the other class as in its own.

Finite Theorem 6. *Every finite graph has an unfriendly partition.*

Proof. Take a partition having maximal number of edges between the classes of the partition. This partition should be unfriendly. ∎

After proving that large classes of infinite graphs have unfriendly partitions it was natural to formulate the following conjecture, [2]

Unfriendly Partition Conjecture. *Every graph has an unfriendly partition.*

However, this conjecture was refuted:

Infinite Theorem 6.1 (Shelah, [18]). *There is an uncountable graph without unfriendly partitions.*

Having disproved the plain generalization what are the other possibilities?

Infinite Theorem 6.2 (Shelah[18]). *Every graph has a partition into three pieces such that every vertex has at least as many neighbor in the two other classes as in its own.*

Or you can get a positive theorem for infinite graphs provided you consider only graphs which are similar to a finite graph. A graph is called *locally finite* iff every vertex has finite degree.

Infinite Theorem 6.3. *Every locally finite graph has an unfriendly partition.*

Proof. We will apply Gödel's Compactness Theorem below.

Gödel's Compactness Theorem. *A first order theory T has a model iff every finite subset of T has a model.*

In many cases (including this one) you can substitute Gödel's Compactness Theorem by other results, e.g. by König's Lemma, but I think that the familiarity with Gödel's Compactness Theorem is very useful if one wants to do infinite combinatorics.

So let $G = (V, E)$ a locally finite graph.

Consider the following first order language $\mathcal{L}$: $\{c_v \colon v \in V\}$ is the set of constant symbols, and R_A and R_B are unary relation symbols.

Define the following formulas:

$$\psi \colon \forall x \, \big(R_A(x) \leftrightarrow \neg R_B(x) \big),$$

for all $v \in V$ write $\mathcal{F}_v = \big\{ F \subset E(v) \colon |F| \geq d(v)/2 \big\}$ and put

$$\varphi_{v,A} \colon R_A(c_v) \to \bigvee_{F \in \mathcal{F}_v} \bigwedge_{x \in F} R_B(c_x),$$

$$\varphi_{v,B} \colon R_B(c_v) \to \bigvee_{F \in \mathcal{F}_v} \bigwedge_{x \in F} R_A(c_x).$$

Now define our theory T as follows: $T = \{\psi, \varphi_{v,A}, \varphi_{v,B} \colon v \in V\}$.

Claim 1. *Every $T' \in [T]^{<\omega}$ has a model.*

Indeed, let $W = \{v \colon c_v \text{ occurs in } T'\}$. Then $G[W]$ has an unfriendly partition (A, B). Let M be the following model: the underlying set M is W, c_v is interpreted as v for $v \in W$, and R_A is interpreted as A and R_B is interpreted as B. Then $M \models T'$.

Using the Claim and Gödel's Compactness Theorem we obtain that T has a model M. Let $A = \big\{ v \in V \colon M \models R_A(c_v) \big\}$ and $B = \big\{ v \in V \colon M \models R_B(c_v) \big\}$.

Then (A, B) is a partition because ψ holds in M. Moreover if $v \in A$ then v has at least as many neighbor in B as in A because $\varphi_{v,A}$ holds.

Hence (A, B) is an unfriendly partition of G. $\blacksquare$

Shelah's counterexample is uncountable which led to the following reformulation of the refuted conjecture:

Unfriendly Partition Conjecture, Revised. *Every countable graph has an unfriendly partition.*

Let us remark that if $G = (V, E)$ is countable and every $v \in V$ has infinite degree then G clearly has an unfriendly partition. We have seen that G has an unfriendly partition if every vertex has finite degree. So the hard case is the "mixed" countable case.

So far the revised unfriendly partition conjecture is completely open.

2.7. Splitting antichains

Given a poset P an element $y \in P$ is a *cutting point* iff $\exists x, z \in P$ such that $x <_P y <_P z$ and $[x, z] = [x, y] \cup [y, z]$. P is *cut-free* if there is no cutting point in P.

A maximal antichain $A \subset P$ *splits* iff A has a partition $A = B \cup^* C$ such that $P = B^\uparrow \cup C^\downarrow$, i.e. for each $p \in P$ we have either $b \leq p$ for some $b \in B$ or $p \leq c$ for some $c \in C$.

Finite Theorem 7 (Ahlswede, R.; Erdős, P. L.; Graham, Niall, [3]). *In a finite cut-free poset every finite maximal antichain splits.*

The plain generalization fails for infinite posets. In fact, in [12] it was proved that if P is an infinite cut-free poset whose order structure is "rich enough" then there are both splitting and non-splitting maximal antichains in P.

As usual, the method of a successful generalization for infinite posets was to keep finite certain key structures as follows.

An antichain A in a poset P is *locally finite* iff every element of P is comparable to only finitely many elements of the antichain.

Infinite Theorem 7 (P. L. Erdős, –, [12]). *In a cut-free poset every locally finite maximal antichain splits.*

The interesting point in this generalization is that we do not know how to prove this infinite theorem from the finite one just using Gödel Compactness Theorem! The problem is that if P is cut-free and $Q \subset P$ is finite then there is no way to find a cut-free finite $Q' \supset Q$.

In the next section we will see a problem when we have theorems for the uncountable infinite case but the finite case is harder than even the countable infinite.

2.8. Chromatic number of product of graphs

Hedetniemi's Conjecture. *If G and H are finite graphs then $\chi(G \times H) = \min\left\{\chi(G), \chi(M)\right\}$.*

There are only partial results, e.g.

Finite Theorem 8 (El-Sahar, Sauer). *If* $\min\{\chi(G), \chi(H)\} \geq 4$ *then* $\chi(G \times H) \geq 4$.

Consider first the countable infinite case.

Infinite Theorem 8.1 (Hajnal). *If* $\chi(G), \chi(H) \geq \omega$ *then* $\chi(G \times H) \geq \omega$.

On the other hand, there are counterexamples for uncountable cardinalities:

Infinite Theorem 8.2 (Hajnal, [16]). *There are two ω_1-chromatic graphs G and H on ω_1 such that* $\chi(G \times H) = \omega$.

The construction is based on the existence of disjoint stationary subsets of ω_1.

Infinite Theorem 8.3 (–, [19]). *It is consistent with GCH that there are two ω_2-chromatic graphs G and H on ω_2 s.t.* $\chi(G \times H) = \omega$.

The proof is a forcing construction.

However, there are open problems even for the uncountable cases, e.g.:

Problem 2.4. Is it consistent with GCH that there are two ω_3-chromatic graphs G and H on ω_3 s.t. $\chi(G \times H) = \omega$?

3. CLASSICAL THEOREMS

In this section we investigate the relation between four classical theorems and their infinite versions.

3.1. Euler trails and Euler circles

In a graph G an *Euler circuit* is a circuit containing all the edges of G. An *Euler trail* is a trail containing all the edges of G.

Finite Theorem 9. (1) *A finite connected graph has an Euler-circle iff the graph is Eulerian, i.e. each vertex has even degree.*

(2) *A finite connected graph has an Euler-trail with end-vertices $v \neq w$ iff v and w are the only vertices of odd degree.*

First one should find an infinite version of the notion of Euler trails (Euler circuits).

A *one-way infinite Euler trail* T in a graph G is a one-way infinite sequence $T = (x_0, x_1 \dots,)$ of vertices such that $E(T) = \{x_i x_{i+1} : i \in \mathbb{N}\}$ is a 1–1 enumeration of the edges of G. x_0 is the *end-vertex* of the trail. A *two-way infinite Euler trail* T in a graph G is a two-way infinite sequence $T = (\dots, x_{-2}, x_{-1}, x_0, x_1 \dots,)$ of vertices such that $\{x_i x_{i+1} : i \in \mathbb{Z}\}$ is a 1–1 enumeration of the edges of G.

Problem 3.1 (König). When does a countable infinite graph G contain a one/two-way infinite Euler trail?

The plain generalizations of the finite theorems fail for infinite graphs (see Figure 1 below): in the first graph G each vertex has even degree, but there is no two-way infinite Euler trail, in the second graph H there is exactly one vertex with odd degree but there is no one-way infinite Euler trail.

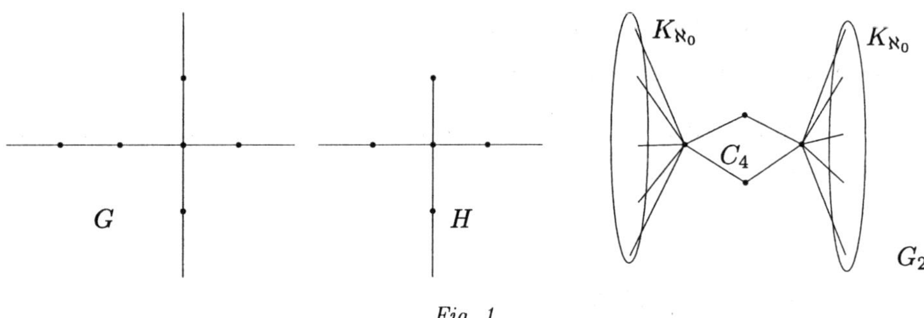

Fig. 1

Infinite Theorem 9 (Erdős, P.; Grünwald, T.; Vázsonyi, E., 1938, [9] and [10]). *A graph $G = (V, E)$ has a one-way infinite Euler trail with end-vertex $v \in V$ iff* (o1)–(o4) *below hold:*
(o1) G *is connected,* $\big| E(G) \big| = \aleph_0$,
(o2) $d_G(v)$ *is odd or infinite,*
(o3) $d_G(v')$ *is even or infinite for each* $v' \in V(G) \setminus \{v\}$,
(o4) $G \setminus E'$ *has one infinite component for each finite* $E' \subset E$.

To simplify our notation we will write $\mathrm{owit}(G, v)$ to mean that (o1)–(o4) above hold for G and v.

If $G = (V, E)$ is a graph and T is a trail in G define the graph $G \setminus T = (V', E')$ as follows: $E' = E \setminus E(T)$ and $V' = \big\{ v \in V : d_{E \setminus E(T)}(v) > 0 \big\}$, i.e. remove the isolated vertices from the graph $\big(V, E \setminus E(T) \big)$.

Proof. The assumptions (o1)–(o4) are clearly necessary.

Assume now that $\mathrm{owit}(G, v)$ holds. The key step of the proof is the following lemma:

Lemma 3.2. *Assume that G is a graph, $v \in V(G)$, $e \in E(G)$ and $\mathrm{owit}(G, v)$ holds. Then there is a trail T with endpoints v and v^* such that $e \in E(T)$ and $\mathrm{owit}(G \setminus T, v^*)$ holds.*

Proof. Since G is connected, there is an endpoint v^* of e and a trail T' in G from v to v^* such that e is the last edge of T'.

Let $G^* = G \setminus T'$. In T' two vertices, v and v^* have odd degree. Hence, by (o2) and (o3), in G^* only one vertex, v^* may have odd degree, and the degree of v^* in G^* is either infinite or odd. So the component G' of v^* in G^* should be infinite because a finite component can not contain exactly one vertex with odd degree. By (o4), all the other components of G^* are finite. Moreover, all these finite components should be Eulerian because in G^* only one vertex, v^* may have odd degree. Let H be the union of T' and the finite components of G^*. This is a connected finite graph in which exactly two vertices, v and v^* have odd degrees. Hence in H there is an Euler-trail T from v to v^*. Then $G \setminus T = G'$. We show that $\mathrm{owit}(G \setminus T, v^*)$ holds. (o1) holds because G' is a component of G^* so it is connected. Since $d_G(x) = d_{G \setminus T}(x) + d_T(x)$ for each $x \in V$, and $d_T(x)$ is odd iff $x = v$ or $x = v^*$, an easy computation gives that (o2) and (o3) also hold for $G \setminus T$. If F is a finite set of edges of $G \setminus T$ then $(G \setminus T) \setminus F = G \setminus (E(T) \cup F)$ so, applying (o4) for G, we obtain that $(G \setminus T) \setminus F$ has only one infinite component. Hence (o4) also holds for $G \setminus T$. Hence T satisfies the requirements. ∎

Using this lemma an easy inductive construction gives a one-way infinite Euler trail in G because G has just countably many edges. ∎

Infinite Theorem 10 (Erdős, P; Grünwald, T.; Vázsonyi, E., 1938, [9] and [10]). *A graph G has a two-way infinite Euler trail iff* (t1)–(t4) *below hold:*
(t1) *G is connected,* $\big| E(G) \big| = \aleph_0$,
(t2) *$d_G(v)$ is even or infinite for each $v' \in V(G)$,*
(t3) *$G \setminus E'$ has at most two infinite components for each finite $E' \subset E$,*
(t4) *$G \setminus E'$ has one infinite component for a finite $E' \subset E$ provided that every degree is even in (V, E').*

We will write $\mathrm{twit}(G)$ to mean that the stipulations (t1)–(t4) above hold for G.

The third graph G_2 on figure 1 shows that we really need to assume (t4): G_2 satisfies (t1)–(t3) but it does not have a two-way infinite Euler trail.

Proof. The assumptions (t1)–(t3) are clearly necessary. To check (t4) assume that $E' \subset E$ is finite such that every degree is even in (V, E'). Let $T = (\ldots, x_{-2}, x_{-1}, x_0, x_1 \ldots,)$ be a two-way infinite Euler line in G. Fix $n \in \mathbb{N}$ such that $E' \subset E_n$, where $E_n = \{x_i x_{i+i}\colon -n \le i < n\}$. Consider the graph $G_n = (V_n, E_n \setminus E')$ where $V_n = \{x_i\colon -n \le i \le n\}$. Then in G_n only the vertices x_{-n} and x_n have odd degree, hence they are in the same connected component. Hence in $G \setminus E'$ the connected component of x_{-n} and x_n contains $V \setminus V_n$, and so there is only one infinite component.

Assume now that $\mathrm{twit}(G)$ holds. We should distinguish two cases.

Case 1.

(∗) *For each finite trail T the graph $G \setminus T$ has one infinite component.*

Lemma 3.3. *Let G be a graph, $v \in V(G)$ and $e \in E(G)$. If $\mathrm{twit}(G)$ and (∗) hold then there is a circuit T in G such that $v \in V(T)$, $e \in E(T)$ and $\mathrm{twit}(G \setminus T)$.*

Proof of the Lemma. Since G is connected, there is a trail T' in G from v to some endpoint v' of e such that e is the last edge of that trail.

Then in $G \setminus T'$ at most two vertices, v and v' may have odd degree. The vertices v and v' can not be in different connected component of $G \setminus T'$. Otherwise one of that components would be finite and would contain exactly one vertex with odd degree, which is impossible. So there is a path S from v to v' in $G \setminus T'$. Then $T'' = S \cup T'$ is a circuit. Let G' be the infinite component of $G \setminus T''$. Clearly all the finite components of $G \setminus T''$ should be Eulerian. Let H is the union of T'' and the finite components of $G \setminus T''$. This is a connected Eulerian finite graph. Let T be an Euler circle of H. Since $G \setminus T'$ had exactly one infinite component, the graph $G \setminus T$ is just that component. Hence $\mathrm{twit}(G \setminus T)$ holds. Hence T satisfies the requirements of the lemma. ∎

By the lemma above there are a sequence $\{v_i\colon i < \omega\}$ of vertices and edge-disjoint circuits $\{T_i\colon i < \omega\}$ in G such that

(a) $x_i, x_{i+1} \in V(T_i)$ for $i < \omega$,

(b) $E(G) = \bigcup \{E(T_i)\colon i < \omega\}$.

Using these circuits we can easily put together a two-way infinite Euler trail.

Case 2.

(∗) *There is a finite trail T such that the graph $G \setminus T$ has two infinite components.*

Let v_1 and v_2 be the endpoints of T. These vertices should be in different components of $G \setminus T$ otherwise there were a circuit $T' \supset T$ containing v_1 and v_2 and so $G \setminus T'$ would have two infinite components, which contradicts (t4).

Let G_1 and G_2 be the components of v_1 and v_2, respectively, in $G \setminus T$. G_1 and G_2 should be infinite since a finite graph can not contain exactly one vertex with odd degree.

Hence all the finite components of $G \setminus T$ should be Eulerian. Let H be the union of T and these finite components. This is a connected finite graph in which exactly two vertices, v and v' have odd degree. Hence in H there is an Euler-trail T' from v_1 to v_2. Then the graphs G_1, G_2 and T' are edge disjoint,

(A) $E(G) = E(G_1) \cup E(G_2) \cup E(T')$,

(B) $\text{owit}(G_1, v_1)$ and $\text{owit}(G_2, v_2)$ hold.

Hence in G_i there is a one way infinite Euler trail T_i with end-vertex v_i, for $i = 1, 2$. Thus the concatenation of T_1, T' and T_2 is a two-way infinite Euler trail in G. ∎

3.2. Covering and matching

Given a graph $G = (V, E)$ a set of edges is *independent* if no two elements are adjacent. If an edge e is incident with a vertex x we say that x *covers* e and e *covers* x. Given a graph $G = (V, E)$ a set $A \subset V$ is *matchable* into $B \subset V$ iff there is a set F of independent edges between A and B such that F covers A, i.e. every $a \in A$ is covered by some $e \in F$.

If G is bipartite with bipartition $V = W \cup^* M$ we will write $G = (M, W, E)$.

Hall's Theorem. *In a finite bipartite graph $G = (M, W, E)$ the set M is matchable into W iff $\left| \Gamma_G(A) \right| \geq |A|$ for each $A \subset M$.*

König's Theorem. *In a finite bipartite graph $G = (M, W, E)$*

$$\max \left\{ |F| \colon F \subset E \text{ is independent } \right\}$$

$$= \min \left\{ |C| \colon C \subset M \cup W \text{ and } C \text{ covers } E \right\}.$$

Menger's Theorem. *If $G = (V, E)$ is a finite graph, and v and w are non-adjacent vertices in G then*

$$\min \big\{ |X| : X \subset V \text{ separates } v \text{ and } w \big\}$$

$$= \max \big\{ |\mathcal{P}| : \mathcal{P} \text{ is a family of vertex-disjoint } v\text{-}w\text{-paths} \big\}.$$

The plain generalizations of König's and Menger's Theorems hold for infinite graphs, but as Erdős observed these "generalizations" says almost nothing about infinite graphs. Indeed, consider the infinite version of König's Theorem: if a maximal independent family F of edges of G is infinite then let C be just the set of end-points of the elements of F. Then C clearly covers all the edges by the maximality of F and $|C| = 2|F| = |F|$ because $|F|$ was infinite.

However, (finite) König's and Menger's Theorems can be reformulated in such a way that the plain infinite versions of the reformulated theorems are deep results.

König's Theorem, reformulated. *In a finite bipartite graph $G = (M, W, E)$ there is an independent set $F \subset E$ and a set $C \subset M \cup W$ which covers E such that $|e \cap C| = 1$ for each $e \in F$.*

In 1984 Aharoni proved the infinite version of this reformulation:

Infinite König's Theorem (R. Aharoni). *Every bipartite graph $G = (M, W, E)$ has an independent set F of edges and a set $C \subset M \cup W$ which covers E such that $|e \cap C| = 1$ for each $e \in F$.*

Menger's Theorem, reformulated. *If $G = (V, E)$ is a finite graph, and v and w are non-adjacent vertices in G, then there is a v-w-separating set X and there is a family $\mathcal{P}$ of vertex-disjoint v-w-paths such that $|P \cap X| = 1$ for each $P \in \mathcal{P}$.*

Based on this reformulation Erdős formulated the Erdős–Menger conjecture, which was proved by Aharoni and Berger in 2005:

Infinite Menger's Theorem (Aharoni, Berger, [1]). *If $G = (V, E)$ is an arbitrary graph, and v and w are non-adjacent vertices in G, then there is a v-w-separating set X and there is a family $\mathcal{P}$ of vertex-disjoint v-w-paths such that $|P \cap X| = 1$ for each $P \in \mathcal{P}$.*

There is a different problem with the plain generalization of Hall's Theorem: namely it fails! Indeed, consider the following "playboy" example: $M = \{m_i : i \geq 0\}$ $W = \{w_i : i \geq 1\}$ $E = \{(m_i, w_i) : i \geq 1\} \cup \{(m_0, w_i) : i \geq 1\}$, and let $G = (M, W, E)$. Then $|\Gamma_G(A)| \geq |A|$ for each $A \subset M$, but M is not matchable into W. The problem is that $A = \{m_i : i \geq 1\} =\subsetneq M$ has the property that every matching of A covers W.

But as it turned out, this is the only possible problem:

Infinite Hall's Theorem (Aharoni, 1984). *If in a bipartite graph $G = (M, W, E)$ the set M does not have a matching then there is $X \subset M$ such that X is unmatchable but $\Gamma_G(X)$ is matchable into X.*

It is worth to note that for finite graphs Aharoni's theorem above is just the classical Hall's Theorem. Indeed, if $\Gamma_G(X)$ is matchable into X then $|\Gamma_G(X)| \leq |X|$. So since the matching is not a bijection and $|X|$ is finite we have $|\Gamma_G(X)| < |X|$.

4. Multi-way Cuts

In this section we will see that some plain generalization holds, the countable case is not harder than the finite. However, the uncountable case will demand a model-theoretic method.

Given a graph $G = (V, E)$ and $S \subset V$ let $G - S = G[V \setminus S]$, i.e. the induced subgraph on $V \setminus S$. An *S-colouring* of G is a function $f : V \longrightarrow S$ with $f \restriction S = \mathrm{id}_S$, i.e. f is the identity on S. The *value $e_G(f)$* of an S-colouring f is the number of *bi-chromatic* edges, i.e. the number of edges whose endpoints have different colours.

If G is finite, let

$$\pi_{G,S} = \min\{e_G(f) : f \text{ is an } S\text{-colouring.}\}$$

Multiway Cut Problem. *Given a finite graph $G = (V, E)$ and a non-empty set $S \subset V$ determine $\pi_{G,S}$!*

This problem is NP-complete, [6]. However, there are some lower bounds for $\pi_{G,S}$.

The lower bound $\nu_{G,S}$ was introduced and studied in [14] and in [11].

Let $\vec{G}$ be a directed graph obtained by an orientation of the edges of G. For each $s \in S$ let $\mathcal{P}_s$ be a family of edge-disjoint directed paths from s into some element of $S \setminus \{s\}$ in $\vec{G}$. Put $\mathcal{P} = \cup\{\mathcal{P}_s : s \in S\}$. Let f be an arbitrary S-colouring. Then

($\bullet$) *there is an injection $e_{\vec{G}}$ from $\mathcal{P}$ into the set of f-bi-chromatic edges.*

Indeed, for each $P \in \mathcal{P}_s \subset \mathcal{P}$ let $e_{\vec{G}}(P)$ be the first f-bi-chromatic edge of the path P.

Hence, if we define $\nu_{G,S}$ as the maximum of $|\mathcal{P}|$ where the maximum is taken over all orientations $\vec{G}$ of G, then we have

$$\nu_{G,S} \leq \pi_{G,S}.$$

Finite Theorem 10 (Erdős, P. L.; Frank, A.; Székely, L. [11]). *If $G = (V, E)$ is a finite graph and $S \subset V$ has at least two elements such that $G - S$ is a tree then $\nu_{G,S} = \pi_{G,S}$.*

If you want to find an infinite version of this theorem you can easily recognize that cardinality is "too coarse" invariant, see just the argument after the first infinite version of König's Theorem.

However, ($\bullet$) holds even for an infinite graph, and the finite theorem can be reformulated as follows: *if $G = (V, E)$ is a finite graph and $S \subset V$ has at least two elements such that $G - S$ is a tree then*

– *there is an orientation $\vec{G}$ of G, and*

– *there is a family $\mathcal{P} = \cup\{\mathcal{P}_s : s \in S\}$, where $\mathcal{P}_s$ is a family of edge-disjoint directed paths from s into some element of $S \setminus \{s\}$ in $\vec{G}$,*

such that $e_{\vec{G}}$ (as it was defined above) is a bijection between $\mathcal{P}$ and the f-bi-chromatic edges.

This is the version of the theorem which is meaningful and non-trivial even for infinite graphs.

Infinite Theorem 11. *Assume that $G = (V, E)$ is an infinite graph and $S \subset V$ is a finite subset having at least two elements such that $G - S$ is a tree which does not contain infinite paths. Then*

– *there is an orientation $\vec{G}$ of G, and*

– *there is a family $\mathcal{P} = \cup\{\mathcal{P}_s; s \in S\}$, where $\mathcal{P}_s$ is a family of edge-disjoint directed paths from s into some element of $S \setminus \{s\}$ in $\vec{G}$,*

such that $e_{\vec{G}}$ is a bijection between $\mathcal{P}$ and the f-bi-chromatic edges.

Proof. The proof tries to imitate the arguments from [11].

Consider the tree $T = G - S$.

We can assume that if ws is an edge for some $s \in S$ and $w \in T$ then w is a leaf of T and $d_G(w) = 2$ because we can subdivide the edge ws by a new node.

We can assume that every leaf of T is connected to some element of S.

Fix a vertex r as the root of T, and let $\vec{T}$ be the rooted tree order of T. Since T does not contain infinite paths, we have that $\vec{T}^*$, the inverse order of $\vec{T}$ is well-founded. Hence we can define a function $L \colon T \longrightarrow P(S) \setminus \{\emptyset\}$ by the following well-founded induction.

Assume that $L(w')$ is defined for $w <_{\vec{T}} w'$. If w is a leaf, let $L(w) = \{s \in S \colon (s, w) \in E\}$. By our assumption, we have $\big|L(w)\big| = 1$.

Assume that w is not a leaf. For each $s \in S$ let

$$K(w, s) = \{w' \colon (w, w') \in E, w <_{\vec{T}} w', s \in L(w')\},$$

then put

$$\kappa_w = \max\{\big|K(w, s)\big| \colon s \in S\}$$

and

$$L(w) = \{s \in S \colon \big|K(w, s)\big| = \kappa_w\}.$$

Since S is finite, κ_w is always defined and so $L(w) \neq \emptyset$.

Since a rooted tree order is always well-founded we can define the S-colouring f of G as follows.

For the root $r \in T$ let $f(r) \in L(r)$ be arbitrary. Assume that $f(w')$ is defined for the immediate $\vec{T}$-predecessor w' of w.

If $f(w') \in L(w)$ then let $f(w) = f(w')$. If $f(w') \notin L(w)$ then let $f(w) \in L(w)$ be arbitrary.

Next we determine the orientation of the edges of G in $\vec{G}$. We will say that an edge uv in $\vec{G}$ is an *up-edge* iff $u <_{\vec{T}} v$, it is a *down-edge* otherwise.

The bi-chromatic edges are defined to be down-edges. Now for each bi-chromatic edge uw, $w <_{\vec{T}} u$, fix an $f(u)$-monochromatic, $<_T$-increasing path Q_u from u to $f(u)$ in G. Let the edges of Q_u be all down edges. So Q_u is a directed path from $f(u)$ into u in $\vec{G}$ and the edges in Q_u are all $f(u)$-monochromatic. All the other edges of T are defined to be up-edges.

If us is an edge in G for some $s \in S$ and $u \in T$ then orient us such that $\text{in}_{\vec{G}}(u) = \text{out}_{\vec{G}}(u) = 1$.

In this way we obtained an orientation $\vec{G}$ of G. Let us denote the families of up-edges and the down-edges by $\vec{E}^{\mathrm{up}}$ and $\vec{E}^{\mathrm{down}}$, respectively.

For each $s \in S$ let

$$F_s = \left\{ uw \in \vec{G} \colon w <_{\vec{T}} u, \ s = f(u) \neq f(w) \right\}$$

and

$$A_s = \{ u \in V \colon \exists w \in V \ uw \in F_s \}.$$

Then $\mathcal{Q}_s = \{Q_u \colon u \in A_s\}$ is a family of edge disjoint directed paths in $\vec{G}$. It is enough to find a family $\mathcal{R}_s = \{R_u \colon u \in A_s\}$ of directed paths in $\vec{G}$ such that

(A) R_u is a directed path from u to some element of $S \setminus \{s\}$ and the first edge in R_u is just $uw \in F_s$,

(B) the paths $\mathcal{R}_s \cup \mathcal{Q}_s$ are pairwise edge-disjoint.

Indeed, let $\mathcal{P}_s = \{Q_u{}^\frown R_u \colon u \in A_s\}$ for $s \in S$. Then $e_{\vec{G}}(Q_u{}^\frown R_u) = uw$ where w is the $<_{\vec{T}}$-predecessor of u. Hence $e''_{\vec{G}}\mathcal{P}_s = F_s$ and so $e''_{\vec{G}}\mathcal{P}$ is just the family of bi-chromatic edges.

Let $V_s = A_s \cup \{ v \in T \colon f(v) \neq s \}$,

$$E_s = \left(\left\{ yw \in \vec{E}^{\mathrm{up}} \colon s \notin L(w) \right\} \cup \left\{ yw \in \vec{E}^{\mathrm{down}} \colon s \in L(y) \wedge f(w) \neq s \right\} \right),$$

and

$$B_s = \left\{ y \in T \colon yt \in \vec{E} \text{ for some } t \in S \setminus \{s\} \right\}.$$

Let $\vec{H}_s = (V_s, E_s)$. We want to use Theorem 4.1 below for $\vec{H}$, A_s and B_s to get the desired family $\mathcal{R}_s$ of directed paths.

Theorem 4.1. *Assume that $G = (V, E)$ is a directed graph which does not contain directed infinite walks, and A and B are disjoint vertex sets such that*

(1) $\mathrm{in}(a) = 0$ *and* $\mathrm{out}(a) = 1$ *for each* $a \in A$,

(2) $\mathrm{in}(b) = 1$ *and* $\mathrm{out}(b) = 0$ *for each* $b \in B$,

(3) $\mathrm{in}(x) \leq \mathrm{out}(x)$ *for each* $x \in V \setminus (A \cup B)$.

Then there is a family $\mathcal{R}$ of edge-disjoint paths such that

(4) $\left\{ \mathrm{first}(p) \colon p \in \mathcal{R} \right\} = A$,

(5) $\left\{ \mathrm{last}(p) \colon p \in \mathcal{R} \right\} \subset B$.

We postpone the proof of this theorem.

It is clear that (1) and (2) hold for $\vec{H}_s$, A_s and B_s. To check (3) let $u \in V_s \setminus (A_s \cup B_s)$.

If $f(u) = s$ then $in_{\vec{H}_s}(u) = 0$.

Assume that $f(u) = t \neq s$.

Let
$$C = K(u, s) \setminus K(u, t) \text{ and } D = K(u, t) \setminus K(u, s).$$

Since $f(u) \in L(u)$ we have

(*) $|D| \geq |C|$ and $|C| = |D|$ implies $s \in L(u)$.

Let x be the predecessor of u in $\vec{T}$ provided $u \neq r$.

Case 1. $u = r$ or $xu \in \vec{G}$, i.e. xu is an up-edge.

Then $\mathrm{In}_{\vec{H}_s}(u) \setminus \{x\} = C$. Indeed, if $y \in K(u, t)$ then $f(y) = t$. So yu can not be an edge from some path Q_z for some $z \in A_t$ because xu is an up-edge.

Hence $\mathrm{In}_{\vec{H}_s}(u) \leq |C| + 1$ and $\mathrm{In}_{\vec{H}_s}(u) = |C|$ if $u = r$.

Moreover $\mathrm{Out}_{\vec{H}_s}(u) \supset D$. Hence $\mathrm{Out}_{\vec{H}_s}(u) \geq |D|$.

If $u = r$ or $|C| < |D|$ then we are done. If $u \neq r$ and $|C| = |D|$ then $s \in L(u)$ and so $xu \notin E_s$. Thus $in_{\vec{H}_s}(u) \leq |C| \leq |D| \leq out_{\vec{H}_s}(u)$.

Case 2. $ux \in \vec{G}$, i.e. ux is a down-edge.

Let us start with an observation:

(*) If $|C| = |D|$ then $ux \in E_s$.

Indeed, if $|C| = |D|$ then $s \in L(u)$. Now $f(x) \neq s$ because $f(u) \neq s$ and $s \in L(u)$. Hence $ux \in E_s$.

Now for some $u <_{\vec{T}} z$ we have that zu is a down-edge from some path $Q \in \mathcal{Q}_t$.

If $z \in D$ then $s \notin L(z)$ hence $zu \notin E_s$. Hence $\mathrm{In}_{\vec{H}_s}(u) = C$. Moreover $D \setminus \{z\} \supset \mathrm{Out}_{\vec{H}_s}(u)$.

If $|D| > |C|$ then $\mathrm{Out}_{\vec{H}_s}(u) \geq |D| - 1 \geq |C| \geq \mathrm{In}_{\vec{H}_s}(u)$.

If $|D| = |C|$ then $ux \in E_s$ by (*) and so $\mathrm{Out}_{\vec{H}_s}(u) \geq (|D| - 1) + 1 = |D| \geq |C| \geq \mathrm{In}_{\vec{H}_s}(u)$.

Finally, if $z \notin D$ then $\mathrm{In}_{\vec{H}_s}(u) \leq |C| + 1$. If $|D| \geq |C| + 1$ then $\mathrm{Out}_{\vec{H}_s}(u) \geq |D| \geq |C| + 1 \geq \mathrm{In}_{\vec{H}_s}(u)$.

If $|D| = |C|$ then $ux \in E_s$ by $(*)$ and so $\mathrm{Out}_{\vec{H}_s}(u) \geq |D|+1 \geq |C|+1 \geq \mathrm{In}_{\vec{H}_s}(u)$.

Hence 4.1(3) also holds, so we apply theorem 4.1 to get a family $\mathcal{R}'_s = \{R'_u : u \in A_s\}$ of edge-disjoint paths from A_s into B_s. The only problem is that the end-points of these paths are leaves of T instead of elements of S. This problem is cured in the next step.

Let $\mathcal{R}_u = \{R'_u {}^\frown f(\,\mathrm{last}(R_u)) : u \in U\}$. Then $\mathcal{R}_u$ satisfies our requirements. ∎

Proof of Theorem 4.1. We prove the theorem by transfinite induction on $|A|$.

Assume first that A is countable, $A = \{a_n : n < \omega\}$.

Let p' be a maximal directed walk from a_0. Since G does not contain infinite walks it follows that p' is finite. Conditions (1)–(3) imply $b_0 = \mathrm{last}(p') \in B$. Thus there is a directed path $p_0 \subset p'$ with $\mathrm{first}(p_0) = a_0$ and $\mathrm{last}(p_0) = b_0$.

Let $V' = V \setminus \{a_0, b_0\}$, $A' = A \setminus \{a_0\}$, $B' = B \setminus \{b_0\}$ and $E' = E \setminus E(p_0)$. Then the directed graph $G' = (V', E')$ and the disjoint vertex sets A' and B' satisfy (1)–(3), so we can repeat the procedure above to find a directed path p_1 in G' with $\mathrm{first}(p_1) = a_1$ and $\mathrm{last}(p_1) \in B'$.

Repeating this procedure we obtain a family $\mathcal{R} = \{p_n : n \in \omega\}$ of edge-disjoint paths with $\mathrm{first}(p_n) = a_n$. Thus $\mathcal{R}$ satisfies the requirements.

Assume now that $|A| = \kappa > \omega$. The natural idea is just to fix an enumeration $\{a_\xi : \xi < \kappa\}$ of A and try to simulate the procedure above.

However, in this case we can stuck even in the case $\kappa = \omega_1$.

Consider the following example: $A = \{a_\xi : \xi < \omega_1\}$, $B = \{b_\xi : \xi < \omega_1\}$, $V = A \cup B \cup \{v\}$, and

$$E = \big\{ (a_n, v), (v, b_n) : n < \omega \big\} \cup \big\{ (a_\omega, v) \big\} \cup \big\{ (a_\xi, b_\xi) : \omega + 1 \leq \xi < \omega_1 \big\}$$

If for each $n < \omega$ in the nth step we pick the path $p_n = a_n v b_n$ then in the ωth step there is no edge-disjoint path from a_ω into B.

So instead of the direct approach we use some induction.

Let $|A| = \kappa > \omega$ and assume that the theorem holds for all triples (G', A', B') with $|A'| < \kappa$.

We will partition A into some pieces and will use the inductive hypothesis inside the pieces. However we need some tool to find the right partition.

Let θ be a large regular cardinal, typically $\theta = \left(2^{|G|}\right)^{+}$ is enough.

The *transitive closure*, $TC(x)$, of a set x is the set

$$x \cup \left(\cup(x) \cup (\cup \cup x) \ldots \right),$$

i.e. the smallest transitive set containing x as a subset.

Let $H(\theta)$ be the family of sets whose transitive closure has cardinality less than θ. Put $\mathcal{H}(\theta) = \left\langle H(\theta), \in, \lhd \right\rangle$, where $\lhd$ is a well-ordering of $H(\theta)$.

Lemma 4.2. *If* $G, A, B \in M \prec N \prec \mathcal{H}(\theta)$, $M \in N$, $|M| \subset M$, $|N| \subset N$ *and* $|N| < \kappa$ *then there is a family* $\mathcal{R}$ *of edge disjoint paths such that*

(i) $\left\{ \mathrm{first}(p) \colon p \in \mathcal{R} \right\} = A \cap (N \setminus M)$,

(ii) $\left\{ \mathrm{last}(p) \colon p \in \mathcal{R} \right\} \subset B \cap (N \setminus M)$,

(iii) $\mathrm{E}(p) \subset N \setminus M$ *for* $p \in \mathcal{R}$.

Proof of the lemma. Let

$$V' = \left(V \cap (N \setminus M) \right) \cup \left\{ w \in (V \cap M) \colon \exists v \in V \cap (N \setminus M) \; vw \in E \right\},$$

$E' = E \cap (N \setminus M)$ and $G' = (V', E')$.

Claim 4.2.1. *The graph* G' *and the vertex sets* $A \cap (N \setminus M)$ *and* $B \cap (N \setminus M)$ *satisfy* (1)–(3).

Proof of the Claim. If $a \in A \cap (N \setminus M)$ then there is exactly one edge $av \in E$ for some $v \in V$. Then $e \in N$ but $e \notin M$ because a is definable from e. Hence (1) holds. Similarly, we can obtain (2).

To check (3) let $x \in V' \setminus (A \cup B)$.

Assume first that $x \in N \setminus M$. If $\mathrm{out}_G(x) > |N|$ then $\mathrm{out}_{G'}(x) = |N|$ because $N \models$ "$\left| \mathrm{Out}_G(x) \setminus (\mathrm{Out}_G(x) \cap M) \right| = \mathrm{out}_G(x)$." Since $\mathrm{in}_{G'}(x) \leq |N|$ we have (3).

If $\mathrm{out}_G(x) \leq |N|$ then $\mathrm{in}_G(x) \leq |N|$ and so $\mathrm{Out}_G(x) \cup \mathrm{In}_G(x) \subset N$. But $xy \notin M$ for $y \in \mathrm{Out}_G(x) \cup \mathrm{In}_G(x)$ because $xy \in M$ implies $x \in M$. Hence $\mathrm{Out}_{G'}(x) = \mathrm{Out}_G(x)$ and $\mathrm{In}_{G'}(x) \subseteq \mathrm{In}_G(x)$ and so (3) holds.

Assume finally that $x \in V' \cap M$. Then $\mathrm{In}_G(x) \not\subset M$ and so $\mathrm{in}_G(x) > |M|$. If $\mathrm{out}_G(x) > |N|$ then $\mathrm{out}_{G'}(x) = |N|$ and so $\mathrm{in}_{G'}(x) \leq |N| = \mathrm{out}_{G'}(x)$. If $\mathrm{out}_G(x) \leq |N|$ then $\mathrm{Out}_G(x) \subset N$. Since $|M| < \mathrm{in}_G(x) \leq \mathrm{out}_G(x) \leq |N|$ we have $\mathrm{out}_{G'}(x) = \mathrm{out}_G(x) \geq \mathrm{in}_G(x) \geq \mathrm{in}_{G'}(x)$. ∎

Since $\big| A \cap (N \setminus M) \big| \leq |N] < \kappa$ we can apply the inductive hypothesis to find a family $\mathcal{R}$ of edge disjoint directed G'-paths satisfying (4) and (5), i.e. (i) and (ii) hold for $\mathcal{R}$. But the elements of $\mathcal{R}$ are G'-paths, so $E(p) \subset E \cap (N \cap M)$ and so (iii) also holds. This completes the proof of Lemma 4.2. ∎

Let $\langle M_\alpha \colon \alpha < \kappa \rangle$ be an increasing continuous chain of elementary submodels of $\langle H(\theta), \in \rangle$, i.e.,

(1) M_α is a elementary submodel of $\mathcal{H}(\theta)$ for $\alpha < \kappa$,

(2) $\langle M_\beta \colon \beta \leq \alpha \rangle \in M_{\alpha+1}$ for $\alpha < \kappa$,

(3) $M_\alpha = \bigcup \{ M_\beta \colon \beta < \alpha \}$ for limit α,

such that $\alpha \subset M_\alpha$, $|M_\alpha| = \alpha + \omega$ and $G, A, B \in M_0$.

For each $\alpha < \kappa$ apply the Lemma above for $M = N_\alpha$ and $N = N_{\alpha+1}$ to obtain a family $\mathcal{R}_\alpha$ of edge-disjoint paths such that

(i) $\big\{ \operatorname{first}(p) \colon p \in \mathcal{R}_\alpha \big\} = A \cap (N_{\alpha+1} \setminus N_\alpha)$,

(ii) $\big\{ \operatorname{last}(p) \colon p \in \mathcal{R}_\alpha \big\} \subset B \cap (N_{\alpha+1} \setminus N_\alpha)$,

(iii) $E(p) \subset N_{\alpha+1} \setminus N_\alpha$ for $p \in \mathcal{R}$.

Then $\mathcal{R} = \cup \{ \mathcal{R}_\alpha \colon \alpha < \kappa \}$ satisfies the requirements. (4) is clear because because $A = \cup \big\{ A \cap \big(N_{\alpha+1} \setminus N_\alpha \big) : \alpha < \kappa \big\}$ by the continuity. (5) is trivial. Finally the elements of $\mathcal{R}$ are edge-disjoint because if $p \in \mathcal{R}_\alpha$ and $q \in \mathcal{R}_\beta$ for some $\alpha < \beta < \kappa$ then $E(p) \subset N_{\alpha+1} \subset N_\beta$ and $E(q) \cap N_\beta = \emptyset$. ∎

REFERENCES

[1] R. Aharoni and E. Berger, Menger's Theorem for Infinite Graphs, *J. Graph Theory*, **50** (2005), 199–211.

[2] R. Aharoni, E. C. Milner and K. Prikry, Unfriendly partitions of a graph, *J. Combin. Theory, Ser. B*, **50** (1990), no. 1, 1–10.

[3] R. Ahlswede, P. L. Erdős and N. Graham, A splitting property of maximal antichains, *Combinatorica*, **15** (1995), no. 4, 475–480.

[4] B. Bollobás, *Modern graph theory*, Graduate Texts in Mathematics, 184, Springer-Verlag (New York, 1998).

[5] V. Chvátal and L. Lovász, *Every directed graph has a semi-kernel*, Hypergraph Seminar (Proc. First Working Sem., Ohio State Univ., Columbus, Ohio, 1972; dedicated to Arnold Ross), pp. 175. Lecture Notes in Math., Vol. 411, Springer (Berlin, 1974).

[6] E. Dalhaus, D. S. Johnson, C. H. Papadimitriou, P. D. Seymour and M. Yannakakis, *The Complexity of Multiway Cuts*, Proc. 24th Annual ACM Symp. on Theory of Computing (1992), pp. 241–251.

[7] R. Diestel, *Graph theory*, Third edition. Graduate Texts in Mathematics, 173. Springer-Verlag (Berlin, 2005).

[8] R. Diestel, *Directions in Infinite Graph Theory and Combinatorics*, Topics in Discrete Mathematics 3, Elsevier – North Holland, 1992.

[9] P. Erdős, T. Grünwald and E. Vázsonyi, Végtelen gráfok Euler vonalairól (On Euler lines of infinite graphs, in Hungarian), *Mat. Fiz. Lapok*, **43** (1936).

[10] P. Erdős, T. Grünwald and E. Vázsonyi, Über Euler-Linien unendlicher Graphen, *J. Math. Phys., Mass. Inst. Techn.*, **17** (1938), 59–75.

[11] P. L. Erdős, A. Frank and L. A. Székely, Minimum multiway cuts in trees, *Discrete Appl. Math.*, **87** (1998), no. 1–3, 67–75.

[12] P. L. Erdős and L. Soukup, How to split antichains in infinite posets, *Combinatorica*, **27** (2007), no. 2, 147–161.

[13] P. L. Erdős and L. Soukup, *Quasi-kernels and quasi-sinks in infinite graphs*, submitted.

[14] P. L. Erdős and L. A. Székely, Evolutionary trees: an integer multicommodity max-flow–min-cut theorem, *Adv. in Appl. Math.*, **13** (1992), no. 4, 375–389.

[15] A. Hajnal, *Infinite combinatorics*, Handbook of combinatorics, Vol. 1, 2, 2085–2116, Elsevier (Amsterdam, 1995).

[16] A. Hajnal, The chromatic number of the product of two $\aleph_1$-chromatic graphs can be countable, *Combinatorica*, **5** (1985), no. 2, 137–139.

[17] P. Komjáth and V. Totik, *Problems and theorems in classical set theory*, Problem Books in Mathematics, Springer (New York, 2006).

[18] S. Shelah and E. C. Milner, *Graphs with no unfriendly partitions*, A tribute to Paul Erdős, Cambridge Univ. Press (Cambridge, 1990), pp. 373–384.

[19] L. Soukup, On chromatic number of product of graphs, *Comment. Math. Univ. Carolin.*, **29** (1988), no. 1, 1–12.

Lajos Soukup

Rényi Institute
Budapest, Hungary

e-mail: soukup@renyi.hu

BOLYAI SOCIETY
MATHEMATICAL STUDIES, 17

Horizons of Combinatorics
Balatonalmádi
pp. 215–234.

The Random Walk Method for Intersecting Families

NORIHIDE TOKUSHIGE*

Let $m(n,k,r,t)$ be the maximum size of $\mathscr{F} \subset \binom{[n]}{k}$ satisfying $|F_1 \cap \cdots \cap F_r| \geq t$ for all $F_1, \ldots, F_r \in \mathscr{F}$. We report some known results about $m(n,k,r,t)$. The random walk method introduced by Frankl is a strong tool to investigate $m(n,k,r,t)$. Using a concrete example, we explain the method and how to use it.

1. Introduction

Let n, k, r and t be positive integers, and let $[n] = \{1, 2, \ldots, n\}$. A family $\mathscr{G} \subset 2^{[n]}$ is called r-wise t-intersecting if $|G_1 \cap \cdots \cap G_r| \geq t$ holds for all $G_1, \ldots, G_r \in \mathscr{G}$. Let us define a typical r-wise t-intersecting family $\mathscr{G}_i(n,r,t)$ and its k-uniform subfamily $\mathscr{F}_i(n,k,r,t)$, where $0 \leq i \leq \left\lfloor \frac{n-t}{r} \right\rfloor$, as follows:

$$\mathscr{G}_i(n,r,t) = \left\{ G \subset [n] \colon \big| G \cap [t+ri] \big| \geq t + (r-1)i \right\},$$

$$\mathscr{F}_i(n,k,r,t) = \mathscr{G}_i(n,r,t) \cap \binom{[n]}{k}.$$

Two families $\mathscr{G}, \mathscr{G}' \subset 2^{[n]}$ are said to be isomorphic, and denoted by $\mathscr{G} \cong \mathscr{G}'$, if there exists a vertex permutation τ on $[n]$ such that $\mathscr{G}' = \big\{ \{ \tau(g) \colon g \in G \} \colon G \in \mathscr{G} \big\}$.

Let $m(n,k,r,t)$ be the maximum size of k-uniform r-wise t-intersecting families on n vertices. To determine $m(n,k,r,t)$ is one of the oldest

*The author was supported by MEXT Grant-in-Aid for Scientific Research (B) 16340027.

problems in extremal set theory, which is still widely open. The case $r = 2$ was observed by Erdős–Ko–Rado [6], Frankl [10], Wilson [30], and then $m(n, k, 2, t) = \max_i \left| \mathscr{F}_i(n, k, 2, t) \right|$ was finally proved by Ahlswede and Khachatrian [2]. Frankl [8] showed $m(n, k, r, 1) = \left| \mathscr{F}_0(n, k, r, 1) \right|$ if $(r - 1)n \geq rk$. Partial results for the cases $r \geq 3$ and $t \geq 2$ are found in [14, 16, 24, 26, 27, 23, 29]. All known results suggest

$$(1) \qquad m(n, k, r, t) = \max_i \left| \mathscr{F}_i(n, k, r, t) \right|.$$

Now we introduce the p-weight version of the Erdős–Ko–Rado theorem. Throughout this paper, p and $q = 1 - p$ denote positive real numbers. For $X \subset [n]$ and a family $\mathscr{G} \subset 2^X$ we define the p-weight of $\mathscr{G}$, denoted by $w_p(\mathscr{G} : X)$, as follows:

$$w_p(\mathscr{G} : X) = \sum_{G \in \mathscr{G}} p^{|G|} q^{|X| - |G|} = \sum_{i=0}^{|X|} \left| \mathscr{G} \cap \binom{X}{i} \right| p^i q^{|X| - i}.$$

We simply write $w_p(\mathscr{G})$ for the case $X = [n]$, for example, we have $w_p\big(\mathscr{G}_0(n, r, t)\big) = p^t$.

Let $w(n, p, r, t)$ be the maximum p-weight of r-wise t-intersecting families on n vertices. It might be natural to expect

$$w(n, p, r, t) = \max_i w_p\big(\mathscr{G}_i(n, r, t)\big).$$

Ahlswede and Khachatrian proved that this is true for $r = 2$ in [3] (cf. [5, 7, 23]). This includes the Katona theorem [19] about $w(n, 1/2, 2, t)$. It is shown in [15] that

$$(2) \qquad w(n, p, r, 1) = p \quad \text{for} \quad p \leq (r - 1)/r.$$

To state some more related results let us define some collections of families as follows.

$$\mathbf{G}(n, r, t) = \left\{ \mathscr{G} \subset 2^{[n]} : \mathscr{G} \text{ is } r\text{-wise } t\text{-intersecting} \right\},$$

$$\mathbf{G}_j(n, r, t) = \left\{ \mathscr{G} \subset 2^{[n]} : \mathscr{G} \subset \mathscr{G}' \text{ for some } \mathscr{G}' \cong \mathscr{G}_j(n, r, t) \right\},$$

$$\mathbf{X}^i(n, r, t) = \mathbf{G}(n, r, t) - \bigcup_{0 \leq j \leq i} \mathbf{G}_j(n, r, t),$$

$$\mathbf{Y}^i(n, k, r, t) = \left\{ \mathscr{F} \subset \binom{[n]}{k} : \mathscr{F} \in \mathbf{X}^i(n, r, t) \right\}.$$

Finally let us define

$$m^i(n, k, r, t) = \max\left\{ |\mathscr{F}| \colon \mathscr{F} \in \mathbf{Y}^i(n, k, r, t) \right\},$$

$$w^i(n, p, r, t) = \max\left\{ w_p(\mathscr{G}) \colon \mathscr{G} \in \mathbf{X}^i(n, r, t) \right\}.$$

Ahlswede and Khachatrian [1] determined $m^0(n, k, 2, t)$ completely, extending the earlier results by Hilton–Milner [18] and Frankl [11]. Brace and Daykin [4] determined $w^0(n, 1/2, r, 1)$ and Frankl determined $w^0(n, 1/2, r, t)$ for $r \geq 5$ and $1 \leq t \leq 2^r - r - 1$; in both cases $\mathscr{G}_1(n, r, t)$ has the maximum p-weight. (But $\mathscr{G}_1$ is not always optimal for w^0, for example, we have $w^0(n, p, r, 1) > w_p(\mathscr{G}_1(n, r, 1))$ if $p > \frac{1}{2}$ and $r \leq 5$, see [28].) More results for $m^0(n, k, r, t)$ with $k/n \approx 1/2$, and $w^0(n, p, r, t)$ with $p \approx 1/2$ are found in [17, 28, 29].

In this article we will introduce the random walk method originated by Frankl, which is a strong tool to investigate $w(n, p, r, t)$. In the next section, we explain the key idea of the method. In Section 3 we prepare some tools to apply the method. Then in Section 4 we illustrate the method by determining $w(n, 1/3, 4, 36)$, and a general setup to get $w(n, p, r, t)$ will be given in Section 5. In the last section we discuss how to derive $m(n, k, r, t)$ from $w(n, p, r, t)$ when $p \approx k/n$. As a consequence, we get the following result (see Theorem 10).

Theorem 1. *Let* $p_0 \in (0, 1)$ *and* $r, t, i \in \mathbb{N}$ *be given. Suppose that* $\max_j \left\{ w_{p_0}(\mathscr{G}_j(n, r, t)) \right\}$ *is attained by* $j = i - 1$ *or* i. *Then* (W) *implies* (M).

(W) *There exist positive constants* $\gamma_0, \varepsilon_0, n_0$ *such that, for all* p *with* $|p - p_0| < \varepsilon_0$ *and all* n *with* $n \geq n_0$, *the following is true: If* $\mathscr{G} \in \mathbf{X}^i(n, r, t)$ *is shifted and* $\bigcap \mathscr{G} = \emptyset$ *then we have* $w_p(\mathscr{G}) < (1 - \gamma_0) \max \left\{ w_p(\mathscr{G}_{i-1}(n, r, t)), w_p(\mathscr{G}_i(n, r, t)) \right\}$.

(M) *There exist positive constants* ε, n_1 *such that, for all* $n > n_1$ *and* k *with* $\left| \frac{k}{n} - p_0 \right| < \varepsilon$, *we have* (1) *with equality holding only if* $\mathscr{F}_{i-1}(n, k, r, t)$ *or* $\mathscr{F}_i(n, k, r, t)$ *(up to isomorphism).*

We can in fact show (W) in some particular choices of p_0, r, t, i by the random walk method. As an example we verify (1) for $r \geq 4$, $t \leq (3^r - 2r - 1)/2$, $k/n \leq 1/3$, and n large enough (Theorem 12). Although it is still beyond our reach to determine $m(n, k, r, t)$ and $w(n, p, r, t)$ completely, we hope that the strategy described in this article will provide a better understanding of multiply intersecting families.

2. The Random Walk Method

In [10] Frankl found a way to connect the number of walks of certain types with an upper bound for the size of intersecting families. He then extended the idea to bound the size of 3-wise 2-intersecting families in [9], where the random walk method was explicitly used for the first time. One of the highlights of the method is [13], where he got many interesting results on multiply intersecting families, and most of them have no alternative proofs so far. A survey [12] by himself is highly recommended.

In this section we explain the key idea of the method. Let p and q be positive reals with $p + q = 1$, and let $\alpha_{r,p} \in (p, 1)$ be the unique root of the equation $qx^r - x + p = 0$. The random walk method is basically to use the following inequality:

$$(3) \qquad w(n, p, r, t) \leq \alpha_{r,p}^t.$$

This inequality itself is not sharp, but we often get sharp upper bounds for the p-weight of intersecting families using (3) with some additional argument.

We outline how to get (3) here. (One can find the proof in [12] (for the case $p = 1/2$) and we also include some more explanation about shifting technique etc. for convenience in the next section.) For $G \subset [n]$ we define the corresponding n-step walk on $\mathbb{Z}^2$, denoted by walk(G), as follows. The walk is from $(0,0)$ to $\big(|G|, n - |G|\big)$, and the i-th step is one unit up ($\uparrow$) if $i \in G$, or one unit to the right ($\rightarrow$) if $i \notin G$. Let $\mathscr{G} \in \mathbf{G}(n, r, t)$. We can find a shifted $\mathscr{G}^* \in \mathbf{G}(n, r, t)$ with $w_p(\mathscr{G}) = w_p(\mathscr{G}^*)$. Then, for each $G \in \mathscr{G}^*$, walk(G) touches the line $L : y = (r - 1)x + t$ (see Lemma 4). Thus we have $\mathscr{G}^* \subset \mathscr{W}_n$, where $\mathscr{W}_n = \big\{ W \subset [n] : \text{walk}(W) \text{ touches } L \big\}$. We note that $\mathscr{W}_n$ is not necessarily r-wise t-intersecting.

Now consider the infinite random walk in $\mathbb{Z}^2$ starting from $(0,0)$, taking $\uparrow$ with probability p and $\rightarrow$ with probability q at each step independently. Suppose that $\mathscr{G}$ has the maximum p-weight. Then it follows that

$$(4) \qquad w(n, p, r, t) = \sum_{G \in \mathscr{G}} p^{|G|} q^{n-|G|} \leq \sum_{W \in \mathscr{W}_n} p^{|W|} q^{n-|W|}$$

$$\leq \lim_{n \to \infty} \sum_{W \in \mathscr{W}_n} p^{|W|} q^{n-|W|}$$

$$= \mathbf{P} \, (\text{the infinite random walk touches } L) = \alpha_{r,p}^t.$$

The last equality (4) can be shown as follows. Let X_s be the probability that the infinite random walk touches the line $y = (r-1)x + s$. After the first step, we are at $(1,0)$ with probability p, or at $(0,1)$ with probability q. Thus we have

$$(5) \qquad\qquad X_t = pX_{t-1} + qX_{t+r-1}.$$

Let a_i be the number of walks from $(0,0)$ to $A_i = \bigl(i, (r-1)i + t\bigr)$ which touch L only at A_i. Then we have $X_t = \sum_{i \geq 0} a_i p^{(r-1)i+t} q^i$. To touch the line $L' \colon y = (r-1)x + t + 1$, we need to hit L somewhere, say, at A_i for the first time. Then the probability that we hit L' starting from A_i is equal to X_1. Thus we have

$$(6) \qquad\qquad X_{t+1} = \sum_{i \geq 0} \bigl(a_i p^{(r-1)i+t} q^i \bigr) X_1 = X_t X_1 = X_1^{t+1}.$$

By (5) and (6) we have $X_1 = p + qX_1^r$. This equation has unique root $X_1 = \alpha_{r,p}$ in $(0,1)$, and then (6) gives $X_t = \alpha_{r,p}^t$, which proves (4). One can also show that $a_i = \frac{t}{ri+t}\binom{ri+t}{i}$ and $\sum_{i \geq 0} a_i p^{(r-1)i+t} q^i = \alpha_{r,p}^t$ in a different way, see e.g., [22].

To consider the k-uniform version problem, let us review the very original idea of the method from [10]. Let $\mathscr{F} \subset \binom{[n]}{k}$ be 2-wise t-intersecting. Then for every $F \in \mathscr{F}$, walk(F) is from $(0,0)$ to $(n-k, k)$, which touches the line $y = x + t$. The total number of walks with this property is, by the reflection principle, equal to the total number of walks from $(-t, t)$ to $(n-k, k)$, which is $\binom{n}{k-t}$. This gives $m(n, k, 2, t) \leq \binom{n}{k-t} \leq \bigl(\frac{k}{n-k} \bigr)^t \binom{n}{k}$. On the other hand, by setting $p = \frac{k}{n}$, we have $\alpha_{2,p} = \frac{p}{q} = \frac{k}{n-k}$, and $m(n, k, 2, t) \leq \alpha_{2,p}^t \binom{n}{k}$. This suggests the following k-uniform version of (3):

$$m(n, k, r, t) \leq \alpha_{r,p}^t \binom{n}{k},$$

where $p = \frac{k}{n}$. This is true if $p < \frac{r-1}{r+1}$ is fixed and n is large enough, see [25]. We will discuss how to get $m(n, k, r, t)$ from $w(n, p, r, t)$ in the last section.

3. Tools

Let us introduce the shifting operation. For integers $1 \leq i < j \leq n$ and a family $\mathscr{G} \subset 2^{[n]}$, we define the (i,j)-shift σ_{ij} as follows:

$$\sigma_{ij}(\mathscr{G}) = \bigl\{ \sigma_{ij}(G) \colon G \in \mathscr{G} \bigr\},$$

where

$$\sigma_{ij}(G) = \begin{cases} (G - \{j\}) \cup \{i\} & \text{if } i \notin G, \ j \in G, \ (G - \{j\}) \cup \{i\} \notin \mathscr{G}, \\ G & \text{otherwise.} \end{cases}$$

This operation preserves r-wise t-intersecting property, namely, if $\mathscr{G}$ is r-wise t-intersecting, then so is $\sigma_{ij}(\mathscr{G})$. Note also that shifting does not change the p-weight, i.e., $w_p\big(\sigma_{ij}(\mathscr{G})\big) = w_p(\mathscr{G})$.

A family $\mathscr{G} \subset 2^{[n]}$ is called *shifted* if $\sigma_{ij}(\mathscr{G}) = \mathscr{G}$ for all $1 \leq i < j \leq n$, and $\mathscr{G}$ is called *tame* if it is shifted and $\bigcap \mathscr{G} = \emptyset$. Starting from a given $\mathscr{G}$ we can always get a shifted $\mathscr{G}'$ by a finite sequence of shifting operations. To see this fact, let $s(\mathscr{G}) = \sum \big\{ \sum \{g : g \in G\} : G \in \mathscr{G} \big\} \in \mathbb{N}$ and observe $s\big(\sigma_{ij}(\mathscr{G})\big) < s(\mathscr{G})$ if $\sigma_{ij}(\mathscr{G}) \neq \mathscr{G}$.

Lemma 2. $\mathbf{X}^0(n, r, t) \subset \mathbf{X}^0(n, r-1, t+1)$ and $w^0(n, p, r, t) \leq w^0(n, p, r-1, t+1)$.

Proof. Let $\mathscr{G} \in \mathbf{X}^0(n, r, t)$. Then clearly we have $\mathscr{G} \notin \mathbf{G}_0(n, r-1, t+1)$. Thus it suffices to show that $\mathscr{G} \in \mathbf{G}(n, r-1, t+1)$. If it is not, then we can find $G_1, \ldots, G_{r-1} \in \mathscr{G}$ such that $|G_1 \cap \cdots \cap G_{r-1}| = t$. But $\mathscr{G}$ is r-wise t-intersecting and so every $G \in \mathscr{G}$ must contain $G_1 \cap \cdots \cap G_{r-1}$. This means $\mathscr{G} \notin \mathbf{X}^0(n, r, t)$, a contradiction. $\blacksquare$

Lemma 3. If $\mathscr{G} \in \mathbf{X}^0(n, r, t)$ has maximum p-weight then we can find a tame $\mathscr{G}' \in \mathbf{X}^0(n, r, t)$ with $w_p(\mathscr{G}') = w_p(\mathscr{G})$.

Proof. If $\mathscr{G} \in \mathbf{X}^0(n, r, t)$ then $\mathscr{G} \in \mathbf{X}^0(n, r-1, t+1)$ by Lemma 2. We apply shifting operations to $\mathscr{G}$ to get a shifted family $\mathscr{G}' \in \mathbf{X}^0(n, r, t) \subset \mathbf{X}^0(n, r-1, t+1)$.

We have to show that $\bigcap \mathscr{G}' = \emptyset$. Otherwise we may assume that $1 \in \bigcap \mathscr{G}'$ and $H = [2, n] \notin \mathscr{G}'$. Since $\mathscr{G}'$ is p-weight maximum we can find $G_1, \ldots, G_{r-1} \in \mathscr{G}'$ such that $|G_1 \cap \cdots \cap G_{r-1} \cap H| < t$. Then we have $|G_1 \cap \cdots \cap G_{r-1}| < t + 1$, which is a contradiction. $\blacksquare$

Lemma 4. Let $\mathscr{G} \in \mathbf{G}(n, r, t)$ be shifted. Then $\mathrm{walk}(G)$ touches the line $L: y = (r-1)x + t$ for all $G \in \mathscr{G}$.

Proof. Let $H = [n] - \{t, t+r, t+2r, t+3r, \ldots\}$. Then $\mathrm{walk}(H)$ does not touch L. Moreover this walk is the maximal one with this property. Namely, if $\mathrm{walk}(F)$ does not touch L, then we can find $F' \supset F$ such that H is obtained from F' by a sequence of shifting operations.

Let $\mathcal{G} \in \mathbf{G}(n, r, t)$. Suppose that we have some $G \in \mathcal{G}$ such that walk(G) does not touch L. We may assume that $\mathcal{G}$ is size maximal, and so $G = H$. For $1 \le i < r$, let $H_i = [n] - \{t + i, t + r + i, t + 2r + i, t + 3r + i, \ldots\}$. We get H_i from H by shifting. Since $\mathcal{G}$ is shifted we have $H, H_1, \ldots, H_{r-1} \in \mathcal{G}$ and $H \cap H_1 \cap \cdots \cap H_{r-1} = [t - 1]$, which is a contradiction. $\blacksquare$

Lemma 5 [28]. *Let p, r, t_0, c be fixed constants, and let $\alpha \in (p, 1)$ be the root of the equation $qx^r - x + p = 0$. Suppose that $w(n, p, r, t_0) \le c$ holds for all $n \ge t_0$. Then we have $w(n, p, r, t) \le c\alpha^{t - t_0}$ for all $t \ge t_0$ and $n \ge t$.*

Proof. If $\mathcal{G} \subset 2^{[n]}$ is trivial r-wise t_0-intersecting, i.e., $|\bigcap \mathcal{G}| \ge t_0$, then we have $\mathcal{G} \subset \{G \subset [n]: [t_0] \subset G\}$ and $w_p(\mathcal{G}) \le p^{t_0}$. Thus we may assume that $c \ge p^{t_0}$. Note also that $p < \alpha$.

We prove the result by double induction on $s = n - t$ and t. One of the initial steps for $t = t_0$ follows from our assumption. For the other initial step for s, we prove the result for the cases $0 \le s \le r - 1$, or equivalently, $t \le n \le t + r - 1$. Suppose that $\mathcal{G} \subset 2^{[n]}$ satisfies $w_p(\mathcal{G}) = w(n, p, r, t)$. We may assume that $\mathcal{G}$ is shifted and size maximal. If $\mathcal{G}$ is trivial, i.e., $|\bigcap \mathcal{G}| \ge t$, then we have $w_p(\mathcal{G}) \le p^t = p^{t_0} p^{t - t_0} < c\alpha^{t - t_0}$ and we are done. Otherwise we have $G \in \mathcal{G}$ such that $[t] \not\subset G$, and we may assume that $G_t = [n] - \{t\} \in \mathcal{G}$ because $\mathcal{G}$ is shifted and maximal. Then again by the shiftedness we have $G_i = [n] - \{i\} \in \mathcal{G}$ for all $t \le i \le n$. This implies $\left|\bigcap_{i=t}^{n} G_i\right| = t - 1$. But this is impossible because $\mathcal{G}$ is r-wise t-intersecting and $n - t + 1 \le r$.

Next we show the induction step. Let $s \ge r$ and $t > t_0$. We show the case (s, t). We assume that the result holds for $\{(s, b): b < t\} \cup \{(a, b): a < s, b \ge t_0\}$. In particular, we can apply induction hypothesis to the case $(s, t - 1)$ and $(s - r, t + r - 1)$.

Let $\mathcal{G} \subset 2^{[n]}$ be r-wise t-intersecting. Define $\mathcal{G}_1, \mathcal{G}_{\bar{1}} \subset 2^{[2,n]}$ as follows:

$$\mathcal{G}_1 = \{G - \{1\}: 1 \in G \in \mathcal{G}\}, \quad \mathcal{G}_{\bar{1}} = \{G: 1 \notin G \in \mathcal{G}\}.$$

Then $\mathcal{G}_1$ is clearly r-wise $(t-1)$-intersecting. On the other hand, $\mathcal{G}_{\bar{1}}$ is r-wise $(t + r - 1)$-intersecting. To see this fact suppose, on the contrary, that there exist $G_2 \ldots G_{r+1} \in \mathcal{G}_{\bar{1}}$ such that $\bigcap_{i=2}^{r+1} G_i = [2, t + r - 1]$. By the shiftedness we have $G'_i = \{1\} \cup (G_i - \{i\}) \in \mathcal{G}$ for all $2 \le i \le r + 1$. But then we have $\bigcap_{i=2}^{r+1} G'_i = [t + r - 1] - [2, r + 1]$, which contradicts r-wise t-intersecting property of $\mathcal{G}$.

Note that s for $\mathscr{G}_1$ is $(n-1)-(t-1) = s$ and s for $\mathscr{G}_{\bar{1}}$ is $(n-1)-(t+r-1) = s - r$. Therefore using the induction hypothesis, we have

$$w_p(\mathscr{G}) = pw_p\big(\mathscr{G}_1 : [2,n]\big) + qw_p\big(\mathscr{G}_{\bar{1}} : [2,n]\big) \leq pc\alpha^{t-t_0-1} + qc\alpha^{t+r-t_0-1}$$

$$= c\alpha^{t-t_0-1}(p + q\alpha^r) = c\alpha^{t-t_0}. \qquad \blacksquare$$

Lemma 6. *For any $i \geq 0$ we have $w^i(n+1,p,r,t) \geq w^i(n,p,r,t)$.*

Proof. Choose $\mathscr{G} \in \mathbf{X}^i(n,r,t)$ with $w_p(\mathscr{G}) = w^i(n,p,r,t)$. Then $\mathscr{G}' := \mathscr{G} \cup \big\{ G \cup \{n+1\} : G \in \mathscr{G} \big\} \in \mathbf{X}^i(n+1,r,t)$ and $w_p\big(\mathscr{G}' : [n+1]\big) = w_p\big(\mathscr{G} : [n]\big)(q+p) = w^i(n,p,r,t)$, which means $w^i(n+1,p,r,t) \geq w^i(n,p,r,t)$. $\blacksquare$

4. An Example

As a toy example, we consider the case $r = 4$ and $t = 36$. Let $p \in (0,1)$ and $q = 1 - p$, and set $\mathscr{G}_j = \mathscr{G}_j(n,4,36)$. Simple computation shows that $w_p(\mathscr{G}_0) \geq w_p(\mathscr{G}_1)$ iff $p \leq 1/3$. To give a feel of the random walk method, we will show that

$$(7) \qquad w(n,p,4,36) = w_p(\mathscr{G}_0) = p^{36}$$

for all $n \geq 40$ and $p \leq 1/3$.

Clearly we have $w(n,p,4,36) \leq w(n,p,2,36)$, and the Ahlswede–Khachatrian result [3] already shows (7) for $p \leq 1/(t+1) = 1/37$. We can easily improve this upper bound for p using (3). Suppose that $\mathscr{G} \in \mathbf{G}(n,4,36)$. If $\mathscr{G} \in \mathbf{G}_0(n,4,36)$ then we have $w_p(\mathscr{G}) \leq p^{36}$. Otherwise we have $\mathscr{G} \in \mathbf{X}^0(n,4,36) \subset \mathbf{X}^0(n,3,37)$ by Lemma 2. Now by (3) we have $w_p(\mathscr{G}) \leq \alpha_{3,p}^{37}$. Then we find that $\alpha_{3,p}^{37} < p^{36}$ if $p \leq 1/5$. In this way we get (7) for $p \leq 1/5$.

To get (7) for $p \leq 1/3$, we will prove the following slightly stronger inequality, that is,

$$(8) \qquad w^1(n,p,4,36) < 0.9999 \max\big\{ w_p(\mathscr{G}_0), w_p(\mathscr{G}_1) \big\}$$

for all $n \geq 40$ and $p \leq 0.34$. This gives $w(n,p,4,36) = \max\big\{ w_p(\mathscr{G}_0), w_p(\mathscr{G}_1) \big\}$ for $p \leq 0.34$, and in particular this implies (7) for $p \leq 1/3$.

Choose $\mathscr{G} \in \mathbf{X}^1(n,4,36)$ with the maximum p-weight, and choose a tame $\mathscr{G}^* \in \mathbf{X}^0(n,4,36)$ with $w_p(\mathscr{G}) = w_p(\mathscr{G}^*)$ by Lemma 3. We will show the following.

(i) If $\mathcal{G}^* \not\subset \mathcal{G}_1$ then $w_p(\mathcal{G}^*) < 0.99\, w_p(\mathcal{G}_0)$ for $p \leq 0.34$.

(ii) If $\mathcal{G}^* \subset \mathcal{G}_1$ then $w_p(\mathcal{G}^*) < 0.9999\, w_p(\mathcal{G}_1)$ for $p \leq 0.34$.

We can show (ii) in a more general setting as we will see in the next section. Here we show (i). So we assume that $\mathcal{G}^* \not\subset \mathcal{G}_1$ and rename it $\mathcal{G}$.

Let $s = \max\{j \colon \mathcal{G} \in \mathbf{G}(n,3,j)\}$. By Lemma 2 we have $s \geq 37$. If $s \geq 40$ then by (3) we have

$$(9) \qquad w_p(\mathcal{G}) \leq w(n,p,3,40) \leq \alpha_{3,p}^{40} < 0.99 p^{36}$$

for $p \leq 0.34$. Thus we may assume that $37 \leq s \leq 39$. After [13] let

$$h = \min\left\{ j \colon \left|G \cap [36+j]\right| \geq 36 \text{ for all } G \in \mathcal{G}\right\}.$$

This is the maximum size of "holes" in $[36+h]$.

Claim 1. $1 \leq h \leq s - 36 \leq 3$.

Proof. Since $\mathcal{G} \in \mathbf{X}^0(n,4,36)$, we have $h \geq 1$. By the definition of s and the shiftedness of $\mathcal{G}$, we have $G_1, G_2, G_3 \in \mathcal{G}$ such that $G_1 \cap G_2 \cap G_3 = [s]$. Since $\mathcal{G} \in \mathbf{G}(n,4,36)$ it follows that $\left|G \cap [s]\right| \geq 36$ for all $G \in \mathcal{G}$, namely, $36 + h \leq s$. ■

Let $b = 36 + (h-1) = 35 + h$ and let $T_i = [b+1-i, b]$ be the right-most i-set in $[b]$ ($T_0 = \emptyset$). For $A \subset [b]$ let

$$\mathcal{G}(A) = \left\{ G \cap [b+1, n] \colon G \in \mathcal{G},\ [b] \setminus G = A \right\}.$$

Since $\mathcal{G}$ is shifted, we have $\mathcal{G}(A) \subset \mathcal{G}(T_i)$ for all $A \in \binom{[b]}{i}$, and thus we have

$$(10) \qquad w_p(\mathcal{G}) \leq \sum_{i=0}^{h} \binom{b}{i} p^{b-i} q^i\, w_p\big(\mathcal{G}(T_i) \colon [b+1, n]\big).$$

To bound $w_p\big(\mathcal{G}(T_i) \colon [b+1, n]\big)$ we use the fact that $\mathcal{G}(T_i)$ is highly-intersecting as we see below.

Claim 2. For $0 \leq i < h$ we have $\mathcal{G}(T_i) \in \mathbf{G}(n,3,3i+1)$.

Proof. Suppose that $\mathcal{G}(T_i) \notin \mathbf{G}(n,3,3i+1)$. Then we can find $G_1, G_2, G_3 \in \mathcal{G}(T_i)$ such that $|G_1 \cap G_2 \cap G_3| \leq 3i$. Since $\mathcal{G}$ is shifted, we may assume that $G_1 \cap G_2 \cap G_3 \subset [b+1, b+3i]$. For $1 \leq \ell \leq 3$, by shifting $\big(G_\ell \cup [b]\big) - T_i \in \mathcal{G}$, we get $G'_\ell := \big(G_\ell \cup [b]\big) - [b+1+(\ell-1)i, b+\ell i] \in \mathcal{G}$.

By the definition of h we have some $H \in \mathcal{G}$ such that $\left|H \cap [b]\right| < 36$ and due to the shiftedness of $\mathcal{G}$ we may assume that $H = [n] - [36, b]$. Then we have $G'_1 \cap G'_2 \cap G'_3 \cap H = [35]$, which contradicts the fact $\mathcal{G} \in \mathbf{G}(n,4,36)$. ■

Claim 3. *If $\mathscr{G} \not\subset \mathscr{G}_h$ then $\mathscr{G}(T_h) \in \mathbf{G}(n, 3, 3h + 2)$.*

Proof. Suppose that $\mathscr{G}(T_h) \notin \mathbf{G}(n, 3, w)$, where $w = 3h + 2$. Then we can find $G_1, G_2, G_3 \in \mathscr{G}(T_h)$ such that $G_1 \cap G_2 \cap G_3 \subset [b+1, b+w-1]$. By shifting $\big(G_\ell \cup [b]\big) - T_h \in \mathscr{G}$ we get $G'_\ell := \big(G_\ell \cup [b]\big) - \big[36 + (\ell - 1)h, 35 + \ell h\big] \in \mathscr{G}$ for $1 \leq \ell \leq 3$. Since $\mathscr{G} \not\subset \mathscr{G}_h$ we have $G'_4 := [n] - [36 + 3h, 36 + 4h] \in \mathscr{G}$. Then we have $|G'_1 \cap \cdots \cap G'_4| < 36$, a contradiction. $\blacksquare$

We may assume that $\mathscr{G} \not\subset \mathscr{G}_h$ for $1 \leq h \leq 3$. In fact, we have already assumed $\mathscr{G} \not\subset \mathscr{G}_1$, and we have $w_p(\mathscr{G}_i) < 0.99 \max\big\{ w_p(\mathscr{G}_0), w_p(\mathscr{G}_1) \big\}$ for $i = 2, 3$ and $p \leq 0.34$.

First we consider the case $h = 1$. In this case, by Claim 2 we have $\mathscr{G}(T_0) \in \mathbf{G}(n, 3, 1)$. Since $\mathscr{G} \not\subset \mathscr{G}_1$ it follows from Claim 3 that $\mathscr{G}(T_1) \in \mathbf{G}(n, 3, 5)$. Thus (3) gives $w_p\big(\mathscr{G}(T_0) \colon [b+1, n]\big) \leq \alpha_{3,p}$ and $w_p\big(\mathscr{G}(T_1) \colon [b+1, n]\big) \leq \alpha_{3,p}^5$. Finally by (10) we have

$$w_p(\mathscr{G}) \leq p^{36}\alpha_{3,p} + 36p^{35}q\alpha_{3,p}^5 < 0.99p^{36}$$

for $p \leq 0.34$.

Next we consider the case $h = 2$. In this case, Claim 2 gives $\mathscr{G}(T_0) \in \mathbf{G}(n, 3, 1)$ and $\mathscr{G}(T_1) \in \mathbf{G}(n, 3, 4)$, and Claim 3 gives $\mathscr{G}(T_2) \in \mathbf{G}(n, 3, 8)$. Thus (3) and (10) imply

$$w_p(\mathscr{G}) \leq p^{37}\alpha_{3,p} + 37p^{36}q\alpha_{3,p}^4 + \binom{37}{2}p^{35}q^2\alpha_{3,p}^8 < 0.99p^{36}.$$

Similarly, in the case $h = 3$, we have

(11)
$$w_p(\mathscr{G}) \leq p^{38}\alpha_{3,p} + 38p^{37}q\alpha_{3,p}^4 + \binom{38}{2}p^{36}q^2\alpha_{3,p}^7 + \binom{38}{3}p^{35}q^3\alpha_{3,p}^{11} < 0.99p^{36}.$$

This completes the proof of (i). $\blacksquare$

If we have more information about $w(n, p, 3, *)$ then we get simpler proof. For example, using a result in [27] we have $w(n, p, 3, 8) \leq p^8$ for $p \leq 0.34$. This together with Lemma 5 gives

$$w(n, p, 3, 39) \leq p^8\alpha_{3,p}^{31} < 0.99p^{36}.$$

By replacing (9) with the above estimation, we can conclude that $37 \leq s \leq 38$ and so $1 \leq h \leq 2$. This means we do not have to deal with (11).

5. A General Setup

Let n, p, r, t be fixed and let $\mathscr{G}_i = \mathscr{G}_i(n,r,t)$. Suppose that $\max \{ w_p(\mathscr{G}_{i-1}), w_p(\mathscr{G}_i) \} > w_p(\mathscr{G}_j)$ for all $j \notin \{i-1, i\}$, and consider the situation that we are trying to show

$$(12) \qquad w(n,p,r,t) = \max \{ w_p(\mathscr{G}_{i-1}), w_p(\mathscr{G}_i) \},$$

with equality holding only if $\mathscr{G} \cong \mathscr{G}_{i-1}$ or $\mathscr{G}_i$. If $\mathscr{G} \notin \mathbf{X}^i(n,r,t)$ then there is nothing to show. So suppose that $\mathscr{G} \in \mathbf{X}^i(n,r,t)$ and we want to show that $w_p(\mathscr{G})$ is much less than $\max \{ w_p(\mathscr{G}_{i-1}), w_p(\mathscr{G}_i) \}$. Let $\mathscr{G}^*$ be a tame family obtained from $\mathscr{G}$ by shifting. Then we have two cases:

(a) $\mathscr{G}^* \not\subset \mathscr{G}_j(n,r,t)$ for all $0 \le j \le i$. This is the essential case we need to estimate $w_p(\mathscr{G}^*)$ by the random walk method. To use the method, it is important that $\mathscr{G}^*$ is shifted. We saw an example in this case in the previous section.

(b) $\mathscr{G}^* \subset \mathscr{G}_j(n,r,t)$ for some $0 \le j \le i$. In this case, we will see that $w_p(\mathscr{G}^*)$ cannot be large by Theorem 7 below.

Consequently, to get (12) with the uniqueness of the optimal configuration, it is enough to consider a tame $\mathscr{G} \in \mathbf{X}^i(n,r,t)$ from the beginning.

Theorem 7. *Let r, t and i be positive integers with $r \ge 4$, and let $p \in \left(0, \frac{r-3}{r-2}\right]$. Then there exists $\gamma > 0$ such that for all $n \ge t + r$ the following is true.*

Let $\mathscr{G} \in \mathbf{X}^i(n,r,t)$, and let $\mathscr{G}^* \in \mathbf{X}^0(n,r,t)$ be a tame family obtained from $\mathscr{G}$ by shifting. If $\mathscr{G}^* \subset \mathscr{G}_i(n,r,t)$ then

$$w_p(\mathscr{G}) < (1 - \gamma) w_p\big(\mathscr{G}_i(n,r,t)\big).$$

Proof. Set $\mathscr{G}_i = \mathscr{G}_i(n,r,t)$. Note that $\mathscr{G}$ is not necessarily shifted. Since $\mathscr{G}^* \subset \mathscr{G}_i$, we may assume (by renaming the starting family if necessary) that $\mathscr{G}^* = \sigma_{xy}(\mathscr{G}) \subset \mathscr{G}_i$, where $x = t+ri$, $y = x+1$. We note that $\big|[x] \backslash G\big| \le i+1$ for all $G \in \mathscr{G}$. Moreover if $\big|[x] \backslash G\big| = i+1$ then $G \cap \{x,y\} = \{y\}$ and $\big(G - \{y\}\big) \cup \{x\} \notin \mathscr{G}$.

For $A \in \binom{[x]}{i}$ set $\mathscr{G}(A) = \{ G \in \mathscr{G} : [y] \backslash G = A \}$, and for $B \in \binom{[x-1]}{i}$ and $z \in \{x, y\}$ let $\mathscr{G}_z(B) = \{ G \in \mathscr{G} : [y] \backslash G = B \cup \{z\} \}$. Since $\sigma_{xy}(\mathscr{G}) \subset \mathscr{G}_i$ we have $\mathscr{G}_x(B) \cap \mathscr{G}_y(B) = \emptyset$ and so $w_p\big(\mathscr{G}_x(B)\big) + w_p\big(\mathscr{G}_y(B)\big) \le p^{x-i}q^{i+1}$.

Set $\mathscr{G}' = \left\{ G \in \mathscr{G} : \left| [x] \setminus G \right| < i \right\}$, $\mathscr{G}'' = \left\{ G \in \mathscr{G} : \left| [x-1] \setminus G \right| = i-1, G \cap \{x,y\} = \emptyset \right\}$ and let $e = \min\left\{ w_p(\mathscr{G}(A)) : A \in \binom{[x]}{i} \right\}$. Then we have

$$(13) \quad w_p(\mathscr{G}) = \sum_{A \in \binom{[x]}{i}} w_p(\mathscr{G}(A)) + \sum_{B \in \binom{[x-1]}{i}} \left(w_p(\mathscr{G}_x(B)) + w_p(\mathscr{G}_y(B)) \right)$$

$$+ w_p(\mathscr{G}') + w_p(\mathscr{G}'')$$

$$\leq e + \left(\binom{x}{i} - 1 \right) p^{x-i+1} q^i + \binom{x-1}{i} p^{x-i} q^{i+1}$$

$$+ \sum_{j=0}^{i-1} \binom{x}{j} p^{x-j} q^j + \binom{x-1}{i-1} p^{x-i} q^{i+1}$$

$$(14) \qquad = e + (\eta - 1) p^{x-i+1} q^i,$$

where $\eta = \sum_{j=0}^{i} \binom{x}{j} p^{i-j-1} q^{-i+j}$. Note that $e \leq p^{x-i+1} q^i$, and (14) coincides $w_p(\mathscr{G}_i) = \eta p^{x-i+1} q^i$ iff $e = p^{x-i+1} q^i$. If there is some $B \in \binom{[x-1]}{i}$ such that $\mathscr{G}_x(B) \cup \mathscr{G}_y(B) = \emptyset$, then by (13) we get $w_p(\mathscr{G}) \leq w_p(\mathscr{G}_i) - p^{x-i} q^{i+1} = \left(1 - q/(\eta p) \right) w_p(\mathscr{G}_i)$, and we are done. Thus we may assume that

$$(15) \qquad \mathscr{G}_x(B) \cup \mathscr{G}_y(B) \neq \emptyset \quad \text{for all} \quad B \in \binom{[x-1]}{i}.$$

To prove $w_p(\mathscr{G}) < (1-\gamma) w_p(\mathscr{G}_i)$ by contradiction, let us assume that for any $\gamma > 0$ and any n_0 there is some $n > n_0$ such that

$$(16) \qquad w_p(\mathscr{G}) > (1-\gamma) w_p(\mathscr{G}_i) = (1-\gamma) \eta p^{x-i+1} q^i.$$

By (14) and (16) we have $e > (1-\gamma\eta) p^{x-i+1} q^i$. This means, letting $\mathscr{H}(A) = \left\{ G \setminus [y] : G \in \mathscr{G}(A) \right\}$ and $Y = [y+1, n]$, we have $w_p\left(\mathscr{H}(A) : Y \right) > 1 - \gamma\eta$, namely,

$$(17) \qquad w_p\left(2^Y - \mathscr{H}(A) : Y \right) > \gamma\eta \quad \text{for all} \quad A \in \binom{[x]}{i}.$$

Since $\mathscr{G} \in \mathbf{X}^i(n, r, t)$ both $\bigcup_{B \in \binom{[x-1]}{i}} \mathscr{G}_x(B)$ and $\bigcup_{B \in \binom{[x-1]}{i}} \mathscr{G}_y(B)$ are non-empty. Using this with (15), we can choose $G \in \mathscr{G}_x(B)$ and $G' \in \mathscr{G}_y(B')$

with $B, B' \in \binom{[x-1]}{i}$ and $B \cap B' = \emptyset$. Let $L = [x-1] - (B \cup B')$ and $\mathscr{H}^* = \bigcap_{A \in \binom{L}{i}} \mathscr{H}(A)$. Then by (17) we have

(18)

$$w_p(\mathscr{H}^*: Y) = 1 - w_p(2^Y - \mathscr{H}^*: Y) = 1 - w_p\left(\bigcup_{A \in \binom{L}{i}} (2^Y - \mathscr{H}(A)): Y \right)$$

$$\geq 1 - \sum_{A \in \binom{L}{i}} w_p\big(2^Y - \mathscr{H}(A): Y\big) > 1 - \binom{|L|}{i}\gamma\eta.$$

If $\mathscr{H}^* \subset 2^Y$ is not $(r-2)$-wise 1-intersecting, then we can find $H_1, \ldots H_{r-2} \in \mathscr{H}^*$ such that $H_1 \cap \cdots \cap H_{r-2} = \emptyset$. Choose disjoint i-sets $B_\ell \subset L$, $1 \leq \ell \leq r - 2$, and set $G_\ell := \big([y] - B_\ell\big) \cup H_\ell \in \mathscr{G}$. Then we have $|G_1 \cap \cdots \cap G_{r-2} \cap G \cap G'| = t - 1$, which contradicts the r-wise t-intersecting property of $\mathscr{G}$. Thus $\mathscr{H}^*$ is $(r - 2)$-wise 1-intersecting and $w_p(\mathscr{H}^*: Y) \leq p$ by (2). (We need $r \geq 4$ and $p \leq \frac{r-3}{r-2}$ here.) But this contradicts (18) because we can choose γ so small that $p \ll 1 - \binom{|L|}{i}\gamma\eta$. $\blacksquare$

This theorem implies (ii) of the previous section by taking $\gamma = 0.0001$. In fact we have $q/(\eta p) > \gamma$ and $p \leq 1 - \binom{|L|}{i}\gamma\eta = 1 - 37\gamma\big(\frac{1}{q} + \frac{40}{p}\big)$ for $p \leq 0.34$. Consequently we have proved (8). It is an easy exercise to get

$$w^1(n, p, 4, t) < (1 - \gamma)\max\big\{w_p\big(\mathscr{G}_0(n, 4, t)\big), w_p\big(\mathscr{G}_1(n, 4, t)\big)\big\}$$

for all $n \geq 40$, $1 \leq t \leq 36$ and $p \leq 0.34$, where $\gamma > 0$ is an absolute constant. Then using induction on r with more careful analysis (but very much in the same way we did for the case $r = 4$ and $t = 36$) one can show the following.

Theorem 8. *For all $r \geq 4$ there exist positive constants ε, γ such that*

$$w^1(n, p, r, t) < (1 - \gamma)\max\big\{w_p\big(\mathscr{G}_0(n, r, t)\big), w_p\big(\mathscr{G}_1(n, r, t)\big)\big\}$$

holds for all $n \geq t + r$, $1 \leq t \leq (3^r - 2r - 1)/2$ and $p \leq \frac{1}{3} + \varepsilon$.

We note that $w_p\big(\mathscr{G}_0(n, r, t)\big) = w_p\big(\mathscr{G}_1(n, r, t)\big)$ if $p = 1/3$ and $t = (3^r - 2r - 1)/2$. As a corollary we get the following.

Corollary 9. *For all $r \geq 4$, $n \geq t + r$, $1 \leq t \leq (3^r - 2r - 1)/2$ and $p \leq 1/3$ we have*

$$w(n, p, r, t) = w_p\big(\mathscr{G}_0(n, r, t)\big) = p^t.$$

Moreover if $t = (3^r - 2r - 1)/2$ and $p = 1/3$ then $\mathscr{G}_0(n, r, t)$ and $\mathscr{G}_1(n, r, t)$ are the only optimal configurations (up to isomorphism). Otherwise $\mathscr{G}_0(n, r, t)$ is the only optimal configuration (up to isomorphism).

6. From p-weight Version to k-uniform Version

In this section, we show that a k-uniform version problem for $m(n,k,r,t)$ can be reduced to a p-weight version problem for $w(n,p,r,t)$ when $k/n \approx p$ (Theorems 10 and 11). Using these results, we will get a k-uniform version (Theorem 12) corresponding to Theorem 8. Theorem 1 in the introduction is an immediate consequence of the following result.

Theorem 10. *Let $p_0 \in (0,1)$ and $r,t,i \in \mathbb{N}$ be given. Then (W) implies (M).*

(W) *There exist positive constants $\gamma_0, \varepsilon_0, n_0$ such that*

$$w^i(n,p,r,t) < (1-\gamma_0)\max\left\{w_p\big(\mathscr{G}_{i-1}(n,r,t)\big), w_p\big(\mathscr{G}_i(n,r,t)\big)\right\}$$

holds for all p with $|p - p_0| < \varepsilon_0$ and all n with $n \geq n_0$.

(M) *There exist positive constants γ, ε, n_1 such that*

$$m^i(n,k,r,t) < (1-\gamma)\max\left\{\big|\mathscr{F}_{i-1}(n,k,r,t)\big|, \big|\mathscr{F}_i(n,k,r,t)\big|\right\}$$

holds for all $n > n_1$ and k with $\left|\frac{k}{n} - p_0\right| < \varepsilon$. (We can choose $\varepsilon = \frac{\varepsilon_0}{2}$, $\gamma = \frac{\gamma_0}{4}$.)

For reals $0 < b < a$ we write $a \pm b$ to mean the open interval $(a-b, a+b)$, and for $n \in \mathbb{N}$, $n(a \pm b)$ means $\big((a-b)n, (a+b)n\big) \cap \mathbb{N}$.

Proof. Assuming the negation of (M), we will construct a counterexample to (W).

For fixed r and t we note that

$$f(p) := \max\left\{w_p\big(\mathscr{G}_{i-1}(n,r,t)\big), w_p\big(\mathscr{G}_i(n,r,t)\big)\right\}$$

is a uniformly continuous function of p on $p_0 \pm \varepsilon_0$. Let $\varepsilon = \frac{\varepsilon_0}{2}$, $\gamma = \frac{\gamma_0}{4}$, and $I = p_0 \pm \varepsilon$.

Choose $\varepsilon_1 \ll \varepsilon$ so that

$$(19) \qquad\qquad (1-3\gamma)f(p) > (1-4\gamma)f(p+\delta)$$

holds for all $p \in I$ and all $0 < \delta \leq \varepsilon_1$. Choose n_2 so that

$$(20) \qquad\qquad \sum_{j \in J}\binom{n}{j}p_1^j(1-p_1)^{n-j} > (1-3\gamma)/(1-2\gamma)$$

holds for all $n > n_2$ and all $p_1 \in I_0 := p_0 \pm \frac{3\varepsilon}{2}$, where $J = n(p_1 \pm \varepsilon_1)$. Choose n_3 so that

$$(21) \quad (1-\gamma)\max\left\{\left|\mathscr{F}_{i-1}(n,k,r,t)\right|, \left|\mathscr{F}_i(n,k,r,t)\right|\right\} > (1-2\gamma)f(k/n)\binom{n}{k}$$

holds for all $n > n_3$ and k with $k/n \in I$. Finally set $n_1 = \max\{n_0, n_2, n_3\}$.

Suppose that (M) fails. Then for our choice of ε, γ and n_1, we can find some n, k and $\mathscr{F} \in \mathbf{Y}^i(n,k,r,t)$ with $|\mathscr{F}| \geq (1-\gamma)\max\left\{\left|\mathscr{F}_{i-1}(n,k,r,t)\right|, \left|\mathscr{F}_i(n,k,r,t)\right|\right\}$, where $n > n_1$ and $\frac{k}{n} \in I$. We fix n, k and $\mathscr{F}$, and let $p = \frac{k}{n}$. By (21) we have $|\mathscr{F}| > c\binom{n}{k}$, where $c = (1-2\gamma)f(p)$. Let $\mathscr{G} = \bigcup_{k \leq j \leq n}\left(\nabla_j(\mathscr{F})\right) \in \mathbf{X}^i(n,r,t)$ be the collection of all upper shadows of $\mathscr{F}$, where $\nabla_j(\mathscr{F}) = \left\{H \in \binom{[n]}{j} : H \supset \exists F \in \mathscr{F}\right\}$. Let $p_1 = p + \varepsilon_1 \in I_0$, and $J = n(p_1 \pm \varepsilon_1) = (k, k + 2\varepsilon_1 n) \cap \mathbb{N}$.

Claim 4. $\left|\nabla_j(\mathscr{F})\right| \geq c\binom{n}{j}$ for $j \in J$.

Proof. Choose a real $x \leq n$ so that $c\binom{n}{k} = \binom{x}{n-k}$. Since $|\mathscr{F}| > c\binom{n}{k} = \binom{x}{n-k}$ the Kruskal–Katona Theorem [21, 20] implies that $\left|\nabla_j(\mathscr{F})\right| \geq \binom{x}{n-j}$. Thus it suffices to show that $\binom{x}{n-j} \geq c\binom{n}{j}$, or equivalently,

$$\frac{\binom{x}{n-j}}{\binom{x}{n-k}} \geq \frac{c\binom{n}{j}}{c\binom{n}{k}}.$$

Using $j \geq k$ this is equivalent to $j \cdots (k+1) \geq (x-n+j) \cdots (x-n+k+1)$, which follows from $x \leq n$. ∎

By the claim we have

$$(22) \quad w_{p_1}(\mathscr{G}) \geq \sum_{j \in J}\left|\nabla_j(\mathscr{F})\right| p_1^j(1-p_1)^{n-j} \geq c\sum_{j \in J}\binom{n}{j}p_1^j(1-p_1)^{n-j}.$$

Using (20) and (19), the RHS of (22) is more than

$$c(1-3\gamma)/(1-2\gamma) = (1-3\gamma)f(p) > (1-4\gamma)f(p+\varepsilon_1) = (1-\gamma_0)f(p_1).$$

This means $w_{p_1}(\mathscr{G}) > (1-\gamma_0)\max\left\{w_{p_1}\left(\mathscr{G}_{i-1}(n,r,t)\right), w_{p_1}\left(\mathscr{G}_i(n,r,t)\right)\right\}$, which contradicts (W) because $p_1 \in I_0 \subset p_0 \pm \varepsilon_0$. ∎

Theorem 11. *Let $r, t \in \mathbb{N}$ with $r \geq 4$, and let $p_0 \in \left(0, \frac{r-3}{r-2}\right]$. Suppose that*

$$p_0^t > (t+r)p_0^{t+r-1}(1 - p_0) + p_0^{t+r},$$

i.e., $w_{p_0}\big(\mathscr{G}_0(n,r,t)\big) > w_{p_0}\big(\mathscr{G}_1(n,r,t)\big)$ for all $n \geq t+r$. Then (W0) implies (M1) and (W1).

(W0) *There exist positive constants $\gamma_0, \varepsilon_0, n_0$ such that $w^0(n,p,r,t) < (1-\gamma_0)p^t$ holds for all p with $|p - p_0| < \varepsilon_0$ and all n with $n \geq n_0$.*

(M1) *There exist positive constants $\gamma_1, \varepsilon_1, n_1$ such that*

$$(23) \qquad m^0(n,k,r,t) < (1-\gamma_1)\binom{n-t}{k-t}$$

holds for all $n > n_1$ and k with $\frac{k}{n} < p_0 + \varepsilon_1$.

(W1) *There exist positive constants γ_2, ε_2 such that $w^0(n,p,r,t) < (1-\gamma_2)p^t$ holds for all p with $p < p_0 + \varepsilon_2$ and all n with $n \geq t$.*

Proof. For simplicity, we write $\mathscr{G}_j$ for $\mathscr{G}_j(n,r,t)$ and $\mathscr{F}_j$ for $\mathscr{F}_j(n,k,r,t)$.

Assume (W0). First we show (M1). Choose ε_0 from (W0). Since $w_{p_0}(\mathscr{G}_0) > w_{p_0}(\mathscr{G}_1)$ we may assume that $w_p(\mathscr{G}_0) > w_p(\mathscr{G}_1)$ for all p with $|p - p_0| < \varepsilon_0$ (if necessary we replace ε_0 so that this property holds). We can choose n_1 so that $|\mathscr{F}_0| > |\mathscr{F}_1|$ holds for all $n > n_1$ and k with $\left|\frac{k}{n} - p_0\right| < \varepsilon_0$. Then for the parameters chosen as above, we have $w^0(n,p,r,t) = w^1(n,p,r,t)$ and $m^0(n,k,r,t) = m^1(n,k,r,t)$. Thus (23) for the case $\left|\frac{k}{n} - p_0\right| < \varepsilon_1 := \frac{\varepsilon_0}{2}$ follows from Theorem 10 by setting $i = 1$. We will show (23) for $\frac{k}{n} \leq p_0 - \varepsilon_1$. Let $p = p_0 - \frac{\varepsilon_1}{2}$. Since $p < p_0$ and $w_p(\mathscr{G}_0) = p^t > w_p(\mathscr{G}_1)$ we can choose $\gamma_1 > 0$ so that

$$(24) \qquad (1 - 2\gamma_1)p^t > w_p\big(\mathscr{G}_1(n,r,t)\big).$$

Then choose n_0 so that

$$(25) \qquad \sum_{i \in J} \binom{n-t}{i-t} p^i (1-p)^{n-i} > p^t (1 - 2\gamma_1)/(1 - \gamma_1)$$

holds for all $n > n_0$, where $J = n\left(p \pm \frac{\varepsilon_1}{2}\right) = \big((p_0 - \varepsilon_1)n, p_0 n\big) \cap \mathbb{N}$.

To show (23), suppose, on the contrary, that we can find some n, k and $\mathscr{F} \in \mathbf{Y}^0(n,k,r,t)$ with $|\mathscr{F}| \geq (1-\gamma_1)\binom{n-t}{k-t}$, where $n > n_1$ and $\frac{k}{n} \leq p_0 - \varepsilon_1$. We fix n, k and $\mathscr{F}$. Let $\mathscr{G} = \bigcup_{k \leq i \leq n}\big(\nabla_i(\mathscr{F})\big) \in \mathbf{X}^0(n,r,t)$ be the collection of all upper shadows of $\mathscr{F}$.

Claim 5. $\left|\nabla_i(\mathcal{F})\right| \geq (1-\gamma_1)\binom{n-t}{i-t}$ *for* $i \in J$.

Proof. Choose a real $x \leq n-t$ so that $(1-\gamma_1)\binom{n-t}{k-t} = \binom{x}{n-k}$. Since $|\mathcal{F}| \geq \binom{x}{n-k}$ the Kruskal–Katona Theorem implies that $\left|\nabla_i(\mathcal{F})\right| \geq \binom{x}{n-i}$. Thus it suffices to show that $\binom{x}{n-i} \geq (1-\gamma_1)\binom{n-t}{i-t}$, or equivalently,

$$\frac{\binom{x}{n-i}}{\binom{x}{n-k}} \geq \frac{(1-\gamma_1)\binom{n-t}{i-t}}{(1-\gamma_1)\binom{n-t}{k-t}}.$$

Using $i > (p_0 - \varepsilon_1)n \geq k$ this is equivalent to $(i-t)\cdots(k-t+1) \geq (x-n+i)\cdots(x-n+k+1)$, which follows from $x \leq n-t$. $\blacksquare$

By the claim we have

$$(26) \quad w_p(\mathcal{G}) \geq \sum_{i \in J}\left|\nabla_i(\mathcal{F})\right|p^i(1-p)^{n-i} \geq (1-\gamma_1)\sum_{i \in J}\binom{n-t}{i-t}p^i(1-p)^{n-i}.$$

By (25) and (24), the RHS of (26) is more than $(1-\gamma_1)\cdot p^t(1-2\gamma_1)/(1-\gamma_1) = p^t(1-2\gamma_1) > w_p\big(\mathcal{G}_1(n,r,t)\big)$, which contradicts (W0). This completes the proof of (M1).

Next we show (W2). Let $\varepsilon_1 = \frac{\varepsilon_0}{2}$ and let $p \leq p_0 - \varepsilon_1$ be given. By (M1) we can find $\gamma_1 > 0$ and n_1 such that $m^0(n,k,r,t) < (1-\gamma_1)\binom{n-t}{k-t}$ holds for all $n > n_1$ and k with $\frac{k}{n} < p_0$. Choose $0 < \delta \ll \varepsilon_1$ so that $p \pm \delta \subset (0,p_0)$. Choose n_2 so that

$$(27) \qquad (1-\gamma_1)\sum_{k \in J}\binom{n-t}{k-t}p^k q^{n-k} + \sum_{k \notin J}\binom{n}{k}p^k q^{n-k} < \left(1-\frac{\gamma_1}{2}\right)p^t$$

holds for all $n > n_2$, where $J = n(p \pm \delta)$. Let $n > \max\{n_1, n_2\}$ and choose $\mathcal{G} \in \mathbf{X}^0(n,r,t)$ with $w_p(\mathcal{G}) = w^0(n,p,r,t)$. Let $\mathcal{G}^{(k)} = \mathcal{G} \cap \binom{[n]}{k}$ for $k \in J$.

If $\mathcal{G}^{(k)} \in \mathbf{Y}^0(n,k,r,t)$ then we have

$$\left|\mathcal{G}^{(k)}\right| \leq m^0(n,k,r,t) < (1-\gamma_1)\binom{n-t}{k-t}.$$

If $\mathcal{G}^{(k)}$ fixes t vertices, say $[t]$, then $\tilde{\mathcal{G}}^{(k)} := \left\{G - [t]\colon G \in \mathcal{G}^{(k)}\right\}$ is $(r-1)$-wise 1-intersecting. (Otherwise $\mathcal{G}$ fixes $[t]$.) Thus we have $\left|\mathcal{G}^{(k)}\right| = \left|\tilde{\mathcal{G}}^{(k)}\right| \leq \binom{n-t-1}{k-t-1} = \frac{k-t}{n-t}\binom{n-t}{k-t} < p_0\binom{n-t}{k-t}$ by (2). Consequently, in both cases, we have

$$(28) \qquad \left|\mathcal{G}^{(k)}\right| < (1-\gamma_1)\binom{n-t}{k-t}.$$

Using (28) and (27), we have

$$w_p(\mathscr{G}) \leq \sum_{k \in J} \left|\mathscr{G}^{(k)}\right| p^k q^{n-k} + \sum_{k \notin J} \binom{n}{k} p^k q^{n-k} < \left(1 - \frac{\gamma_1}{2}\right) p^t,$$

and this is true for all $n \geq t$ by Lemma 6. This completes the proof of (W1).
■

By Theorems 8, 10 and 11, we have the following.

Theorem 12. *Let $r \geq 4$. There exists n_1 such that*

$$m(n, k, r, t) = \max\left\{\left|\mathscr{F}_0(n, k, r, t)\right|, \left|\mathscr{F}_1(n, k, r, t)\right|\right\}$$

holds for all t with $1 \leq t \leq (3^r - 2r - 1)/2$, and for all $n > n_1$ and k with $\frac{k}{n} < \frac{1}{3} + \varepsilon$. Moreover $\mathscr{F}_0(n, k, r, t)$ and $\mathscr{F}_1(n, k, r, t)$ are the only possible optimal configurations (up to isomorphism).

REFERENCES

[1] R. Ahlswede and L. H. Khachatrian, The complete nontrivial-intersection theorem for systems of finite sets, *J. Combin. Theory (A)*, **76** (1996), 121–138.

[2] R. Ahlswede and L. H. Khachatrian, The complete intersection theorem for systems of finite sets, *European J. Combin.*, **18** (1997), 125–136.

[3] R. Ahlswede and L. H. Khachatrian, The diametric theorem in Hamming spaces – Optimal anticodes, *Adv. in Appl. Math.*, **20** (1998), 429–449.

[4] A. Brace and D. E. Daykin, A finite set covering theorem, *Bull. Austral. Math. Soc.*, **5** (1971), 197–202.

[5] C. Bey and K. Engel, Old and new results for the weighted t-intersection problem via AK-methods, *Numbers, Information and Complexity, Althofer, Ingo, Eds. et al., Dordrecht,* Kluwer Academic Publishers (2000), pp. 45–74.

[6] P. Erdős, C. Ko and R. Rado, Intersection theorems for systems of finite sets, *Quart. J. Math. Oxford (2)*, **12** (1961), 313–320.

[7] I. Dinur and S. Safra, On the Hardness of Approximating Minimum Vertex-Cover, *Annals of Mathematics*, **162** (2005), 439–485.

[8] P. Frankl and On Sperner families satisfying an additional condition, *J. Combin. Theory (A)*, **20** (1976), 1–11.

[9] P. Frankl, Families of finite sets satisfying an intersection condition, *Bull. Austral. Math. Soc.*, **15** (1976), 73–79.

[10] P. Frankl, The Erdős–Ko–Rado theorem is true for $n = ckt$, *Combinatorics (Proc. Fifth Hungarian Colloq., Keszthely, 1976), Vol. I*, 365–375, Colloq. Math. Soc. János Bolyai, 18, North–Holland (1978).

[11] P. Frankl, On intersecting families of finite sets, *J. Combin. Theory (A)*, **24** (1978), 141–161.

[12] P. Frankl, The shifting technique in extremal set theory, "Surveys in Combinatorics 1987" (C. Whitehead, Ed. LMS Lecture Note Series 123), 81–110, Cambridge Univ. Press (1987).

[13] P. Frankl, Multiply-intersecting families, *J. Combin. Theory (B)*, **53** (1991), 195–234.

[14] P. Frankl and N. Tokushige, Weighted 3-wise 2-intersecting families, *J. Combin. Theory (A)*, **100** (2002), 94–115.

[15] P. Frankl and N. Tokushige, Weighted multiply intersecting families, *Studia Sci. Math. Hungarica*, **40** (2003), 287–291.

[16] P. Frankl and N. Tokushige, Random walks and multiply intersecting families, *J. Combin. Theory (A)*, **109** (2005), 121–134.

[17] P. Frankl and N. Tokushige, Weighted non-trivial multiply intersecting families, *Combinatorica*, **26** (2006), 37–46.

[18] A. J. W. Hilton and E. C. Milner, Some intersection theorems for systems of finite sets, *Quart. J. Math. Oxford*, **18** (1967), 369–384.

[19] G. O. H. Katona, Intersection theorems for systems of finite sets, *Acta Math. Acad. Sci. Hung.*, **15** (1964), 329–337.

[20] G. O. H. Katona, A theorem of finite sets, in: *Theory of Graphs*, Proc. Colloq. Tihany, 1966 (Akadémiai Kiadó, 1968), pp. 187–207.

[21] J. B. Kruskal, The number of simplices in a complex, in: *Math. Opt. Techniques* (Univ. of Calif. Press, 1963), pp. 251–278.

[22] N. Tokushige, A frog's random jump and the Pólya identity, *Ryukyu Math. Journal*, **17** (2004), 89–103.

[23] N. Tokushige, Intersecting families – uniform versus weighted, *Ryukyu Math. J.*, **18** (2005), 89–103.

[24] N. Tokushige, Extending the Erdős–Ko–Rado theorem, *J. Combin. Designs*, **14** (2006), 52–55.

[25] N. Tokushige, The maximum size of 4-wise 2-intersecting and 4-wise 2-union families, *European J. of Comb.*, **27** (2006), 814–825.

[26] N. Tokushige, The maximum size of 3-wise t-intersecting families, *European J. Combin*, **28** (2007), 152–166.

[27] N. Tokushige, EKR type inequalities for 4-wise intersecting families, *J. Combin. Theory (A)*, **114** (2007), 575–596.

[28] N. Tokushige, Brace–Daykin type inequalities for intersecting families, *European J. Combin*, **29** (2008), 273–285.

[29] N. Tokushige, Multiply-intersecting families revisited, *J. Combin. Theory (B),* **97** (2007), 929–948.

[30] R. M. Wilson, The exact bound in the Erdős–Ko–Rado theorem, *Combinatorica,* **4** (1984), 247–257.

Norihide Tokushige

College of Education
Ryukyu University
Nishihara, Okinawa
903-0213 Japan

e-mail: hide@edu.u-ryukyu.ac.jp

BOLYAI SOCIETY
MATHEMATICAL STUDIES, 17

Horizons of Combinatorics
Balatonalmádi
pp. 235–255.

PROBLEMS AND RESULTS ON COLORINGS OF MIXED HYPERGRAPHS

ZSOLT TUZA[*] and VITALY VOLOSHIN[†]

We survey results and open problems on 'mixed hypergraphs' that are hypergraphs with two types of edges. In a proper vertex coloring the edges of the first type must not be monochromatic, while the edges of the second type must not be completely multicolored. Though the first condition just means 'classical' hypergraph coloring, its combination with the second one causes rather unusual behavior. For instance, hypergraphs occur that are uncolorable, or that admit colorings with certain numbers k' and k'' of colors but no colorings with exactly k colors for any $k' < k < k''$.

1. INTRODUCTION

In the classical theory of hypergraph coloring, colors have to be assigned to the vertices in such a way that no edge is colored completely with the same color. In this paper we consider a fruitful generalization of proper colorings, introduced in [45, 46]. It turns out that under the more complex conditions involved, a more powerful model is obtained and, on the other hand, rather unusual phenomena arise. The goal of this paper is to provide an overview of results and open problems in this area.

Basic definitions. A *mixed hypergraph* is a triple $\mathcal{H} = (X, \mathcal{C}, \mathcal{D})$, where X is the *vertex set*, and $\mathcal{C}$ and $\mathcal{D}$ are families of subsets of X, called the $\mathcal{C}$-*edges*

[*]Research supported in part by the Hungarian Scientific Research Fund, OTKA grant T-049613.
[†]Partially supported by Troy University Research Grant.

and $\mathcal{D}$-*edges*, respectively. A *proper coloring* of $\mathcal{H}$ is a mapping from X into a set of k *colors* so that each $\mathcal{C}$-edge has two vertices with a $\mathcal{C}$ommon color and each $\mathcal{D}$-edge has two vertices with $\mathcal{D}$istinct colors. A *bi-edge* is a vertex subset that is both a $\mathcal{C}$-edge and $\mathcal{D}$-edge.

A coloring may also be viewed as a *partition* of X, where the *color classes* (partition classes) are the sets of vertices assigned to the same color. Then the condition is that no class may contain a $\mathcal{D}$-edge, and each $\mathcal{C}$-edge has to meet some class in more than one vertex. A mixed hypergraph is k-*colorable* if it has a proper coloring with at most k colors. A *strict k-coloring* is a proper k-coloring using all of the k colors.

We obtain classical hypergraph coloring in the special case of $\mathcal{H} = (X, \emptyset, \mathcal{D})$, which is denoted by $\mathcal{H}_{\mathcal{D}}$ and called a $\mathcal{D}$-*hypergraph*. A hypergraph $\mathcal{H} = (X, \mathcal{C}, \emptyset)$ will be denoted by $\mathcal{H}_{\mathcal{C}}$ and called a $\mathcal{C}$-*hypergraph*. Mixed hypergraphs with $\mathcal{C} = \mathcal{D}$ are called *bi-hypergraphs*.

The maximum number of colors in a strict coloring of $\mathcal{H} = (X, \mathcal{C}, \mathcal{D})$ is the *upper chromatic number* $\bar{\chi}(\mathcal{H})$; and the minimum number of colors is the *lower chromatic number* $\chi(\mathcal{H})$. Thus, general mixed hypergraphs represent structures where problems on both the minimum and maximum number of colors occur.

Some types of (mixed) hypergraphs. For a mixed hypergraph $\mathcal{H}$, a *host graph* is a graph G on the same vertex set as $\mathcal{H}$, and such that every $\mathcal{C}$-edge and every $\mathcal{D}$-edge induces a *connected* subgraph in G. For a given $\mathcal{H}$, there can be many host graphs G. Depending on the type of G, particular terminology is used for $\mathcal{H}$:

- If G is a path, then $\mathcal{H}$ is called an *interval hypergraph*.
- If G is a tree, then $\mathcal{H}$ is called a *hypertree*.
- If G is a cycle, then $\mathcal{H}$ is called a *circular hypergraph*.

Let us mention further that $\mathcal{H}$ is called r-*uniform* if each of its $\mathcal{C}$- and $\mathcal{D}$-edges has exactly r vertices.

For general information and concepts not defined here, we refer to [3] on 'classical' hypergraphs and [47] on mixed hypergraphs; see also the regularly updated web site [48].

2. COLORABILITY PROBLEM

Although colorings are always possible in the classical or $\mathcal{C}$-hypergraph setting, the mixed hypergraphs in general may have no colorings at all. Hence, mixed hypergraphs can model not only extremal problems but also *existence problems*. In this way, the first problem that appears in mixed hypergraphs is to find out if they admit *any* coloring. This problem is called the *colorability problem*. A mixed hypergraph with no colorings at all is *uncolorable* [46]; otherwise it is called *colorable*.

The colorability problem represents a new type of problems in the theory of coloring. The structure of uncolorable mixed hypergraphs is unknown, and there is not much hope for a general description since the recognition problem of colorable mixed hypergraphs is already NP-complete for 3-uniform mixed hypergraphs (Tuza, Voloshin, Zhou [44]).

The first results concerning uncolorability were obtained by Tuza and Voloshin in [43]. There the existence of uncolorable mixed hypergraphs $\mathcal{H} = (X, \mathcal{C}, \mathcal{D})$ having an arbitrarily large difference between $\bar{\chi}(\mathcal{H}_\mathcal{C})$ and $\chi(\mathcal{H}_\mathcal{D})$ was proven. More precisely, it was shown that for any $k = \bar{\chi}(\mathcal{H}_\mathcal{C}) - \chi(\mathcal{H}_\mathcal{D}) > 0$ the minimum number of vertices of an inclusionwise minimal uncolorable mixed hypergraph is exactly $k+4$. (Minimality has to be assumed, since the trivial example of a 2-element bi-edge together with $k + 1$ isolated vertices would yield a smaller bound.) This result also provides a clear evidence that, though $\bar{\chi}(\mathcal{H}_\mathcal{C}) < \chi(\mathcal{H}_\mathcal{D})$ is a sufficient condition for uncolorability, it is very far from being necessary.

A measure of uncolorability, called the *vertex uncolorability number*, has been introduced [43]. It is the minimum number of vertices to be deleted in such a way that the mixed hypergraph obtained becomes colorable. A greedy algorithm – which is the first greedy mixed hypergraph coloring algorithm – to determine an estimate on the vertex uncolorability number has also been developed.

It has been shown that the colorability problem can be expressed as an integer linear programming problem. There are two different formulations, the first one [43] involving the maximal independent sets of $\mathcal{H}_\mathcal{C}$ (hence, the description of its condition set may be exponentially large in terms of the input size), while the other one [34] reduces colorability to a polynomial-size integer linear program accepting $(0, 1)$ solutions only. This latter model has been elaborated further to determine whether the input hypergraph admits a *strict k-coloring* ([34]).

A further connection to other fields is that the list colorability problem of graphs – where the existence of a proper coloring is a central issue – represents a special case of the colorability problem on mixed hypergraphs *without lists* [31, 42, 43]. In fact, a large number of coloring problems on graphs can be modeled using mixed hypergraph coloring (Kráľ [27]).

Some particular classes of uncolorable mixed hypergraphs, too, were considered in [43], including e.g. the so-called complete (l, m)-uniform mixed hypergraphs; it means that all l-element subsets are viewed as C-edges and all m-element subsets are D-edges. The running parameter of the structure is the number n of vertices. It was observed that for any *fixed* pair (l, m), almost all complete (l, m)-uniform mixed hypergraphs are uncolorable, the largest colorable one having as few as $(l - 1)(m - 1)$ vertices. On the other hand, if we do not fix l and m, generally *almost all* complete mixed hypergraphs *are colorable*, as n gets large.

A subclass of uncolorable mixed hypergraphs can be derived from the class of *k-chromatic graphs* (graphs whose chromatic number is equal to k) [43], for any $k \geq 3$. In this construction the edges of a graph G are taken as D-edges of the hypergraph; and the C-edges are the vertex sets of the k-vertex paths in G. Since k-critical graphs are hard to recognize, this example is a further indication that uncolorable mixed hypergraphs may have a rather complex structure in general.

Criteria of uncolorability for some special classes of mixed hypergraphs have been obtained, too. For example, a mixed hypertree is uncolorable if and only if it contains an 'evidently uncolorable' D-edge, that is a D-edge inside which each edge of the host tree is a 2-element C-edge of the hypergraph [43]. In spite of the fact that the presence of an evidently uncolorable edge is 'evidently' sufficient for uncolorability, very little is known about classes of mixed hypergraphs where this condition is necessary, too.

Open problems.
1. Describe further structural properties and further subclasses of uncolorable mixed hypergraphs.
2. Search for the criteria of colorability for mixed hypergraphs derived from important classes, especially for 3-uniform bi-hypergraphs.
3. Investigate uncolorable mixed hypergraphs that are critical with respect to the deletion of vertices and/or edges.

4. Determine the minimum value of $|\mathcal{C}|$, of $|\mathcal{D}|$, and of $|\mathcal{C}| + |\mathcal{D}|$, in inclusionwise minimal uncolorable r-uniform (bi-) hypergraphs on n vertices, as a function of n and r.

5. Develop algorithms for testing colorability and estimating the vertex uncolorability number. Results on the threshold concerning running time vs. the precision of approximation would be of great interest.

6. Characterize uncolorable mixed hypergraphs with vertex degree k, for $k \geq 2$ fixed.

3. Uniquely Colorable Mixed Hypergraphs

A mixed hypergraph is called *uniquely colorable* (UC, for short) if all of its proper colorings induce the same partition of the vertex set. The only uniquely colorable hypergraphs in classic coloring theory are the complete graphs (cliques). Thus, the uniquely colorable mixed hypergraphs are in this sense generalizations of cliques, and so they represent new 'absolutely rigid' combinatorial objects. But their structure is rather general. The first paper about uniquely colorable mixed hypergraphs is by Tuza, Voloshin and Zhou [44]. It is shown there, for example, that every mixed hypergraph having at least one coloring is an induced subhypergraph of some uniquely colorable mixed hypergraph. (Analogously to graph theory, the subhypergraph induced by a subset of vertices consists of all the $\mathcal{C}$-edges and $\mathcal{D}$-edges contained in the subset.)

Recursive formulas can be derived by introducing the concept of uniquely colorable separator (Voloshin, Zhou [50]). In this setting, $\mathcal{H}_0$, $\mathcal{H}_1$, $\mathcal{H}_2$ are assumed to be induced subhypergraphs of a mixed hypergraph $\mathcal{H}$, such that $\mathcal{H}_1 \cup \mathcal{H}_2 = \mathcal{H}$, $\mathcal{H}_1 \cap \mathcal{H}_2 = \mathcal{H}_0$, and $\mathcal{H}_0$ is uniquely colorable. Formulas have been presented that relate the upper and lower chromatic numbers of $\mathcal{H}_0$, $\mathcal{H}_1$, $\mathcal{H}_2$ with the upper and lower chromatic number of the original mixed hypergraph $\mathcal{H}$, and also the numbers of colorings of those three subhypergraphs with that of $\mathcal{H}$. The uniquely colorable separators open the way to build up new structures with interesting coloring properties, analogously to the way as chordal graphs are constructed from cliques.

It is easy to decide whether or not a *graph* is uniquely colorable, because in that case it is a clique. In general, however, it is algorithmically hard to determine if a mixed hypergraph is uniquely colorable. More precisely, given

$\mathcal{H}$ together with one of its proper colorings, it is co-NP-complete to decide whether $\mathcal{H}$ is uniquely colorable (Tuza, Voloshin, Zhou [44]).

A weaker property has also been studied, where the mixed hypergraph has precisely one color partition with both $\bar{\chi}$ and χ colors, for $\bar{\chi} > \chi$ [44]. The class of these 'weakly uniquely colorable' mixed hypergraphs contains all uniquely colorable graphs in the usual graph-theoretic sense that requires a unique coloring with just χ colors (apart from the permutations of colors).

The following classes of uniquely colorable mixed hypergraphs have been characterized: those with $\chi = n-1$ and $\chi = n-2$ (Niculitsa, Voss [40]); UC mixed hypertrees (Niculitsa, Voloshin [39]); and UC circular mixed hypergraphs (Voloshin, Voss [49]). Moreover, pseudo-chordal mixed hypergraphs as a generalization of chordal graphs have been introduced and described by Voloshin and Zhou in [50]. Further, the possible *size distributions* of color classes in uniquely colorable *r-uniform bi-hypergraphs* have been characterized by Bacsó, Tuza, and Voloshin in [2].

A mixed hypergraph is *UC-orderable* [44] if there exists an ordering $x_1, x_2, \ldots, x_n$ of the vertex set such that every subhypergraph induced by an initial segment $\{x_1, \ldots, x_i\}$ of this ordering is uniquely colorable. It had been expected that this class of mixed hypergraphs could be efficiently recognized; but this has turned out to be false, as disproved by Bujtás and Tuza in [4]. As a matter of fact, testing UC-orderability is NP-complete. On the other hand, the possible *color sequences* of *uniquely* UC-orderable mixed hypergraphs have been characterized, and a linear-time algorithm for their recognition has been given [4]. Moreover, the UC-orderable mixed *hypertrees* have been completely characterized by Niculitsa and Voloshin in [39].

Open problems.

1. Search for conditions that are necessary or sufficient for a mixed hypergraph to be uniquely colorable.

2. Find the characterization of unique colorability for mixed hypergraphs derived from important classes, such as pseudo-chordal and planar mixed hypergraphs and those discussed in [3].

3. Search for conditions that are necessary or sufficient for a mixed hypergraph to be UC-orderable.

4. Determine the minimum value of $|\mathcal{C}|$, of $|\mathcal{D}|$, and of $|\mathcal{C}|+|\mathcal{D}|$, in uniquely colorable or UC-orderable or uniquely UC-orderable r-uniform (bi-) hypergraphs on n vertices, as a function of n and r.

5. Develop algorithms for testing the unique colorability of mixed hyper-graphs.

6. Characterize UC mixed hypergraphs with vertex degree k, for $k \geq 2$ fixed.

7. Characterize the structure of mixed hypergraphs whose UC-ordering is unique apart from transposing the first two vertices, and develop algorithms to recognize them.

4. CHROMATIC SPECTRUM

The set of values of k for which $\mathcal{H}$ has a strict k-coloring is the *feasible set*. For each k, let r_k denote the number of *partitions* of the vertex set into k nonempty parts (color classes) such that the coloring constraint is satisfied on each $\mathcal{C}$- and $\mathcal{D}$-edge. Such partitions are called *feasible partitions*. In fact, r_k is strongly related to the number of strict k-colorings, with the only difference that in r_k we do not count the permutations (re-labeling) of colors as distinct colorings. The vector $R(\mathcal{H}) = (r_1, \ldots, r_n)$ is termed the *chromatic spectrum* [45, 46].

Given a colorable mixed hypergraph $\mathcal{H}$, it is natural and important to ask whether $\mathcal{H}$ has strict k-colorings for all k such that $\chi(\mathcal{H}) \leq k \leq \bar{\chi}(\mathcal{H})$. Open since the introduction of mixed hypergraphs, this question has been solved in [23]. The answer is surprisingly negative: there may indeed be gaps in the chromatic spectrum.

A mixed hypergraph has a *gap at k* if its feasible set contains elements larger and smaller than k but omits k. A *gap of size g* means g consecutive gaps. If some gaps occur, the feasible set and the chromatic spectrum of $\mathcal{H}$ are said to be *broken*, and if there are no gaps then they are called *continuous* or *gap-free*.

A mixed hypergraph $\mathcal{H}_{s,t}$ has been constructed in [23] for all $2 \leq s \leq t-2$, that has feasible set $\{s, t\}$. Furthermore, it has been verified that $\mathcal{H}_{s,t}$ has the fewest vertices among all s-colorable mixed hypergraphs that have a gap at $t - 1$; this minimum number of vertices is $2t - s$. It follows, in particular, that $2g + 4$ is a tight lower bound on the number of vertices in mixed hypergraphs having a gap of size g. It has also been proven that a finite set of positive integers is a feasible set if and only if it is an initial interval $\{1, \ldots, t\}$ or does not contain the element 1. This remains valid for r-*uniform* mixed- or bi-hypergraphs as well, except that no gaps smaller than r can occur if all edges are required to have the same size r (Bujtás, Tuza [6]). Constructions also show (Kráľ [26]) that in the chromatic spectra of non-1-colorable mixed hypergraphs the sequence $(r_2, r_3, \ldots, r_{\bar{\chi}})$ can take all combinations of nonnegative integers in its entries, except for the natural condition $r_{\bar{\chi}} > 0$.

It is obvious that the feasible sets of all r-uniform $\mathcal{C}$-hypergraphs contain $\{1, \ldots, r-1\}$ as a subset. On n vertices, the minimum number of r-element $\mathcal{C}$-edges to generate this smallest possible feasible set is $\lceil n(n-2)/3 \rceil$ in the particular case of $r = 3$, but only some lower and upper estimates of the order $\Theta(n^{r-1})$ are known if $r \geq 4$ (Diao et al. [14, 15]).

The chromatic spectrum of some special families of mixed hypergraphs has been investigated, too. In particular, it has been shown that no gaps can occur in mixed interval hypergraphs (Jiang et al. [23]), or more generally in hypertrees (Kráľ et al. [28]), and in mixed hypergraphs over a host graph in which any two cycles are vertex-disjoint (Kráľ, Kratochvíl, Voss [30]). Gaps cannot arise either, if the mixed hypergraph has a uniquely colorable separator and the derived subhypergraphs do not have any gaps. This result implies that the so-called pseudo-chordal mixed hypergraphs [50] have gap-free chromatic spectra.

One of the important results states that mixed hypergraphs with maximum vertex degree 2 have no gaps (Kráľ, Kratochvíl, Voss [29]). If we consider hypergraphs dual to these structures, then they are just *multigraphs*. In such multigraphs, we have *vertices* of two types: $\mathcal{C}$- and $\mathcal{D}$-vertices. Now, in a proper coloring of the *edges*, we require that every $\mathcal{C}$-vertex be incident with at least two edges of the same color, and every $\mathcal{D}$-vertex be incident with at least two edges of different colors. Hence, in (multi)graphs, there is a natural way to introduce the concept of *lower and upper chromatic indices*, and the result states that proper colorings exist using every intermediate number of colors between minimum and maximum.

It is interesting to note that in the special case with just $\mathcal{C}$-vertices, an exact formula can be given: if such a multigraph has n vertices, m edges and p pendant vertices, and c denotes the maximum number of vertex-disjoint cycles in it, then the upper chromatic index is equal to $c + m - n + p$ (M. Gionfriddo, Milazzo, Voloshin [22]). This result may be viewed as a dual version of Vizing's celebrated theorem, with the remarkable aspect that in this 'mixed' case the graphs of 'class two' do not occur.

Further notable facts in this direction are that some *planar* mixed hypergraphs have gaps (Kobler, Kündgen [24]), and mixed hypergraphs derived from some block designs may have gaps in their chromatic spectrum as well (L. Gionfriddo [20]); we return to these structures in a greater detail in Sections 6 and 7, respectively.

The discovery of gaps in the chromatic spectrum has far-reaching consequences in general coloring theory and its applications. As a matter of fact, mixed hypergraph colorings can model many combinatorial problems in a much more general context than graph colorings do. For example, they can model list colorings without lists, resource allocation, Ramsey-type problems, graph homomorphisms, etc., see [47, Chapter 12] and [27]. The presence or absence of gaps in the chromatic spectrum is vitally important in many of such applications.

Open problems.

1. Investigate the (non-) existence of gaps in the chromatic spectrum of further types of mixed hypergraphs.

2. Develop efficient algorithms to determine if the chromatic spectrum of mixed hypergraphs from some classes is gap-free.

3. Given a finite set S of positive integers, determine or estimate the minimum and maximum numbers of ($\mathcal{C}$-, $\mathcal{D}$-, bi-) edges in a mixed (bi-) hypergraph whose feasible set is S.

4. Is there a mixed hypergraph with gaps in the coefficients of its chromatic polynomial? ([47])

5. Investigate 'gap-critical' mixed hypergraphs; i.e., those with gaps in their chromatic spectrum, such that after deleting any edge some gaps disappear.

5. PERFECT MIXED HYPERGRAPHS

Every graph G satisfies the trivial inequality $\chi(G) \geq \omega(G)$, where $\omega(G)$ is the size of the largest clique and $\chi(G)$ is the chromatic number in the usual sense. The *perfect graphs* are the graphs such that the equality $\chi(G') = \omega(G')$ holds for every induced subgraph G' of G. Many subclasses of perfect graphs have beautiful structural properties and admit fast algorithms for various optimization problems.

A natural analogue of perfection for the upper chromatic number was introduced in [46]. In a mixed hypergraph, a set of vertices is *C-stable* if it contains no C-edges. The *C-stability number* $\alpha_C(\mathcal{H})$ is the maximum cardinality of a C-stable set in $\mathcal{H}$. It is easy to see that the upper bound $\bar{\chi}(\mathcal{H}) \leq \alpha_C(\mathcal{H})$ always is valid, because a set with more colors than $\alpha_C(\mathcal{H})$ would assign distinct colors to all the vertices of some C-edge. A mixed hypergraph $\mathcal{H}$ is called *perfect* [46] if $\bar{\chi}(\mathcal{H}') = \alpha_C(\mathcal{H}')$ holds for every induced subhypergraph $\mathcal{H}'$. Notice that the perfection of *graphs* is related to the *lower* chromatic number, while the perfection of *hypergraphs* involves the *upper* chromatic number. In this setting, every $\mathcal{D}$-hypergraph (that is, hypergraph in the classical sense, without C-edges) is perfect, because – in each of its subhypergraphs – the upper chromatic number is equal to the number of vertices.

Several classes of perfect and minimal imperfect mixed hypergraphs have been found. A *cycloid* [46] is an r-uniform C-hypergraph on n vertices, denoted by C_n^r, which has n C-edges and admits a simple cycle on n vertices as a host graph. A *polystar* is a mixed hypergraph with at least two C-edges, in which the set Y of vertices common to all C-edges (center) is nonempty, and every vertex pair in Y forms a $\mathcal{D}$-edge. When the center consists of just one vertex, the polystar is also called *monostar*. Hence, *every* polystar in a C-hypergraph is a monostar. A *bistar* (called co-bistar in [46]) is a mixed hypergraph in which there exists a pair of distinct vertices common to all C-edges but not forming a $\mathcal{D}$-edge.

Bistars are perfect, while polystars are not [46]. Also, cycloids of the form C_{2r-1}^r are not perfect [46]. Indeed, when $n = 2r - 1$, we have $\alpha_C(C_n^r) = 2r - 3$ and $\bar{\chi}(C_n^r) = 2r - 4$. These cycloids are analogous to the well-known minimal imperfect graphs. Polystars and cycloids of the form C_{2r-1}^r, $r \geq 3$, are minimal imperfect mixed hypergraphs in the sense that every proper induced subhypergraph of such a cycloid is perfect, and every subhypergraph of a polystar that is not a polystar is perfect. (A cycloid on fewer than $2r - 1$ vertices is perfect, whereas on more than $2r - 1$ vertices it contains a monostar, hence it is imperfect but *not minimally* imperfect.)

It was conjectured in [46] that an r-uniform C-hypergraph is perfect if and only if it has no induced monostar or cycloid of the form C_{2r-1}^r, $r \geq 3$. These two natural imperfect families served as an analogue of Berge's Strong Perfect Graph Conjecture, which stated that a graph G is perfect if and only if no odd cycle of length at least 5 occurs as an induced subgraph of G or its complement $\overline{G}$ (proved recently by Chudnovsky et al. [13]).

There are some classes of $\mathcal{C}$-hypergraphs for which the conjecture is true. For example, Bulgaru and Voloshin [12] proved that a mixed interval hypergraph is perfect if and only if it has no induced polystars. In [47] it was proved that if a mixed hypertree does not contain polystars as partial subhypergraphs, then it is perfect. It is clear that the situation is more complex than in case of graphs. (We are not formulating any guess about the complexity of a possible proof...)

Kráľ [25] has disproved the mixed hypergraph perfection conjecture for each $r \geq 3$, by constructing a new family of minimal imperfect $\mathcal{C}$-hypergraphs (one $\mathcal{C}$-hypergraph for each r, on $2r$ vertices) different from monostars and cycloids. Recently, Bujtás and Tuza [11] have found a larger family of counterexamples for $r \geq 4$, an increasing number of minimally imperfect r-uniform $\mathcal{C}$-hypergraphs as r gets large.

Regarding the 3-uniform case, up to isomorphism the following six examples of minimal imperfect $\mathcal{C}$-hypergraphs are known:

- $V_1 = \big(\{1, 2, 3, 4\}, \{\{1, 2, 3\}, \{1, 3, 4\}, \{1, 2, 4\}\}\big)$ – monostar,
- $V_2 = \big(\{1, 2, 3, 4, 5\}, \{\{1, 2, 3\}, \{1, 4, 5\}\}\big)$ – monostar,
- $V_3 = \big(\{1, 2, 3, 4, 5\}, \{\{1, 2, 3\}, \{1, 3, 4\}, \{1, 4, 5\}\}\big)$ – monostar,
- $V_4 = \big(\{1, 2, 3, 4, 5\}, \{\{1, 2, 3\}, \{1, 3, 4\}, \{1, 4, 5\}, \{1, 2, 5\}\}\big)$ – monostar,
- $V_5 = \big(\{1, 2, 3, 4, 5\}, \{\{1, 2, 3\}, \{2, 3, 4\}, \{3, 4, 5\}, \{4, 5, 1\}, \{5, 1, 2\}\}\big)$ – cycloid C_5^3,
- $K_1 = \big(\{1, 2, 3, 4, 5, 6\}, \{\{1, 2, 4\}, \{2, 3, 5\}, \{3, 4, 6\}, \{4, 5, 1\}, \{5, 6, 2\}, \{6, 1, 3\}, \{1, 3, 5\}, \{2, 4, 6\}\}\big)$ – Kráľ's construction.

There may be other minimal imperfect 3-uniform mixed hypergraphs, as it is the case for $r \geq 4$. On the other hand, polystars generally are not uniform and they already indicate that the family of non-uniform minimal imperfect mixed hypergraphs may be complex. All these results and investigations will lead sooner or later to a more general conjecture about perfect mixed hypergraphs.

Algorithmic complexity aspects of perfection. It is well-known that perfection on graphs has led to efficient polynomial- and linear-time algorithms for solving several problems (not only coloring, but also maximum-weight clique, minimum vertex cover, etc.) that are NP-complete in general.

One can already see that a similar situation occurs with the perfection of hypergraphs.

All classes of perfect mixed hypergraphs known so far can be upper-colored efficiently. These include bistars, mixed interval hypergraphs (Bulgaru, Voloshin [12]) and quasi-interval C-hypergraphs (Prisakaru [41]). These were the simplest cases of perfect mixed hypergraphs. As we have already mentioned, if a mixed hypertree does not contain monostars, then it is perfect. In this case there is an efficient polynomial algorithm (and it is even possible to develop a linear-time algorithm, too) for finding the upper chromatic number and a respective coloring. When monostars are allowed in C-hypertrees, the problem is already NP-complete; it is NP-complete even for monostars themselves [18]. One may expect that perfection will lead to efficient polynomial-time algorithms for finding $\bar{\chi}$ and properly coloring the given hypergraph. In addition, perfection may serve as a hint for the search of efficient polynomial algorithms for other hard combinatorial problems on discrete structures.

Open problems.

1. Search for new classes of perfect and minimal imperfect mixed hypergraphs.

2. Describe classes of uniform C-hypergraphs in which the exclusion of monostars and cycloids of certain lengths implies perfectness.

3. Develop efficient algorithms for computing the upper chromatic number, and for finding maximum colorings, for various classes of perfect mixed hypergraphs.

4. *Prove or disprove:* A 3-uniform C-hypergraph is perfect if and only if it does not contain any of the families V_1–V_5 and K_1 above as an induced subhypergraph.

6. PLANAR MIXED HYPERGRAPHS

Let $\mathcal{H} = (X, \mathcal{E})$ be a hypergraph. The *bipartite representation* of $\mathcal{H}$, denoted by $B(\mathcal{H})$, is the bipartite graph with vertex set $X \cup \mathcal{E}$, where $x \in X$ is adjacent to $E \in \mathcal{E}$ in $B(\mathcal{H})$ if and only if $x \in E$ in $\mathcal{H}$. The following definition is due to Zykov [51]: a hypergraph $\mathcal{H}$ is called *planar* if $B(\mathcal{H})$ is a planar graph.

In this way, planar graphs are the special cases of planar hypergraphs, in which all edges have size 2. As one can see, a planar hypergraph admits an embedding in the plane in such a way that each vertex corresponds to a point in the plane, and each edge corresponds to a closed region homeomorphic to a disk such that its boundary contains the points corresponding to its vertices, and it does not contain any points corresponding to the other vertices. Furthermore, two such regions intersect exactly in the points that correspond to the vertices in the intersection of the corresponding edges. In this way, the *faces of the embedding* of the planar hypergraph are formed by those connected regions of the plane which do not correspond to the edges.

Using properties of the bipartite representation $B(\mathcal{H})$, one can derive many properties of a planar embedding of the hypergraph $\mathcal{H}$. For example, denoting the degree of vertex $x \in X$ in $\mathcal{H}$ by $d_{\mathcal{H}}(x)$, we obtain the following generalization [32] of Euler's formula: for any planar embedding of $\mathcal{H} = (X, \mathcal{E})$ with f faces,

$$|X| + |\mathcal{E}| - \sum_{E \in \mathcal{E}} |E| + f = |X| + |\mathcal{E}| - \sum_{x \in X} d_{\mathcal{H}}(x) + f = 2.$$

In particular, the number of faces is independent of the embedding.

An embedding of a planar hypergraph is called *maximal* if every face (including the unbounded face, too) contains precisely two vertices; or, equivalently, if in the corresponding embedding of $B(\mathcal{H})$ every face is a cycle of length 4. A planar hypergraph is maximal if it has a maximal embedding in the plane. This maximality is relative in the sense that in every such face one can always insert a new edge of size 2. However, if a planar hypergraph $\mathcal{H}$ is not maximal, then there is at least one face of size at least 3, and therefore one can insert a new edge of size at least 3 inside that face.

If we draw the faces of a maximal planar hypergraph as curves connecting respective vertices, then we obtain a plane graph whose faces correspond to the edges of the initial hypergraph. In this way, we may look at a plane graph as a planar embedding of a maximal hypergraph such that the faces of the graph correspond to the edges of the hypergraph.

Let $\mathcal{H} = (X, \mathcal{C}, \mathcal{D})$ be a mixed hypergraph. Denote the underlying edge set of $\mathcal{H}$ by $\mathcal{E} = \mathcal{C} \cup \mathcal{D}$; if a $\mathcal{C}$-edge and a $\mathcal{D}$-edge consist of the same set of vertices (i.e., it is a bi-edge), then this set appears only once in $\mathcal{E}$. We say that $\mathcal{H}' = (X, \mathcal{E})$ is the *underlying hypergraph* of $\mathcal{H}$.

A mixed hypergraph $\mathcal{H} = (X, \mathcal{C}, \mathcal{D})$ is planar if and only if its underlying hypergraph $\mathcal{H}'$ is planar. This can be verified by embedding $\mathcal{H}'$ in the plane and labeling each hyperedge with B, C, or D appropriately, according to whether it is a bi-edge, C-edge, or D-edge. Note that C-edges of size 2 can be contracted, and bi-edges of size 2 lead to uncolorability, so that in general it suffices to restrict our attention to mixed hypergraphs containing neither.

The question of coloring properties of general planar mixed hypergraphs was first raised in [46] (Problem 8, p. 43). This class already contains uncolorable members. The smallest non-trivial (reduced) uncolorable planar mixed hypergraph $\mathcal{H} = (X, C, D)$ has three vertices and four edges: $X = \{1, 2, 3\}$, $C = \big\{ \{1, 2, 3\} \big\}$, $D = \big\{ \{1, 2\}, \{1, 3\}, \{2, 3\} \big\}$. One can easily embed it in the plane with 4 faces (3 of them containing 2 vertices each, and 1 containing 3 vertices). It is not difficult to extend this example to an infinite family of minimal uncolorable planar mixed hypergraphs.

The structure of uncolorable planar mixed hypergraphs is unknown. In general, allowing D-edges of size 2 implies that the four-color problem is a special case of the theory of planar mixed hypergraphs. Therefore, it is reasonable to distinguish planar mixed hypergraphs without edges of size 2 from those containing edges of size 2.

The first interesting case is where $\mathcal{H} = (X, C, D)$ is a *3-uniform maximal planar bi-hypergraph*. Since maximality means that every face is of size 2, we can associate a graph $G(\mathcal{H})$ with $\mathcal{H}$, on the same vertex set: replace each face in $\mathcal{H}$ by an edge in G, so that every edge of $\mathcal{H}$ becomes a face of G. Since $\mathcal{H}$ is maximal 3-uniform, G must be a triangulation in the usual sense. We call both $\mathcal{H}$ and G *bi-triangulations* because every edge of $\mathcal{H}$ is a bi-edge.

Colorings of bi-triangulations have been investigated (Kündgen, Mendelsohn, Voloshin [32]). It has been proved that they are always colorable, the chromatic spectrum is gap-free, and, moreover, their chromatic polynomial has a very special form.

An important discovery in the coloring of planar mixed hypergraphs is that their chromatic spectrum may have gaps. Kobler and Kündgen's smallest example [24] has 6 vertices and its feasible set is $\{2, 4\}$. Moreover, it is proved in [24] that a nonempty set S of positive integers is the feasible set of some planar mixed hypergraph if and only if S is an interval $\{s, s + 1, \ldots, t\}$ with $1 \leq s \leq 4$ or is of the form $\{2, 4, 5, \ldots, t\}$. In other words, planar mixed hypergraphs may have gaps, but the gap can only occur at 3.

Open problems.

1. Characterize the chromatic spectra of planar mixed hypergraphs.

2. Search for the criteria of uncolorability and unique colorability for various subclasses of planar mixed hypergraphs (e.g., r-uniform, containing a fixed number of edges of size 2, pseudo-chordal, etc.)

3. Characterize perfect planar mixed hypergraphs.

7. Coloring Block Designs

For integers $v \geq k > t \geq 2$, a Steiner system $S(t, k, v)$ (of index 1) is a family $\mathcal{B}$ of k-element subsets – called *blocks* – over a v-element vertex set X, such that every t-subset of X is contained in precisely one block. For the particular cases of $(t, k) = (2, 3)$ and $(t, k) = (3, 4)$, the notation $STS(v)$ and $SQS(v)$ are commonly used (Steiner triple systems and Steiner quadruple systems of order v, respectively).

When looking at Steiner systems as mixed hypergraphs, in principle one might decide on each block independently whether it is a $\mathcal{C}$-edge, or a $\mathcal{D}$-edge, or a bi-edge. Homogeneous conditions, however, are of primary interest. To distinguish between the three basic types, we use different notation.

- When all the blocks are regarded as $\mathcal{D}$-edges, we have a $\mathcal{D}$-hypergraph of the type $\mathcal{H} = (X, \emptyset, \mathcal{B})$. In this case, we keep the classic notation $S(t, k, v)$ or $STS(v)$ or $SQS(v)$.

- When all the blocks are regarded as $\mathcal{C}$-edges, we deal with a $\mathcal{C}$-hypergraph of the type $\mathcal{H} = (X, \mathcal{B}, \emptyset)$. In this case, we will use the notation $CS(t, k, v)$ or $CSTS(v)$ or $CSQS(v)$.

- Finally, when all the blocks are bi-edges, we consider a bi-hypergraph of the type $\mathcal{H} = (X, \mathcal{B}, \mathcal{B})$. In this case, we will use the notation $BS(t, k, v)$ or $BSTS(v)$ or $BSQS(v)$.

Evidently, all STS, SQS, and all $CSTS$, $CSQS$ are colorable. The upper chromatic number of $CSTS(v)$ is fairly well understood; for instance, if $v \leq 2^k - 1$, then $\bar{\chi}(CSTS(v)) \leq k$, and the cases of equality are structurally characterized (Milazzo, Tuza [35, 36]). More generally, $\bar{\chi}(CS(k-1, k, v)) = O(\ln v)$ holds for every fixed k; but if $t \leq k - 2$, then $\bar{\chi}(CS(k-1, k, v))$ grows at least as fast as v^c for some constant $c > 0$ (Milazzo, Tuza [37]).

However, some $BSTS, BSQS$ may be uncolorable. More specifically, there are uncolorable systems $BSTS$ for all orders at least 15, see [19, 35]. This information and discussions with Alex Rosa (1998) gave rise to the idea that it may very well be that almost all triple systems $BSTS$ are uncolorable. On the other hand, no uncolorable $BSQS$ has been found yet. For small orders it is known [33] that all $BSQS$ on $v \le 16$ vertices are colorable.

An interesting discovery was that some P_3-designs have gaps in their chromatic spectra, (L. Gionfriddo [20]). It raises questions about the existence of other designs with or without gaps in their chromatic spectra. M. Gionfriddo conjectured in [21] that every $BSTS$ has a gap-free chromatic spectrum.

A survey on colorings of mixed Steiner systems can be found in [38]. Very recently, it has been proved by Bacsó and Tuza in [1] that the best possible general upper bound for the upper chromatic number of finite projective planes of order q is equal to $q^2 - q - \Theta(\sqrt{q})$ as q tends to infinity, both when considered as $\mathcal{C}$- and bi-hypergraphs.

Open problems.
1. Prove or disprove that almost all $BSTS$ are uncolorable.
2. Prove or disprove that all $BSTS$ have gap-free chromatic spectrum.
3. Find the order of the smallest uncolorable $BSQS$ (if there are any).
4. Do there exist $BSQS$ with arbitrarily large upper chromatic number?
5. Do there exist $BSQS$ with arbitrarily large lower chromatic number?
6. Determine $\limsup_{v \to \infty} \bar{\chi}(CSQS(v)) / \log_2 v$ taken over Steiner quadruple systems as $\mathcal{C}$-hypergraphs. Is this limit equal to 1?
7. For $k \ge t + 2 \ge 4$, determine the largest possible exponent $c = c(k, t)$ such that $\bar{\chi}(S) \ge \Omega(v^c)$ for all Steiner systems $S = CS(t, k, v)$.
8. Estimate the upper chromatic number of lines and of higher subspaces in the finite projective and affine spaces viewed as $\mathcal{C}$- or bi-hypergraphs, as a function of order and dimension.

8. MORE GENERAL STRUCTURES

In this concluding section we mention some recent concepts that are generalizations of, or closely related to mixed hypergraphs. We are going to proceed with those structure classes according to a decreasing generality. At the time of writing this survey, all related manuscripts are still in the process of publication. Only few selected results will be mentioned from them, just to illustrate the flavor of those studies.

To a wide extent, the most general model is that of *pattern hypergraphs*, introduced by Dvořák et al. [17]. Each edge is associated with the family of its feasible color partitions – without any *a priori* restrictions on those feasible families – and a coloring of the hypergraph is proper if its color classes induce a feasible partition on each edge. Despite that the model is extremely general, a necessary and sufficient condition can be given for the presence of gaps in the chromatic spectrum. The characterization is established by identifying four special classes of edge patterns.

In the more specified but still fairly general concept of *stably bounded hypergraphs* [9, 10] the patterns are restricted by four functions s, t, a b on the edges, with the following meaning. A coloring is proper if, in each edge E, the largest cardinality of a *monochromatic* subset is at least $a(E)$ and at most $b(E)$, whereas the largest cardinality of a *totally multicolored* subset is at least $s(E)$ and at most $t(E)$. The subclass of *color-bounded hypergraphs* (introduced in [5] under a different name) assumes the functions s and t only, but it still includes the mixed hypergraphs by $s(E) = 2$ (with the trivial bound $t(E) = |E|$) for $\mathcal{D}$-edges, $t(E) = |E| - 1$ (with $s(E) = 1$) for $\mathcal{C}$-edges, and both $s(E) = 2$ and $t(E) = |E| - 1$ for bi-edges.

Concerning the various combinations of the four possible color-bound functions there have been investigations into their interrelations with respect to the possible gaps in feasible sets, the hierarchy of the sets of chromatic polynomials, and substantial differences in the time complexity of unique $(n - 1)$-colorability ([7, 9]); moreover, the unexpectedly rich family of feasible sets for color-bounded hypertrees and efficient algorithms on various subclasses of stably bounded interval hypergraphs and hypertrees have been studied ([8, 10]).

Drgas-Burchardt and Łazuka [16] have investigated the chromatic polynomials for the class of hypergraphs that can be expressed, using the notation above, with the function s alone. That is, the number of colors is lower-bounded in each edge, but no other types of restrictions are considered. In the paper [16], which was a substantial motivation also for introducing the generalizations above, the chromatic polynomial $P(\mathcal{H}, \lambda)$ has been investigated. It has been proved that there exists a strict connection between the coefficients of $P(\mathcal{H}, \lambda)$ and the sizes of the edges of $\mathcal{H}$. Consequences concerning the combinatorial meaning of the first three coefficients have been derived. (It is easy to see that the chromatic polynomial has order n, and the leading coefficient always equals 1. These properties remain valid in the more general model involving the two color-bound functions s and b together, but fail already in the class of mixed hypergraphs.)

Note Added in Proof. Recently, Cs. Bujtás and Zs. Tuza solved asymptotically the extremal problem discussed in the fifth paragraph of Section 4, moreover they characterized several further classes of perfect C-hypergraphs.

REFERENCES

[1] G. Bacsó and Zs. Tuza, Upper chromatic number of finite projective planes, *J. Combin. Designs*, to appear.

[2] G. Bacsó, Zs. Tuza and V. Voloshin, Unique colorings of bi-hypergraphs, *Australasian J. Combin.*, **27** (2003), 33–45.

[3] C. Berge, *Hypergraphs: Combinatorics of Finite Sets*, North Holland (1989).

[4] Cs. Bujtás and Zs. Tuza, Orderings of uniquely colorable mixed hypergraphs, *Discrete Applied Math.*, **155** (2007), 1395–1407.

[5] Cs. Bujtás and Zs. Tuza, Mixed colorings of hypergraphs, *Electronic Notes in Discrete Mathematics*, **24** (2006), 273–275.

[6] Cs. Bujtás and Zs. Tuza, Uniform mixed hypergraphs: The possible numbers of colors, *Graphs Combin.*, to appear.

[7] Cs. Bujtás and Zs. Tuza, Color-bounded hypergraphs, I: General results. Manuscript (2006).

[8] Cs. Bujtás and Zs. Tuza, Color-bounded hypergraphs, II: Interval hypergraphs and hypertrees. Manuscript (2006).

[9] Cs. Bujtás and Zs. Tuza, Color-bounded hypergraphs, III: Model comparison, *Applicable Analysis and Discrete Math.*, **1** (2007), 36–55.

[10] Cs. Bujtás and Zs. Tuza, Color-bounded hypergraphs, IV: Stable colorings of hypertrees (in preparation).

[11] Cs. Bujtás and Zs. Tuza, C-perfect hypergraphs (in preparation).

[12] E. Bulgaru and V. Voloshin, Mixed interval hypergraphs, *Discrete Applied Math.*, **77(1)** (1997), 24–41.

[13] M. Chudnovsky, N. Robertson, P. Seymour and R. Thomas, The strong perfect graph theorem, *Annals of Math.*, **164(1)** (2006), 51–229.

[14] K. Diao, P. Zhao and H. Zhou, About the upper chromatic number of a co-hypergraph, *Discrete Math.*, **220** (2000), 67–73.

[15] K. Diao, G. Liu, D. Rautenbach and P. Zhao, A note on the least number of edges of 3-uniform hypergraphs with upper chromatic number 2, *Discrete Math.*, **306** (2006), 670–672.

[16]　E. Drgas-Burchardt and E. Łazuka, On chromatic polynomials of hypergraphs, *Applied Math. Letters,* to appear.

[17]　Dvořák, Z., J. Kára, D. Král' and O. Pangrác, Pattern hypergraphs. Manuscript (2004).

[18]　E. Flocos, Proprietati cromatice ale co-monostelelor, *Buletinul Academiei de Stiinte a RM, Matematica,* Chisinau (1997), No. 3, pp. 8–19. (in Romanian).

[19]　B. Ganter, private communication (1997).

[20]　L. Gionfriddo, Voloshin's colourings of P_3-designs, *Discrete Math.,* **275** (2004), 137–149.

[21]　M. Gionfriddo, Colourings of hypergraphs and mixed hypergraphs, *Rendiconti del Seminario Matematico di Messina. Serie II,* Tomo XXV, Vol. 9 (2003), 87–98.

[22]　M. Gionfriddo, L. Milazzo and V. Voloshin, On the upper chromatic index of a multigraph, *Computer Sci. J. Moldova,* **10(1)** [28] (2002), 81–91.

[23]　T. Jiang, D. Mubayi, V. Voloshin, Zs. Tuza and D.B. West, The chromatic spectrum of mixed hypergraphs, *Graphs Combin.,* **18** (2002), 309–318.

[24]　D. Kobler and A. Kündgen, Gaps in the chromatic spectrum of face-constrained plane graphs, *Electron. J. Combin.,* **8(1)** (2001), #N3.

[25]　D. Král', A counter-example to Voloshin's hypergraph co-perfectness conjecture, *Australasian J. Combin.,* **27** (2003), 25–41.

[26]　D. Král', On feasible sets of mixed hypergraphs, *Electron. J. Combin.,* **11(1)** (2004), #R19.

[27]　D. Král', Mixed Hypergraphs and other coloring problems, *Discrete Math.,* **307(7–8)** (2007), 923–938.

[28]　D. Král', J. Kratochvíl, A. Proskurowski and H.-J. Voss, Coloring mixed hypertrees, *Discrete Applied Mathematics,* **154(4)** (2006), 660–672.

[29]　D. Král', J. Kratochvíl and H.-J. Voss, Mixed hypergraphs with bounded degree: edge-coloring of mixed multigraphs, *Theoretical Computer Science,* **295(1–3)** (2003), 263–278.

[30]　D. Král', J. Kratochvíl and H.-J. Voss, Mixed hypercacti, *Discrete Math.,* **286(1–2)** (2004), 99–113.

[31]　J. Kratochvíl, Zs. Tuza and M. Voigt, New trends in the theory of graph colorings: Choosability and list coloring, in: *Contemporary Trends in Discrete Mathematics* (R. L. Graham et al., eds.), *DIMACS Series in Discrete Mathematics and Theoretical Computer Science,* **49**, Amer. Math. Soc. (1999), pp. 183–197.

[32]　A. Kündgen, E. Mendelsohn and V. I. Voloshin, Coloring of planar mixed hypergraphs, *Electron. J. Combin.,* **7** (2000), #R60.

[33]　G. Lo Faro, L. Milazzo and A. Tripodi, The first *BSTS* with different lower and upper chromatic numbers, *Australasian J. Combin.,* **22** (2000), 123–133.

[34]　D. Lozovanu and V. Voloshin, Integer programming models for colorings of mixed hypergraphs, *Computer Sci. J. Moldova,* **8** (2000), 64–74.

[35] L. Milazzo and Zs. Tuza, Upper chromatic number of Steiner triple and quadruple systems, *Discrete Math.*, **174** (1997), 247–259.

[36] L. Milazzo and Zs. Tuza, Strict colourings for classes of Steiner triple systems, *Discrete Math.*, **182** (1998), 233–243.

[37] L. Milazzo and Zs. Tuza, Logarithmic upper bound for the upper chromatic number of $S(t, t + 1, v)$ systems. Manuscript (2004).

[38] L. Milazzo, Zs. Tuza and V. I. Voloshin, Strict colorings of Steiner triple and quadruple systems: a survey, *Discrete Math.*, **261** (2003), 399–411.

[39] A. Niculitsa and V. Voloshin, About uniquely colorable mixed hypertrees, *Discuss. Math. Graph Theory,* **20(1)** (2000), 81–91.

[40] A. Niculitsa and H.-J. Voss, A characterization of uniquely colorable mixed hypergraphs of order n with upper chromatic numbers $n - 1$ and $n - 2$, *Australasian J. Combin.*, **21** (2000), 167–177.

[41] V. Prisakaru, The upper chromatic number of quasi-interval co-hypergraphs, *Le Mathematiche,* LII (1997) – Fasc. II, 237–260.

[42] Zs. Tuza, Graph colorings with local constraints – A survey, *Discuss. Math. Graph Theory,* **17(2)** (1997), 161–228.

[43] Zs. Tuza and V. I. Voloshin, Uncolorable mixed hypergraphs, *Discrete Applied Math.*, **99** (2000), 209–227.

[44] Zs. Tuza, V. I. Voloshin and H. Zhou, Uniquely colorable mixed hypergraphs, *Discrete Math.*, **248** (2002), 221–236.

[45] V. I. Voloshin, The mixed hypergraphs, *Computer Sci. J. Moldova,* **1** (1993), 45–52.

[46] V. I. Voloshin, On the upper chromatic number of a hypergraph, *Australasian J. Combin.*, **11** (1995), 25–45.

[47] V. I. Voloshin, *Coloring Mixed Hypergraphs: Theory, Algorithms and Applications,* Fields Institute Monograph, Amer. Math. Soc. (2002).

[48] V. Voloshin, Mixed Hypergraph Coloring Web Site:
http://spectrum.troy.edu/~voloshin/mh.html

[49] V. Voloshin and H.-J. Voss, Circular mixed hypergraphs I: colorability and unique colorability, in: *Proceedings of the Thirty-first Southeastern International Conference on Combinatorics, Graph Theory and Computing,* Boca Raton, FL, 2000, *Congr. Numer.,* **144** (2000), pp. 207–219.

[50] V. I. Voloshin and H. Zhou, Pseudo-chordal mixed hypergraphs, *Discrete Math.*, **202** (1999), 239–248.

[51] A. A. Zykov, Hypergraphs, *Uspekhi Mat. Nauk,* **29** (1974), 89–154. (in Russian)

Zsolt Tuza

Computer and Automation Institute
Hungarian Academy of Sciences
H–1111 Budapest, Kende u. 13–17
Hungary

and

Department of Computer Science
University of Pannonia
H–8200 Veszprém
Hungary

Vitaly Voloshin

Department of Mathematics and
Physics
Troy University
Troy, AL 36082
U.S.A

BOLYAI SOCIETY
MATHEMATICAL STUDIES, 17

Horizons of Combinatorics
Balatonalmádi
pp. 257–280.

RANDOM DISCRETE MATRICES

VAN VU*

In this survey, we discuss some basic problems concerning random matrices with discrete distributions. Several new results, tools and conjectures will be presented.

1. INTRODUCTION

Random matrices is an important area of mathematics, with strong connections to many other areas (mathematical physics, combinatorics, theoretical computer science, to mention a few).

There are two types of random matrices: continuous and discrete. The continuous models have an established theory (see [41], for instance). On the other hand, the discrete models are still not very well understood. In this survey, we discuss a few basic problems concerning these models. The topics to be discussed are:

- The limiting distribution of the spectrum (Section 3).
- The spectral norm and the second largest eigenvalue (Sections 4, 5).
- Determinant (Section 6).
- Rank and Singular probability (Sections 7, 8).
- The condition number (Section 9).
- Tools from additive combinatorics (Sections 10, 11, 12).

---

*V. Vu is an A. Sloan Fellow and is supported by NSF Career Grant 0635606.

Notations. We denote by M_n the n by n random matrix whose entries are i.i.d. Bernoulli random variables (taking values 1 and -1 with probability $1/2$). This matrix is not symmetric. Symmetric matrices often come from graphs. We denote by $Q(n,p)$ the adjacency matrix of the Erdős–Rényi random graph $G(n,p)$. Thus $Q(n,p)$ is a random symmetric matrix whose upper diagonal entries are i.i.d. random variables taking value 1 with probability p and 0 with probability $q = 1 - p$. Another popular model for random graphs is that of random regular graphs. A random regular graph $G_{n,d}$ is obtained by sampling uniformly over the set of all simple d-regular graphs on the vertex set $\{1, \dots, n\}$. The adjacency matrix of this graph is denoted by $Q_{n,d}$.

In the whole paper, we assume that n is large. The asymptotic notation is used under the assumption that $n \to \infty$. We write $A \ll B$ if $A = o(B)$. c denotes a universal constant. All logarithms have natural base, if not specified otherwise.

2. The universality principle

Intuitively, one would expect a universal behavior among random models of the same object. For random matrices in particular, one would expect the distributions of specific eigenvalues to be the same (after a proper normalization), regardless of the model. Thus, given a theorem for continuous models, it is often simple to come up with a reasonable conjecture for discrete ones. For instance, there are fairly accurate tail estimates for the smallest singular value of a random matrix whose entries are i.i.d. Gaussians (see for example Theorem 9.2 in Section 9). It would be natural to try to prove similar estimates for a random matrix whose entries are i.i.d. Bernoulli. However, this kind of task is usually challenging, as the tools developed for continuous models are typically not applicable in a discrete setting. In the last few sections (Sections 10, 11, 12) of this survey we will present new tools developed recently in order to treat the discrete models. These tools, among others, reveal an intriguing connection between the theory of random matrices and additive combinatorics.

For random graphs, there is a specific conjecture which establishes the universality between the two models $G(n,p)$ and $G_{n,d}$ (Erdős–Rényi graphs and random regular graphs).

Conjecture 2.1 (Sandwich Conjecture) [31]. For $d \gg \log n$, there is a joint distribution (or coupling) on random graphs H, $G_{n,d}$, G such that

- H has the same distribution as $G(n, p_1)$ where $p_1 = \frac{d}{n}\left(1 - \frac{c\sqrt{d\log n}}{n}\right)$ and G has the same distribution as $G(n, p_2)$ where $p_2 = \frac{d}{n}\left(1 + \frac{c\sqrt{d\log n}}{n}\right)$.
- $\mathbf{P}(H \subset G_d) = 1 - o(1)$.
- $\mathbf{P}(G_{n,d} \subset G) = 1 - o(1)$.

The conjecture asserts that a random regular graph can be approximated from both below and above by Erdős–Rényi graphs of approximately the same densities. The conjecture has been proved for $d \leq n^{1/3-o(1)}$ [31].

Theorem 2.2. *The sandwich conjecture holds for* $\log n \ll d \ll n^{1/3}/\log^2 n$.

The main difficulty when dealing with the random regular graph $G_{n,d}$ is that the (upper diagonal) entries of its adjacency matrix are not independent variables. But using Theorem 2.2, one can often deduce information about the spectrum of $G_{n,d}$ using information about the spectrum of $G(n,p)$.

3. Limiting Distributions

One of the cornerstones of the theory of random matrices is Wigner's semicircle law, which established the limiting distribution of a certain class of random symmetric matrices [64]. We present here a more general version, due to Arnold [5].

Let a_{ij}, $1 \leq i \leq j \leq n$, be i.i.d. random variables with common variance one and distribution function $F(x)$ such that $\int_0^\infty |x|^k \, dF < \infty$ for all $k = 1, 2, \ldots$. Let A_n be the random symmetric matrix of size n whose upper diagonal entries are $\xi_{ij} = a_{ij}/2\sqrt{n}$. Let $\lambda_1 \geq \cdots \geq \lambda_n$ be the (real) eigenvalues of A_n. Define

$$W_n(x) := \frac{1}{n}\big|\{i \mid \lambda_i \leq x\}\big|.$$

Let $s(x)$ denote the semi-circle density function

$$s(x) := \frac{2}{\pi}\sqrt{1 - x^2}$$

for $|x| \leq 1$ and $s(x) := 0$ otherwise. Define

$$W(x) := \int_{-\infty}^{x} s(x)\partial x.$$

Theorem 3.1 (Semi-circle law). *With probability one,*

$$\lim_{n \to \infty} W_n(x) = W(x).$$

In order to prove the semi-circle law, Wigner introduced the so-called trace method, the heart of which is the calculation of the expectation of $Trace(A_n^k)$ for $k = 1, 2, \ldots$. This method is useful for many other problems (see Section 4 for example).

Let us now turn to the special matrix $Q(n, p)$. The entries of $Q(n, p)$ have variance $\sigma^2 = p(1 - p)$. Dividing each entry of $Q(n, p)$ by σ, we obtain a matrix $Q'(n, p)$ whose entries have common variance one. However, one cannot apply Theorem 3.1 directly as the entries of $Q'(n, p)$ do not have bounded moments when p tends to zero with n. On the other hand, by applying Wigner's trace method, one can prove the following theorem

Theorem 3.2 [23, 63]. *There is a constant c such that the following holds. Let $\lambda_1 \geq \cdots \geq \lambda_n$ be the eigenvalues of $Q(n, p)$ where $p \geq n^{-1} \log^c n$ and define*

$$R_n^1(x) := \frac{1}{n} \left| \left\{ i | \lambda_i \leq 2x \sqrt{np(1 - p)} \right\} \right|.$$

Then with probability one,

$$\lim_{n \to \infty} R_n^1(x) = W(x).$$

A corollary of a general theorem by Guionet and Zeitouni [27] shows that $R_n^1(x)$ and many other quantities concerning the spectrum of $Q(n, p)$ are highly concentrated.

Next we discuss the situation with the random regular graph $G_{n,d}$. Define

$$R_n^2(x) := \frac{1}{n} \left| \left\{ i | \lambda_i \leq 2x \sqrt{d - 1} \right\} \right|$$

and

$$s(d, x) := \frac{d^2 - d}{d^2 - 4(d - 1)x^2} \frac{2}{\pi} \sqrt{1 - x^2}$$

for $|x| \leq 1$ and $s(d, x) := 0$ otherwise. Define $W(d, x) := \int_{-\infty}^{x} s(d, x)\partial x$. A theorem of McKay [42] on the spectrum of regular graphs (not necessarily random) implies

Theorem 3.3 (Distribution of the eigenvalues in random regular graphs with fixed degree). *For any fixed d the following holds with probability one*

$$\lim_{n\to\infty} R_n^2(x) = W(d,x).$$

Observe that the limiting distribution $W(d,x)$ in this theorem is not semi-circular because of the extra term $\frac{d^2-d}{d^2-4(d-1)x^2}$. On the other hand, it becomes arbitrarily close to the semi-circle distribution if d is sufficiently large. Thus it is reasonable to conjecture that if d tends to infinity with n, $Q_{n,d}$ follows the semi-circle law. However, McKay's proof used Wigner's trace method and relied on the crucial fact that the graph has few small cycles. Theorem 3.3 still holds for $d = n^{o(1)}$. But for $d = n^\varepsilon$ with any constant $\varepsilon > 0$ the graph has too many small cycles and it seems very hard to apply this method. On the other hand, using the sandwiching theorem (Theorem 2.2), Vu and Wu [63] proved that if $\log n \ll d \ll n^{1/3}/\log^2 n$ then with probability one

$$\lim_{n\to\infty} R_n^2(x) = W(x).$$

If the sandwich conjecture holds for all $d \gg \log n$, then this statement can be extended for all $d \gg \log n$. Recently, Zeitouni (private communication) suggested to the author another approach that also seems to work for a wide range of d. The details will appear in [63]. For results concerning more general models of random graphs, see [13, 63].

To conclude this section, let us briefly discuss the case when A_n is not symmetric. Let ξ be a complex random variable with mean zero and bounded variance σ^2. Let N_n be a random matrix of order n with entries being i.i.d. copies of ξ. Let $\lambda_1, \ldots, \lambda_n$ be the eigenvalues of $\frac{1}{\sigma\sqrt{n}}N_n$. Define the *empirical spectral distribution* μ_n of N_n by the formula

$$\mu_n(s,t) := \frac{1}{n}\#\big\{k \le n \mid \Re(\lambda_k) \le s;\ \Im(\lambda_k) \le t\big\}.$$

The following famous conjecture has been open since the 1950's:

Circular law conjecture: μ_n converges to the uniform distribution μ_∞ over the unit disk as n tends to infinity.

Girko [24] and Bai [6] obtained important partial results concerning this conjecture. These results and some refinements are carefully discussed in the book [7]. There have been a rapid developments very recently by Götze-Tikhomirov [26], and Pan-Zhou [46], and Tao-Vu [60]. In particular, Tao and Vu [60] confirmed the conjecture under the slightly stronger assumption that the $(2 + \eta)$th-moment of ξ is bounded, for any $\eta > 0$.

4. The Spectral Norm

The spectral norm of an n by n matrix A is defined as

$$\|A\| = \sup_{v \in \mathbf{R}^n, \, \|v\|=1} |Av|.$$

If A is symmetric, then $\|A\|$ is the largest eigenvalue of A (in absolute value).

We consider the following general model of random symmetric matrices. Let ξ_{ij}, $1 \le i \le j \le n$, be independent (but not necessarily identical) random variables with the following properties

- $|\xi_{ij}| \le K$ for all $1 \le i \le j \le n$.
- $\mathbf{E}(\xi_{ij}) = 0$, for all $1 \le i < j \le n$.
- $\mathbf{Var}(\xi_{ij}) = \sigma^2$, for all $1 \le i < j \le n$.

For a moment, let us assume that σ and K are positive constants.

Define $\xi_{ji} = \xi_{ij}$ and consider the symmetric random matrix $A_n = (\xi_{ij})_1^n$. Notice that for any matrix A,

$$\|A\| = \lim_{k \to \infty} Tr(A^k)^{1/k}.$$

This suggests that the trace method would be an effective tool for bounding $\|A_n\|$. Indeed, all upper bounds mentioned below are based on this method. Füredi and Komlós [23] proved that a.s.

$$\|A_n\| \le 2\sigma\sqrt{n} + cn^{1/3}\ln n.$$

The error term $cn^{1/3}\ln n$ was improved to $cn^{1/4}\ln n$ by Vu [62], and further to $n^{-1/132+\varepsilon}$ by Peche and Soshnyikov [51]. From below, Alon, Krivelevich and Vu [4] showed that a.s.

$$2\sigma\sqrt{n} - c\ln n \le \|A_n\|,$$

for some constant c. Putting these bounds together, we have

Theorem 4.1. *For a random matrix A as above there is a positive constant $c = c(\sigma, K)$ such that*

$$2\sigma\sqrt{n} - c\ln n \le \|A\| \le 2\sigma\sqrt{n} + cn^{-1/132+\varepsilon},$$

holds almost surely.

In many situations σ and K may depend on n. A typical example is when A is the "normalized" adjacency matrix of $G(n,p)$ (1 and 0 are replaced by $1 - p$ and $-p$, respectively; this forces all entries to have mean zero) where p is decreasing with n ($p = n^{-\varepsilon}$, say). In this case, by following the proof of the upper bound in the previous theorem, one can obtain

Theorem 4.2 [62]. *There are constants c and c' such that the following holds. Let ξ_{ij}, $1 \le i \le j \le n$ be independent random variables, each of which has mean 0 and variance at most σ^2 and is bounded in absolute value by K, where $\sigma \ge c'n^{-1/2}K\ln^2 n$. Then almost surely*

$$\|A_n\| \le 2\sigma\sqrt{n} + c(K\sigma)^{1/2}n^{1/4}\ln n.$$

One can also obtain a somewhat weaker bound by following the proof from [23].

If one assumes more about the distributions of the entries ξ_{ij}, one can obtain a sharper bound via computing very high traces. See, for instance [52].

In certain algorithmic applications, it is useful to have a tail distribution for the spectral norm (see [1, 35], for instance). Using Talagrand's inequality, one can show

Theorem 4.3 [35, 4] (Concentration of the spectral norm). *There is a positive constant $c = c(K)$ such that for any $t > 0$*

$$\mathbf{P}\big(\big|\|A_n\| - \mathbf{E}\big(\|A_n\|\big)\big| \ge ct\big) \le 4e^{-t^2/32}.$$

Similar results hold for larger classes of random matrices and also for other eigenvalues (see [4, 43]).

In the case of $Q(n,p)$, if p is sufficiently large, then all rows have a.s. roughly np ones and the norm of $Q(n,p)$ is (a.s.) $(1 + o(1))np$. But if p is relatively small, this is no longer true. Krivelevich and Sudakov [33] proved that $\|Q(n,p)\|$ is almost surely

$$(1 + o(1))\max\big\{np, \sqrt{D}\big\}$$

where D denotes the maximum degree of the (random) graph. See [53, 29] for more results of this type.

5. THE SECOND EIGENVALUE OF RANDOM REGULAR GRAPHS

Let G be a graph on n points and A its adjacency matrix. Let $\lambda_1 \geq \lambda_2 \geq \cdots \geq \lambda_n$ be the eigenvalues of A. If G is d-regular, then $\lambda_1 = d$. In this case, a critical parameter of the graph is

$$\lambda(G) := \max\left\{ |\lambda_2|, |\lambda_n| \right\}.$$

In the literature, $\lambda(G)$ is frequently called the second eigenvalue of G. (This name is inaccurate but somewhat convenient.) A good way to think of $\lambda(G)$ is

$$\lambda(G) = \max_{\|v\|=1,\ v\cdot\mathbf{1}=0} |v^T \cdot Av|,$$

where $\mathbf{1}$ is the all ones vector. One can also think of $\lambda(G)$ as the spectral norm of the "normalized" adjacency matrix of G, where 1 and 0 are replaced by $(n-d)/n$ and $-d/n$, respectively.

One can derive many interesting properties of the graph G from the value of $\lambda(G)$. The general phenomenon here is that if $\lambda(G)$ is significantly less than d, then the edges of G distribute like those of a random graph with edge density d/n [3, 55, 12]. One can use this information to derive various properties of the graphs (see [34] for many results of this kind). The whole concept can be generalized to non-regular graphs. In this case, one needs to consider the Laplacian rather than the adjacency matrix (see, for example, [11]).

Estimating $\lambda(G)$ for a random regular graph is a well known problem in the discrete math/theoretical computer science community. A consequence of the well known Alon-Boppana bound [2] asserts that if d is fixed and n tends to infinity, a.s.

$$\lambda_2(G_{n,d}) \geq 2\sqrt{d-1} - o(1).$$

Since $\lambda(G) \geq \big|\lambda_2(G)\big|$ it follows that a.s.

$$\lambda(G_{n,d}) \geq 2\sqrt{d-1} - o(1).$$

Alon [2] conjectured that for any fixed d, a.s.

$$\lambda_2(G_{n,d}) = 2\sqrt{d-1} + o(1).$$

Friedman [20] and Kahn and Szemerédi [22] showed that if d is fixed and n tends to infinity, then a.s. $\lambda(G_{n,d}) = O(\sqrt{d})$. Recently, Friedman, in a highly technical paper [21], proved Alon's conjecture. In fact, he proved the stronger statement that a.s. $\lambda(G_{n,d}) = 2\sqrt{d-1} + o(1)$. This, together with the lower bound above, determines the asymptotics of $\lambda(G_{n,d})$.

Theorem 5.1 [21] (Second eigenvalue of random regular graphs with fixed degree). *For any fixed d and n tending to infinity, a.s.*

$$\lambda(G_{n,d}) = \big(2 + o(1)\big)\sqrt{d-1}.$$

A d-regular graph G is *Ramanujan* if $\lambda(G) \le 2\sqrt{d-1}$. Explicit constructions of Ramanujan graphs are highly non-trivial and usually come from deep results in number theory (see [38] or [39], for example). On the other hand, the following conjecture has been circulated in the last few years (mentioned to the author by Sarnak)

Conjecture 5.2. For d fixed and n tends to infinity, $G_{n,d}$ is Ramanujan with positive constant probability.

So far, we discussed the case when d is a constant. What happens if d also tends to infinity with n? It is not clear (at least to the author) that Friedman's proof of Alon's conjecture in [21] can be extended to this case. On the other hand, it is not hard to show that $\lambda\big(G(n,p)\big)$, where $G(n,p)$ is the Erdős–Rényi random graph, is $\big(2 + o(1)\big)\sqrt{np(1-p)}$ for sufficiently large p (e.g., $p \ge n^{-1+\varepsilon}$ for any fixed $0 < \varepsilon < 1$). Motivated by the universality principle, we make the following conjecture

Conjecture 5.3. Assume that $d \le n/2$ and both d and n tend to infinity. Then a.s.
$$\lambda(G_{n,d}) = \big(2 + o(1)\big)\sqrt{d(1 - d/n)}.$$

Nilli [44] showed that for any d-regular graph G having two edges with distance at least $2k + 2$ between them $\lambda_2(G) \ge 2\sqrt{d-1} - 2\sqrt{d-1}/(k+1)$. If $d = n^{o(1)}$ then $G_{n,d}$ has diameter $\omega(1)$ with probability $1 - o(1)$. Thus in this case
$$\lambda(G_{n,d}) \ge \lambda_2(G_{n,d}) \ge \big(2 + o(1)\big)\sqrt{d}$$

with probability a.s. This proves the lower bound in Conjecture 5.3. For a general d, it is easy to show (by computing the trace of the square of the adjacency matrix) that any d-regular graph G on n vertices satisfies

$$\lambda(G) \ge \sqrt{d(n-d)/(n-1)} \approx \sqrt{d(1 - d/n)}.$$

(We would like to thank N. Alon for pointing out this bound.)

Let us now turn to the upper bound. For $d = o(n^{1/2})$, one can follow the Kahn-Szemerédi approach to show that $\lambda(G_{n,d}) = O(\sqrt{d})$ a.s. For larger d, there is a weaker bound $o(d)$ [36, Theorem 2.8] proved by the trace method. The following two approaches look promising:

- (Suggested by Krivelevich) Combine the sharp concentration result in the previous section with the probability that a random graph is regular. Using this, one can show for example that $\lambda(G_{n,d}) = O(\sqrt{d \log n})$ for d close to n ($d = n/2$, for instance).

- The Sandwich Theorem (Theorem 2.2) implies

$$\lambda(G_{n,d}) = \lambda\big(G(n, d/n)\big) + O\big(\sqrt{d \log n}\big).$$

For most values of d, $\lambda\big(G(n, d/n)\big) = O(\sqrt{d})$. Thus, if the Sandwich conjecture holds, it would imply an upper bound of $O\big(\sqrt{d \log n}\big)$ for most values of d.

The author feels confident that one can prove that

$$\lambda(G_{n,d}) = O\big(\sqrt{d \log n}\big)$$

for all d using these approaches. However, removing the log term seems non-trivial. In fact, even the following special and weakened case already looks challenging

Problem. Prove that $\lambda(G_{n,n/2}) = O(\sqrt{n})$ almost surely.

6. Determinant

The problem of determining the determinant of M_n has been considered by various researchers for at least 40 years. It was proved by Komlós in 1967 [32] that almost surely $\det M_n$ is not zero. In fact, it is easy to see that $\det M_n$ is divisible by 2^{n-1}, thus it follows that a.s. $|\det M_n| \geq 2^{n-1}$. From above, Hadamard's inequality implies that $|\det M_n| \leq n^{n/2}$ (notice that all row vectors of M_n have length $\sqrt{n}$. It was often conjectured that with probability close to 1, $|\det M_n|$ is close to this upper bound.

Conjecture 6.1. Almost surely $|\det M_n| = n^{(1/2 - o(1))n}$.

This conjecture is supported by the following observation of Turán, whose proof is a simple application of the linearity of expectation.

Fact 6.2.
$$\mathbf{E}\big((\det M_n)^2\big) = n!\,.$$

It follows immediately by Markov's bound that for any function $\omega(n)$ tending to infinity with n, almost surely

$$|\det M_n| \leq \omega(n)\sqrt{n!}\,.$$

Tao and Vu [56] established the matching lower bound, which confirms Conjecture 6.1.

Theorem 6.3. *Almost surely*

$$|\det M_n| \geq \sqrt{n!}\exp\big(-29\sqrt{n\log n}\,\big).$$

We are going to sketch the proof very briefly as it contains a useful lemma. For a more detailed proof, we refer to [56].

Proof. We view $|\det M_n|$ as the volume of the parallelepiped spanned by n random $\{-1,1\}$ vectors. This volume is the product of the distances from the $(d+1)$st vector to the subspace spanned by the first d vectors, where d runs from 0 to $n-1$. We are able to obtain very tight control of this distance (as a random variable), thanks to the following lemma.

Lemma 6.4. *Let W be a fixed subspace of dimension $1 \leq d \leq n-4$ and X a random ± 1 vector. Then*

$$(1) \qquad \mathbf{E}\big(\operatorname{dist}(X,W)^2\big) = n - d.$$

Furthermore, for any $t > 0$

$$(2) \qquad \mathbf{P}\big(\big|\operatorname{dist}(X,W) - \sqrt{n-d}\big| \geq t+1\big) \leq 4\exp(-t^2/16).$$

Observe that in this lemma, we do not need to assume that W is spanned by random vectors. The lemma, however, is not applicable when d is very close to n as it does not imply that the distance is positive almost surely. In this case, we do need to use the assumption that W is random. This assumption allows us to derive information about the normal vector of W, which, combined with Erdős–Littlewood–Offord bound (see Theorem 10.1), provides control of the last few distances. ∎

Remark 6.5. After having written [56], Tao and the author discovered that Girko (Section 6 of [25]) claimed a very general theorem which implies Theorem 6.3. We are not able to understand his proof and have not found anyone who does.

Theorem 6.3 can be extended for much more general models of random matrices. Let ξ_{ij}, $1 \le i, j \le n$, be independent (but not necessarily i.i.d.) r.v.'s with the following two properties:

- Each ξ_{ij} has mean zero and variance one.
- There is a constant K that $|\xi_{ij}| \le K$ with probability one.

Theorem 6.6. *Consider the random matrix M_n' with entries ξ_{ij} as above. Let ε be an arbitrary positive constant. With probability $1 - o(1)$,*

$$| \det M_n'| \ge \sqrt{n!} \exp(-n^{1/2+\varepsilon}).$$

In certain situations, the assumption that $|\xi_{ij}|$ are bounded from above by a constant is too strong. We are going to consider the following less restricted model. Let ξ_{ij}, $1 \le i, j \le n$ be i.i.d. random variables with mean zero and variance one. Assume furthermore that their fourth moment is finite. Consider the random matrix M_n'' with ξ_{ij} as its entries.

Theorem 6.7 [56]. *We have, with probability $1 - o(1)$, that*

$$| \det M_n''| \ge n^{(1/2 - o(1))n}.$$

An open problem concerning determinants is to extend Theorem 6.3 to symmetric matrices. Let Q_n denote the random symmetric matrix whose upper diagonal entries are i.i.d. Bernoulli random variables.

Conjecture 6.8. Almost surely, $|\det Q_n| = n^{(1/2 + o(1))n}$.

It was proved only very recently [15] that a.s. $|\det Q_n|$ is positive (which can be seen as the symmetric version of Komlós' theorem mentioned above). The main difficulty here is that the row vectors of Q_n are no longer independent and so Lemma 6.4 is not applicable as there is a correlation between the subspace W and the vector X.

Finally, let us briefly discuss the situation with the permanent. Notice that the estimate in Fact 6.2 is still valid for the permanent of M_n. Thus, one would expect that, like the determinant, the permanent of M_n is typically of order $n^{(1/2 + o(1))n}$ (in absolute value). However, the following problem is still open.

Problem. Show that the permanent of M_n is a.s. not zero.

7. RANK AND SINGULAR PROBABILITY: NON-SYMMETRIC MODELS

Let us consider the basic model M_n and let p_n be the probability that M_n is singular. Estimating p_n is well known problem in discrete probability. From below it is clear that $p_n \geq \left(1/2 + o(1)\right)^n$, as a matrix is singular if it has two equal rows. A famous conjecture in the field asserts that this trivial lower bound is sharp.

Conjecture 7.1. $p_n = \left(1/2 + o(1)\right)^n$.

There is a refined version of the above conjecture where the right hand side is more precise (see [30]). However, Conjecture 7.1, as formulated, is still open.

It is already non-trivial to show that $p_n = o(1)$. As mentioned in the previous section, this was done by Komlós almost fourty years ago [32]. The bound on p_n in his original proof tends very slowly to zero with n. Later, he found a new proof which showed $p_n = O(n^{-1/2})$. In 1995, a breakthrough by Kahn, Komlós and Szemerédi [30] yielded the first exponential bound $p_n = O(.999^n)$. Their arguments were simplified by Tao and Vu in 2004 [56], resulting in a slightly better bound $O(.958^n)$ and a somewhat simpler proof. Shortly afterwards, Tao and Vu [57] combined the approach from [30] with ideas from additive combinatorics to obtain the following more significant improvement

Theorem 7.2 [57]. $p(n) \leq \left(3/4 + o(1)\right)^n$.

The proof in [57] is highly technical and requires many tools from discrete Fourier analysis, additive combinatorics and the geometry of numbers. On the other hand, the proving scheme is flexible and can be adapted to other models, sometime yielding (surprisingly) sharp bounds. Let us present one such result. Instead of M_n we consider the following "lazy" model M_n^{lazy}. The entries of M_n^{lazy} are i.i.d. random variables which equal zero with probability one half and 1 and -1 with probability one quarter. (If one thinks of the entries of M_n as fair coin flips, then in the "lazy" model about half of the time we are lazy and simply write zero instead flipping a coin.) It is clear that for the lazy model the singular probability p_n^{lazy} is again at least $\left(1/2 + o(1)\right)^n$ (which is the probability that there is a zero row). We are able to show that this bound is actually sharp

Theorem 7.3 [10]. $p_n^{\text{lazy}} = \left(1/2 + o(1)\right)^n$.

Let us conclude this section by two conjectures motivated by studies from random graphs. These questions concern the resilience of a structure, introduced in [31, 54]. Roughly speaking, the resilience of a structure S with respect to a property $\mathcal{P}$ measures how much we have to change S in order to destroy $\mathcal{P}$. We would like to measure the resilience of M_n with respect to the property of being non-singular.

Given $\{-1, 1\}$ matrix M, we denote by $\text{Res}(M)$ the minimum number of entries we need to switch (from 1 to -1 and vice versa) in order to make M singular. If M is a sample of M_n, it is easy to show that $\text{Res}(M)$ is, a.s., at most $\left(1/2 + o(1)\right) n$, as we can, a.s., change that many entries in the first row to make the first two rows equal. We conjecture that this is the best one can do.

Conjecture 7.4. Almost surely $\text{Res}(M_n) = \left(1/2 + o(1)\right) n$.

A closely related question (motivated by the notion of local resilience from [54]) is the following. Call a $\{-1, 1\}$ (n by n) matrix M *good* if all matrices obtained by switching (from 1 to -1 and vice versa) the diagonal entries of M are non-singular (there are 2^n such matrices).

Conjecture 7.5. Almost surely M_n is good.

8. Rank and Singular Probability: Symmetric Models

Let us now consider symmetric matrices. The symmetric counterpart of M_n is Q_n. In fact it is more convenient to consider $Q(n, 1/2)$ instead of Q_n, as the graph terminology is more convenient and leads to natural extensions. (It is easy to show that if $Q(n, 1/2)$ is a.s. non-singular then Q_n is and vice versa.)

While the non-singularity of M_n has been known for forty years since [32], that of $G(n, 1/2)$ was established only recently by Costello, Tao and

Vu [15]. This confirmed a conjecture of B. Weiss (this conjecture was communicated to the author by G. Kalai and N. Linial).

Theorem 8.1. $Q(n, 1/2)$ *is a.s. non-singular.*

As pointed out earlier in Section 6, the main difficulty in going from the non-symmetric setting to the symmetric one is that the row vectors in a random symmetric matrix are not independent. The key tool that helped us to overcome this difficulty was the so-called quadratic Littlewood–Offord inequality (Theorem 10.2), discussed in Section 10.

It is natural to ask if Theorem 8.1 still holds for a smaller density p. The answer is negative after a certain threshold. Indeed, if $p < (1-\varepsilon)\log n/n$ for some positive constant ε, then $G(n, p)$ has a.s. isolated vertices which means that its adjacency matrix has all zero rows and so is singular. Costello and Vu proved that $\log n/n$ is the right threshold.

Theorem 8.2 [14]. *For any constant* $\varepsilon > 0$, $Q\big(n, (1+\varepsilon)\log n/n\big)$ *is a.s. non-singular.*

It remains an interesting problem to estimate the rank of $Q(n, p)$ for $p < \log n/n$. The answer here is not yet conclusive, but some partial results are known. For instance, it was shown in [14] that if $p > (1+\varepsilon)\log n/2n$, then the rank of $Q(n, p)$ is a.s. equal n minus the number of isolated vertices in the graph. (The upper bound is trivial.)

Let us conclude this section by stating two conjectures. The first is a variant of Conjecture 7.1. We denote by p_n^{sym} the probability that $Q(n, 1/2)$ is singular. It is easy to show that this probability is at least $\big(1/2 + o(1)\big)^n$.

Conjecture 8.3. $p_n^{\mathrm{sym}} = \big(1/2 + o(1)\big)^n$.

This conjecture is perhaps very hard. The current best upper bound on p_n^{sym} is $n^{-1/4+o(1)}$. It seems already non-trivial to replace $1/4$ by an arbitrary constant C.

The second conjecture concerns random regular graphs. If $d = 2$, $G_{n,d}$ is a union of disjoint cycles and it is easy to show that its adjacency matrix $Q_{n,d}$ is a.s. singular, as many of these cycles have length divisible by 4. We conjecture that this is the only case.

Conjecture 8.4. For all $d \geq 3$, $Q_{n,d}$ is a.s. non-singular.

9. The Condition Number

For a matrix M the *condition number* $c(M)$ is defined as

$$c(M) := \| (M) \| \cdot \| (M^{-1}) \|.$$

We adopt the convention that $c(M)$ is infinite if M is not invertible.

The condition number plays a crucial role in applied linear algebra. In particular, the complexity of any algorithm which requires solving a system of linear equations usually involves the condition number of a matrix [8]. Another area of mathematics where this parameter is important is the theory of probability in Banach spaces (see [47] and the references therein).

The condition number of a random matrix is a well-studied object (see [16] and the references therein). In the case when the entries of M are i.i.d. Gaussian random variables (with mean zero and variance one), Edelman [16] proved

Theorem 9.1. *Let N_n be a $n \times n$ random matrix, whose entries are i.i.d. Gaussian random variables (with mean zero and variance one). Then* $\mathbf{E}\big(\ln c(N_n)\big) = \ln n + c + o(1)$, *where $c > 0$ is an explicit constant.*

In applications, it is usually useful to have a tail estimate. It was shown by Edelman and Sutton [17] that

Theorem 9.2. *Let N_n be a n by n random matrix, whose entries are i.i.d. Gaussian random variables (with mean zero and variance one). Then for any constant $A > 0$,*

$$\mathbf{P}\big(c(N_n) \geq n^{A+1}\big) = O_A\big(n^{-A}\big).$$

With the universality principle, it is reasonable to conjecture that this estimate holds for the random Bernoulli matrix M_n as well (see [49] for an even more precise conjecture). This was confirmed recently by Rudelson and Vershynin [50], following weaker results from [47, 58, 59].

Theorem 9.3. *For any constant $A > 0$,*

$$\mathbf{P}\big(c(M_n) \geq n^{A+1}\big) = O\big(n^{-A}\big).$$

In fact, the Rudelson and Vershyinin result holds for the general class of random sub-Gaussian matrix (see [50] for more details).

Theorem 9.3 still holds, in a somewhat weaker sense, if we replace M_n by $M + M_n$ where M is an arbitrary matrix with polynomially bounded norm.

Theorem 9.4 [59]. *For any positive constants A and C, there is a positive constant B such that the following holds. For any n by n matrix M where $\|M\| \leq n^C$,*

$$\mathbf{P}\big(c(M + M_n) \geq n^B\big) \leq n^{-A}.$$

The point here is that M itself can have very large condition number. (In fact if M is singular then its condition number is infinity.) Theorem 9.4 asserts that a Bernoulli perturbation of M has small condition number with high probability. The Gaussian version of Theorem 9.4 was proved by Spielman and Teng in [49]. For the connection of these theorems to probability, numerical linear algebra and theoretical computer science, we refer to [49] and [59, 60]. In particular, a variant of this result is critical in the proof of the Circular Law, mentioned in Section 3.

10. Littlewood–Offord and quadratic Littlewood–Offord

Let $\mathbf{v} = \{v_1, \ldots, v_n\}$ be a set of n integers and $\xi_1, \ldots, \xi_n$ be i.i.d. random Bernoulli variables. Define $S := \sum_{i=1}^n \xi_i v_i$ and $p_\mathbf{v}(a) := \mathbf{P}(S = a)$ and $p_\mathbf{v} := \sup_{a \in \mathbf{Z}} p_\mathbf{v}(a)$.

Erdős, answering a question of Littlewood and Offord, proved the following theorem, which we are referring to as the Erdős–Littlewood–Offord inequality.

Theorem 10.1. *Let $v_1, \ldots, v_n$ be non-zero numbers and ξ_i be i.i.d. Bernoulli random variables. Then*

$$p_\mathbf{v} \leq \frac{\binom{n}{\lfloor n/2 \rfloor}}{2^n} = O\big(\sqrt{n}\big).$$

Theorem 10.1 is a classical result in combinatorics and has many non-trivial extensions (see [61, Chapter 7] or [28] and the references therein).

The random sum $S := \sum_{i=1}^n \xi_i v_i$ plays a central role in the study of random Bernoulli matrices. In many problems (such as that of the determinant, rank or condition number), the critical parameter is the distance from a random Bernoulli vector to the hyperplane spanned by another $n-1$ random Bernoulli vectors (these are the row vectors of M_n). If $(v_1, \ldots, v_n)$ is the (unit) normal vector of the hyperplane, then this distance is exactly the absolute value of the random sum $S = \sum_{i=1}^n \xi_i v_i$.

Bounding the above mentioned distance is one of the main difficulties when one goes from Gaussian matrices to Bernoulli matrices. If the entries of the matrix are Gaussian, then the distance in question is a simple object. Thanks to symmetry, the position of the hyperplane does not really matter and so we can condition on it. Furthermore, the distribution of the distance from a random Guassian vector to a fixed hyperplane is well understood. The situation in the Bernoulli case is very different. In this case, the random vectors are chosen from the vertices of the n-dimensional $\{-1,1\}$ cube. Very little is known about the hyperplanes spanned by $n-1$ such vectors. Let us point out, however, that there are planes that contain a constant fraction of the vertices of the $\{-1,1\}$ cube and in this case the distance in question is zero with constant probability.

In order to treat random symmetric matrices (in particular $Q(n, 1/2)$), Costello, Tao and Vu [15] introduced the following quadratic version of Theorem 10.1

Theorem 10.2. *Let c_{ij}, $1 \le i,j \le n$ be non-zero numbers. Consider the quadratic form $F = \sum_{1 \le i \le j \le n} c_{ij}\xi_i\xi_j$, where ξ_i, $1 \le i \le n$ are i.i.d. Bernoulli random variables. Then for any a*

$$\mathbf{P}(F = a) = O(n^{-1/8}).$$

The exponent $-1/8$ was improved to $-1/4$ (see [14]). It is conjectured [14] that the sharp exponent would be $-1/2$. The lower bound is given by the quadratic form $F = \left(\sum_{i=1}^{n} \xi_i \right)^2$.

11. Random walks and Lazy Random Walks

Let $\mathbf{v} = \{v_1, v_1, \ldots, v_n\}$ be the set of n non-zero numbers and consider the random walk W on the real line (starting at 0) which at step i goes to the left by v_i with probability one half and to the right by v_i with probability one half. The probability $p_{\mathbf{v}}(0) = \mathbf{P}(\sum_{i=1}^{n} \xi_i v_i = 0)$ is exactly the probability that W returns to the origin after n steps.

Let μ be a constant between 0 and 1 and consider the "lazy" random walk W^{μ} (starting at 0) which at step i stays with probability $1-\mu$ and goes to the left by v_i with probability $\mu/2$ and to the right by v_i with probability $\mu/2$. Let $p_{\mathbf{v}}^{\mu}(0)$ be the probability that the lazy walk return to zero after n steps.

Intuitively, one would expect that $p_{\mathbf{v}}^{\mu}(0)$ is larger than $p_{\mathbf{v}}(0)$ (especially when μ is small), as the lazy walk has a stronger tendency to stay near the starting point. Quantitatively, one can show (using Fourier analysis and the elementary fact that $|\cos x| \leq 3/4 + 1/4 \cos 2x$) that for any $\mathbf{v}$

$$p_{\mathbf{v}}(0) \leq p_{\mathbf{v}}^{1/4}(0).$$

The next question is: Can one improve this to

$$(3) \qquad\qquad p_{\mathbf{v}}(0) \leq \varepsilon p_{\mathbf{v}}^{1/4}(0),$$

for any positive constant ε? The answer is negative. If we take $\mathbf{v} = \{1, 1, \ldots, 1\}$, then it is easy to show that

$$p_{\mathbf{v}}(0) = \big(c + o(1)\big) p_{\mathbf{v}}^{1/4}(0)$$

where c is a positive constant (depends on $1/4$). However, it is possible to classify all sets $\mathbf{v}$ where (3) fails (under some slight assumptions). This classification is the heart of the proof of Theorem 7.2. The precise statement is somewhat technical (we refer to [57] for details), but roughly it says that if (3) fails then $\mathbf{v}$ is contained in a generalized arithmetic progression with constant rank and small volume.

12. Inverse Littlewood–Offord theorems

A set
$$P = \{c + m_1 a_1 + \cdots + m_d a_d \mid M_i \leq m_i \leq M_i'\}$$

is called a *generalized arithmetic progression* (GAP) of rank d. It is convenient to think of P as the image of an integer box $B := \big\{ (m_1, \ldots, m_d) \mid M_i \leq m_i \leq M_i' \big\}$ in $\mathbf{Z}^d$ under the linear map

$$\Phi : (m_1, \ldots, m_d) \mapsto c + m_1 a_1 + \cdots + m_d a_d.$$

The numbers a_i are the *generators* of P. For a set A of reals and a positive integer k, we define the iterated sumset

$$kA := \{a_1 + \cdots + a_k \mid a_i \in A\}.$$

Let us take another look at Theorem 10.1. This theorem is sharp, as is shown by taking $v_1 = v_2 = \cdots = v_n = 1$. However, the bound changes significantly if one forbids this special case. Erdős and Moser [19] showed that under the stronger assumption that the v_i are non-zero and different,

$$p_{\mathbf{v}} = O(n^{-3/2} \ln n).$$

They conjectured that the logarithmic term is not necessary and this was confirmed by Sárközy and Szemerédi [52] (see also [28]). Again, the bound is sharp (up to a constant factor), as can be seen by taking $v_1, \ldots, v_n$ to be a proper arithmetic progression such as $1, \ldots, n$.

In the above two examples, we observe that in order to make $p_{\mathbf{v}}$ large, we have to impose a very strong additive structure on $\mathbf{v}$ (in one case we set the v_i's to be the same, while in the other we set them to be elements of an arithmetic progression). A more general example is the following

Example. Let P be a GAP of rank d and volume V. Let $v_1, \ldots, v_n$ be (not necessarily different) elements of P. Then the random variable $S = \sum_{i=1}^{n} \xi_i v_i$ takes values in the GAP nP which has volume $n^d V$. From the pigeonhole principle and the definition of $p_{\mathbf{v}}$, it follows that

$$p_{\mathbf{v}} \geq n^{-d} V^{-1}.$$

This example shows that if the elements of $\mathbf{v}$ belong to a GAP with small rank and small volume then $p_{\mathbf{v}}$ is large. One may conjecture that the inverse also holds, namely,

If $p_{\mathbf{v}}$ is large, then (most of) the elements of $\mathbf{v}$ belong to a GAP with small rank and small volume.

Tao and Vu [58] have managed to quantify this statement.

Theorem 12.1 [58]. *Let $A, \alpha > 0$ be arbitrary constants. There are constants d and B depending on A and α such that the following holds. Assume that $\mathbf{v} = \{v_1, \ldots, v_n\}$ is a multiset of integers satisfying $\mathbf{P}(S = 0) \geq n^{-A}$. Then there is a GAP Q of rank at most d and volume at most n^B which contains all but at most n^α elements of $\mathbf{v}$ (counting multiplicity).*

Notice that the small set of exceptional elements is not avoidable. For instance, one can add $O(\log n)$ completely arbitrary elements to $\mathbf{v}$, and only decrease $p_{\mathbf{v}}$ by a factor of $n^{-O(1)}$ at worst.

Theorem 12.1 is one of the main ingredients in the proofs of Theorems 9.3 and 9.4. For many other theorems of this type, see [58].

Acknowledgement. We would like to thank K. Costello and P. Wood for proofreading the manuscript. We also thank the referee for useful comments.

REFERENCES

[1] D. Achlioptas and F. McSherry, Fast computation of low rank matrix approximations, in: *Proceedings of the Thirty-Third Annual ACM Symposium on Theory of Computing*, 611–618 (electronic), ACM, New York (2001).

[2] N. Alon, Eigenvalues and expanders, *Combinatorica*, **6** 1986), 83–96.

[3] N. Alon and V. Milman, λ_1, isoperimetric inequalities for graphs, and superconcentrators, *J. Combin. Theory Ser. B*, **38** (1985), no. 1, 73–88.

[4] N. Alon, M. Krevelevich and V. Vu, On the concentration of eigenvalues of random symmetric matrices, *Israel J. Math.*, **131** (2002), 259–267.

[5] L. Arnold, On the asymptotic distribution of the eigenvalues of random matrices, *J. Math. Anal. Appl.*, **20** (1967), 262–268.

[6] Z. D. Bai, Circular law, *Ann. Probab.*, **25** (1997), no. 1, 494–529.

[7] Z. D. Bai and J. Silverstein, Spectral analysis of large dimensional random matrices, Mathematics Monograph Series **2**, Science Press (Beijing, 2006).

[8] D. Bau and L. Trefethen, *Numerical linear algebra*, Society for Industrial and Applied Mathematics (SIAM), Philadelphia, PA (1997).

[9] B. Bollobás, *Random graphs*, Second edition, Cambridge Studies in Advanced Mathematics, 73. Cambridge University Press, Cambridge (2001).

[10] J. Bourgain, V. Vu and P. Wood, *paper in preparation*.

[11] F. Chung, *Spectral graph theory*, CBMS series, no. 92 (1997).

[12] F. Chung, R. Graham and R. Wilson, Quasi-random graphs, *Combinatorica*, **9** (1989), no. 4, 345–362.

[13] F. Chung, L. Lu and V. Vu, The spectra of random graphs with expected degrees, *Proceedings of National Academy of Sciences*, **100**, no. 11, (2003).

[14] K. Costello and V. Vu, The ranks of random graphs, *Random Structures and Algorithms*, to appear.

[15] K. Costello, T. Tao and V. Vu, Random symmetric matrices are almost surely non-singular, *Duke Math. Journal*, **135**, number 2 (2006), 395–413.

[16] A. Edelman, Eigenvalues and condition numbers of random matrices, *SIAM J. Matrix Anal. Appl.*, **9** (1988), no. 4, 543–560.

[17] A. Edelman and B. Sutton, Tails of condition number distributions, *SIAM J. Matrix Anal. Appl.*, **27** (2005), no. 2

[18] P. Erdős, On a lemma of Littlewood and Offord, *Bull. Amer. Math. Soc.*, **51** (1945), 898–902.

[19] P. Erdős, Extremal problems in number theory, 1965. *Proc. Sympos. Pure Math.*, Vol. VIII, pp. 181–189., Amer. Math. Soc., Providence, R.I.

[20] J. Friedman, On the second eigenvalue and random walks in random d-regular graphs, *Combinatorica*, **11** (1991), no. 4, 331–362

[21] J. Friedman, A proof of Alon's second eigenvalue conjecture, in: *Proceedings of the Thirty-Fifth Annual ACM Symposium on Theory of Computing*, 720–724, ACM, New York (2003).

[22] J. Friedman, J. Kahn, and E. Szemere di, On the second eigenvalue in random regular graphs, *Proc. of 21th ACM STOC* (1989), 587–598.

[23] Z. Füredi and J. Komlós, The eigenvalues of random symmetric matrices, *Combinatorica*, **1** (1981), no. 3, 233–241.

[24] V. Girko, Circle law (Russian), *Teor. Veroyatnost. i Primenen.* **29** (1984), no. 4, 669–679.

[25] V. Girko, Theory of random determinants. Translated from the Russian. Mathematics and its Applications (Sovict Scrics), 45. Kluwer Academic Publishers Group, Dordrecht (1990).

[26] F. Götze and A. N. Tikhomirov, *On the circular law*, preprint.

[27] A. Guionnet and O. Zeitouni, Concentration of the spectral measure for large matrices, *Electron. Comm. Probab.*, **5** (2000), 119–136 (electronic).

[28] G. Halász, Estimates for the concentration function of combinatorial number theory and probability, *Period. Math. Hungar.*, **8** (1977), no. 3–4, 197–211.

[29] S. Janson, The first eigenvalue of random graphs, *Combin. Probab. Comput.*, **14** (2005), no. 5–6, 815–828.

[30] J. Kahn, J. Komlós and E. Szemerédi, On the probability that a random ± 1 matrix is singular, *J. Amer. Math. Soc.*, **8** (1995), 223–240.

[31] J. H. Kim and V. Vu, Sandwiching random graphs: Universality between random models, *Advances in Mathematics*, **188** (2004), 444–469.

[32] J. Komlós, On the determinant of $(0, 1)$ matrices, *Studia Sci. Math. Hungar.*, **2** (1967), 7–22.

[33] M. Krivelevich and B. Sudakov, The largest eigenvalue of sparse random graphs, *Combin. Probab. Comput.*, **12** (2003), no. 1, 61–72.

[34] M. Krivelevich and B. Sudakov, Pseudo-random graphs, in: *More sets, graphs and numbers*, 199–262, Bolyai Soc. Math. Stud., 15, Springer (Berlin, 2006).

[35] M. Krivelevich and V. Vu, Approximating the independence number and the chromatic number in expected polynomial time, *J. Comb. Optim.*, **6** (2002), no. 2, 143–155.

[36] M. Krivelevich, B. Sudakov, V. Vu and N. Wormald, Random regular graphs of high degree, *Random Structures and Algorithms,* **18** (2001), no. 4, 346–363.

[37] J. E. Littlewood and A. C. Offord, On the number of real roots of a random algebraic equation. III. *Rec. Math. [Mat. Sbornik] N.S.,* **12** (1943), 277–286.

[38] A. Lubotzky, R. Phillips and P. Sarnak, Ramanujan graphs, *Combinatorica,* **8** (1988), no. 3, 261–277.

[39] G. Margulis, Explicit group-theoretic constructions of combinatorial schemes and their applications in the construction of expanders and concentrators (Russian), *Problemy Peredachi Informatsii,* **24** (1988), no. 1, 51–60; translation in *Problems Inform. Transmission,* **24** (1988), no. 1, 39–46.

[40] J. E. Littlewood and A. C. Offord, On the number of real roots of a random algebraic equation. III. *Rec. Math. [Mat. Sbornik] N.S.,* **12** (1943), 277–286.

[41] M. L. Mehta, Random matrices, Third edition, Pure and Applied Mathematics, 142, Elsevier/Academic Press (Amsterdam, 2004).

[42] B. McKay, The expected eigenvalue distribution of a large regular graph, *Linear Algebra Appl.,* **40** (1981), 203–216.

[43] M. Meckes, Concentration of norms and eigenvalues of random matrices, *J. Funct. Anal.,* **211** (2004), no. 2, 508–524.

[44] A. Nilli, On the second eigenvalue of a graph, *Discrete Mathematics,* **91** (1991), 207–210.

[45] A. Nilli, Tight estimates for eigenvalues of regular graphs, *Electronic J. Combinatorics,* **11** (2004), N9, 4pp.

[46] G. Pan and W. Zhou, Circular law, *Extreme singular values and potential theory,* preprint.

[47] M. Rudelson, Invertibility of random matrices: Norm of the inverse, to appear in *Annals of Mathematics.*

[48] A. Sárközy and E. Szemerédi, Uber ein Problem von Erdős und Moser, *Acta Arithmetica,* **11** (1965), 205–208.

[49] D. Spielman and S. H. Teng, Smoothed analysis of algorithms, Proceedings of the International Congress of Mathematicians, Vol. I (Beijing, 2002), 597–606, Higher Ed. Press (Beijing, 2002).

[50] M. Rudelson and R. Vershynin, *The Littlewood–Offord problem and invertibility of random matrices,* preprint.

[51] S. Peche and A. Soshnikov, *Wigner random matrices with non-symmetrically distributed entries,* preprint.

[52] Y. Sinai and A. Soshnikov, A refinement of Wigner's semicircle law in a neighborhood of the spectrum edge for random symmetric matrices, (Russian) *Funktsional. Anal. i Prilozhen.,* **32** (1998), no. 2, 56–79, 96; translation in *Funct. Anal. Appl.,* **32** (1998), no. 2, 114–131.

[53] A. Soshnikov and B. Sudakov, On the largest eigenvalue of a random subgraph of the hypercube, *Comm. Math. Phys.,* **239** (2003), no. 1–2, 53–63.

[54] B. Sudakov and V. Vu, *Resilience of graphs*, submitted.

[55] A. Thomason, *Pseudorandom graphs,* Random graphs '85 (Poznań, 1985), 307–331, North-Holland Math. Stud., 144, North-Holland (Amsterdam, 1987).

[56] T. Tao and V. Vu, On random 1 matrices: singularity and determinant, *Random Structures Algorithms,* **28** (2006), no. 1, 1–23.

[57] T. Tao and V. Vu, On the singularity probability of random Bernoulli matrices, *J.A.M.S.,* **20** (2007), no. 3, 603–628.

[58] T. Tao and V. Vu, Inverse Littlewood–Offord theorems and the condition number of random matrices, *to appear in Annals of Mathematics.*

[59] T. Tao and V. Vu, The condition number of a randomly perturbed matrix, *STOC 2007.*

[60] T. Tao and V. Vu, *Random matrices: The circular law,* submitted.

[61] T. Tao and V. Vu, *Additive Combinatorics*, Cambridge Univ. Press (2006).

[62] V. Vu, Spectral Norm of Random Matrices, to appear in *Combinatorica* (extended abstract appeared in STOC 2005).

[63] V. Vu and L. Wu, *paper in preparation.*

[64] Wigner, On the distribution of the roots of certain symmetric matrices, *Annals of Mathematics* (2), **67** (1958), 325–327.

Van Vu

Department of Mathematics
Rutgers, Piscataway
NJ 08854

e-mail: vanvu@math.rutgers.edu